Lecture Notes in Physics
Monographs

Editorial Board

R. Beig, Wien, Austria
J. Ehlers, Potsdam, Germany
U. Frisch, Nice, France
K. Hepp, Zürich, Switzerland
W. Hillebrandt, Garching, Germany
D. Imboden, Zürich, Switzerland
R. L. Jaffe, Cambridge, MA, USA
R. Kippenhahn, Göttingen, Germany
R. Lipowsky, Golm, Germany
H. v. Löhneysen, Karlsruhe, Germany
I. Ojima, Kyoto, Japan
H. A. Weidenmüller, Heidelberg, Germany
J. Wess, München, Germany
J. Zittartz, Köln, Germany

Managing Editor

W. Beiglböck
c/o Springer-Verlag, Physics Editorial Department II
Tiergartenstrasse 17, D-69121 Heidelberg, Germany

Springer
*Berlin
Heidelberg
New York
Barcelona
Hong Kong
London
Milan
Paris
Singapore
Tokyo*

Physics and Astronomy ONLINE LIBRARY

http://www.springer.de/phys/

The Editorial Policy for Monographs

The series Lecture Notes in Physics reports new developments in physical research and teaching - quickly, informally, and at a high level. The type of material considered for publication in the monograph Series includes monographs presenting original research or new angles in a classical field. The timeliness of a manuscript is more important than its form, which may be preliminary or tentative. Manuscripts should be reasonably self-contained. They will often present not only results of the author(s) but also related work by other people and will provide sufficient motivation, examples, and applications.

The manuscripts or a detailed description thereof should be submitted either to one of the series editors or to the managing editor. The proposal is then carefully refereed. A final decision concerning publication can often only be made on the basis of the complete manuscript, but otherwise the editors will try to make a preliminary decision as definite as they can on the basis of the available information.

Manuscripts should be no less than 100 and preferably no more than 400 pages in length. Final manuscripts should be in English. They should include a table of contents and an informative introduction accessible also to readers not particularly familiar with the topic treated. Authors are free to use the material in other publications. However, if extensive use is made elsewhere, the publisher should be informed. Authors receive jointly 30 complimentary copies of their book. They are entitled to purchase further copies of their book at a reduced rate. No reprints of individual contributions can be supplied. No royalty is paid on Lecture Notes in Physics volumes. Commitment to publish is made by letter of interest rather than by signing a formal contract. Springer-Verlag secures the copyright for each volume.

The Production Process

The books are hardbound, and quality paper appropriate to the needs of the author(s) is used. Publication time is about ten weeks. More than twenty years of experience guarantee authors the best possible service. To reach the goal of rapid publication at a low price the technique of photographic reproduction from a camera-ready manuscript was chosen. This process shifts the main responsibility for the technical quality considerably from the publisher to the author. We therefore urge all authors to observe very carefully our guidelines for the preparation of camera-ready manuscripts, which we will supply on request. This applies especially to the quality of figures and halftones submitted for publication. Figures should be submitted as originals or glossy prints, as very often Xerox copies are not suitable for reproduction. For the same reason, any writing within figures should not be smaller than 2.5 mm. It might be useful to look at some of the volumes already published or, especially if some atypical text is planned, to write to the Physics Editorial Department of Springer-Verlag direct. This avoids mistakes and time-consuming correspondence during the production period.

As a special service, we offer free of charge LATEX and TEX macro packages to format the text according to Springer-Verlag's quality requirements. We strongly recommend authors to make use of this offer, as the result will be a book of considerably improved technical quality.

For further information please contact Springer-Verlag, Physics Editorial Department II, Tiergartenstrasse 17, D-69121 Heidelberg, Germany.

Series homepage – http://www.springer.de/phys/books/lnpm

Richard J. Szabo

Equivariant Cohomology and Localization of Path Integrals

 Springer

Author

Richard J. Szabo
The Niels Bohr Institute
Blegdamsvej 17
2100 Copenhagen Ø, Denmark
and
Department of Physics - Theoretical Physics
University of Oxford
1 Keble Road
Oxford OX1 3NP, United Kingdom

Library of Congress Cataloging-in-Publication Data.

Die Deutsche Bibliothek - CIP-Einheitsaufnahme

Szabo, Richard J.:
Equivariant cohomology and localization of path integrals / Richard J. Szabo.
- Berlin ; Heidelberg ; New York ; Barcelona ; Hong Kong ; London ;
Milan ; Paris ; Singapore ; Tokyo : Springer, 2000
(Lecture notes in physics : N.s. M, Monographs ; 63)
ISBN 3-540-67126-9

ISSN 0940-7677 (Lecture Notes in Physics. Monographs)
ISBN 3-540-67126-9 Springer-Verlag Berlin Heidelberg New York

This work is subject to copyright. All rights are reserved, whether the whole or part of the material is concerned, specifically the rights of translation, reprinting, reuse of illustrations, recitation, broadcasting, reproduction on microfilm or in any other way, and storage in data banks. Duplication of this publication or parts thereof is permitted only under the provisions of the German Copyright Law of September 9, 1965, in its current version, and permission for use must always be obtained from Springer-Verlag. Violations are liable for prosecution under the German Copyright Law.

Springer-Verlag is a company in the specialist publishing group BertelsmannSpringer
© Springer-Verlag Berlin Heidelberg 2000
Printed in Germany

The use of general descriptive names, registered names, trademarks, etc. in this publication does not imply, even in the absence of a specific statement, that such names are exempt from the relevant protective laws and regulations and therefore free for general use.

Typesetting: Camera-ready by the author
Cover design: *design & production*, Heidelberg

Printed on acid-free paper
SPIN: 10644521   55/3144/du - 5 4 3 2 1 0

In Memory of Robert Martin

Preface

This book reviews equivariant localization techniques for the evaluation of Feynman path integrals. It develops systematic geometric methods for studying the semi-classical properties of phase space path integrals for dynamical systems, emphasizing the relations with integrable and topological quantum field theories. Beginning with a detailed review of the relevant mathematical background – equivariant cohomology and the Duistermaat-Heckman theorem, it demonstrates how the localization ideas are related to classical integrability and how they can be formally extended to derive explicit localization formulas for path integrals in special instances using BRST quantization techniques. Various loop space localizations are presented and related to notions in quantum integrability and topological field theory. The book emphasizes the common symmetries that such localizable models always possess and uses these symmetries to discuss the range of applicability of the localization formulas. A number of physical and ma thematical applications are presented in connection with elementary quantum mechanics, Morse theory, index theorems, character formulas for semi-simple Lie groups, quantization of spin systems, unitary integrations in matrix models, modular invariants of Riemann surfaces, supersymmetric quantum field theories, two-dimensional Yang-Mills theory, conformal field theory, cohomological field theories and the loop expansion in quantum field theory. Some modern techniques of path integral quantization, such as coherent state methods, are also discussed. The relations between equivariant localization and other ideas in topological field theory, such as the Batalin-Fradkin-Vilkovisky and Mathai-Quillen formalisms, are presented and used to discuss the general relationship between topological field theories and more conventional physical models.

Copenhagen, September 1999   *Richard J. Szabo*

Acknowledgements: I would like to thank I. Kogan, G. Landi, F. Lizzi and G. Semenoff for advice and encouragement during various stages of the writing of this book and for helpful discussions. I am grateful to L. Paniak for his participation in the various calculational and conceptual aspects of Section 7, to E. Gozzi for important historical remarks on the manuscript, and to O. Tirkkonen for illuminating discussions. I would also like to thank D. Austin and R. Douglas for comments and suggestions on some of the more mathematical aspects of this book. This work was supported in part by the Natural Sciences and Engineering Research Council of Canada.

Contents

1. **Introduction** 1
   1.1 Path Integrals in Quantum Mechanics,
       Integrable Models and Topological Field Theory 1
   1.2 Equivariant Localization Theory 5
   1.3 Outline 7

2. **Equivariant Cohomology
   and the Localization Principle** 11
   2.1 Example: The Height Function of the Sphere 11
   2.2 A Brief Review of DeRham Cohomology 13
   2.3 The Cartan Model of Equivariant Cohomology 19
   2.4 Fiber Bundles
       and Equivariant Characteristic Classes 24
   2.5 The Equivariant Localization Principle 33
   2.6 The Berline-Vergne Theorem 38

3. **Finite-Dimensional Localization Theory
   for Dynamical Systems** 43
   3.1 Symplectic Geometry 44
   3.2 Equivariant Cohomology on Symplectic Manifolds 47
   3.3 Stationary-Phase Approximation
       and the Duistermaat-Heckman Theorem 51
   3.4 Morse Theory and Kirwan's Theorem 56
   3.5 Examples: The Height Function
       of a Riemann Surface 58
   3.6 Equivariant Localization and Classical Integrability 62
   3.7 Degenerate Version
       of the Duistermaat-Heckman Theorem 67
   3.8 The Witten Localization Formula 70
   3.9 The Wu Localization Formula 74

4. **Quantum Localization Theory
   for Phase Space Path Integrals** 77
   4.1 Phase Space Path Integrals 78

- 4.2 Example: Path Integral Derivation
  of the Atiyah-Singer Index Theorem 84
- 4.3 Loop Space Symplectic Geometry
  and Equivariant Cohomology 95
- 4.4 Hidden Supersymmetry
  and the Loop Space Localization Principle 99
- 4.5 The WKB Localization Formula 105
- 4.6 Degenerate Path Integrals
  and the Niemi-Tirkkonen Localization Formula 107
- 4.7 Connections with the Duistermaat-Heckman
  Integration Formula 112
- 4.8 Equivariant Localization and Quantum Integrability 114
- 4.9 Localization for Functionals of Isometry Generators 117
- 4.10 Topological Quantum Field Theories 121

5. **Equivariant Localization on Simply Connected Phase Spaces: Applications to Quantum Mechanics, Group Theory and Spin Systems** 127
   - 5.1 Coadjoint Orbit Quantization
     and Character Formulas 130
   - 5.2 Isometry Groups
     of Simply Connected Riemannian Spaces 137
   - 5.3 Euclidean Phase Spaces
     and Holomorphic Quantization 146
   - 5.4 Coherent States on Homogeneous Kähler Manifolds
     and Holomorphic Localization Formulas 153
   - 5.5 Spherical Phase Spaces and Quantization
     of Spin Systems 158
   - 5.6 Hyperbolic Phase Spaces 169
   - 5.7 Localization of Generalized Spin Models
     and Hamiltonian Reduction 171
   - 5.8 Quantization of Isospin Systems 180
   - 5.9 Quantization on Non-Homogeneous Phase Spaces 191

6. **Equivariant Localization on Multiply Connected Phase Spaces: Applications to Homology and Modular Representations** 203
   - 6.1 Isometry Groups of Multiply Connected Spaces 205
   - 6.2 Equivariant Hamiltonian Systems in Genus One 207
   - 6.3 Homology Representations
     and Topological Quantum Field Theory 210
   - 6.4 Integrability Properties and Localization Formulas 213
   - 6.5 Holomorphic Quantization
     and Non-Symmetric Coadjoint Orbits 217
   - 6.6 Generalization to Hyperbolic Riemann Surfaces 226

7. Beyond the Semi-Classical Approximation 233
  7.1 Geometrical Characterizations
      of the Loop Expansion 234
  7.2 Conformal and Geodetic Localization Symmetries 244
  7.3 Corrections to the Duistermaat-Heckman Formula:
      A Geometric Approach 253
  7.4 Examples 259
  7.5 Heuristic Generalizations to Path Integrals:
      Supersymmetry Breaking 266

8. Equivariant Localization
   in Cohomological Field Theory 269
  8.1 Two-Dimensional Yang-Mills Theory: Equivalences
      Between Physical and Topological Gauge Theories 270
  8.2 Symplectic Geometry of Poincaré Supersymmetric Quantum
      Field Theories 276
  8.3 Supergeometry
      and the Batalin-Fradkin-Vilkovisky Formalism 281
  8.4 Equivariant Euler Numbers, Thom Classes
      and the Mathai-Quillen Formalism 287
  8.5 The Mathai-Quillen Formalism
      for Infinite-Dimensional Vector Bundles 291

9. Appendix A: BRST Quantization 295

10. Appendix B: Other Models
   of Equivariant Cohomology 299
  10.1 The Topological Definition 299
  10.2 The Weil Model 300
  10.3 The BRST Model 304
  10.4 Loop Space Extensions 306

References 309

1. Introduction

In this book we shall review the systematic approaches and applications of a theory which investigates situations where Feynman path integrals of physical systems can be evaluated exactly leading to a complete understanding of the quantum physics. These mathematical formalisms are in large part motivated by the symmetries present in integrable systems and topological quantum field theories which make these latter examples exactly solvable problems. Besides providing conceptual understandings of the solvability features of these special classes of problems, this framework yields geometric approaches to evaluating the quantum spectrum of generic quantum mechanical and quantum field theoretical partition functions. The techniques that we shall present here in fact motivate an approach to studying generic physical problems by relating their properties to those of integrable and topological field theories. In doing so, we shall therefore also review some of the more modern quantum field theoretical and mathematical ideas which have been at the forefront of theoretical physics over the past two decades.

1.1 Path Integrals in Quantum Mechanics, Integrable Models and Topological Field Theory

The idea of path integration was introduced by Feynman [46] in the 1940's as a novel approach to quantum theory. Symbolically, the fundamental path integral formula is

$$\mathcal{K}(q',q;T) = \sum_{C_{qq'}} e^{iTL[C_{qq'}]} \qquad (1.1)$$

where the 'sum' is over all paths $C_{qq'}$ between the points q and q' on the configuration space of a physical system, and $L[C_{qq'}]$ is the length of the path. The quantity on the left-hand side of (1.1) represents the probability amplitude for the system to evolve from a state with configuration q to one with configuration q' in a time span T. One of the great advantages of the path integral formulation is that it gives a global (integral) solution of the quantum problem in question, in contrast to the standard approach to quantum mechanics based on the Schrödinger equation which gives a local (differential) formulation of the problem. Of utmost significance at the time was Feynman's generalization of the path integral to quantum electrodynamics from which a systematic derivation of the famous Feynman rules, and hence the basis of most perturbative calculations in quantum field theory, can be carried out [75].

The problem of quantum integrability, i.e. the possibility of solving analytically for the spectrum of a quantum Hamiltonian and the corresponding energy eigenfunctions, is a non-trivial problem. This is even apparent from the point of view of the path integral, which describes the time evolution of wavefunctions. Relatively few quantum systems have been solved exactly and even fewer have had an exactly solvable path integral. At the time that the functional integration (1.1) was introduced, the only known examples where it could be evaluated exactly were the harmonic oscillator and the free particle. The path integrals for these 2 examples can be evaluated using the formal functional analog of the classical Gaussian integration formula [176]

$$\int_{-\infty}^{\infty} \prod_{k=1}^{n} dx^k \; e^{\frac{i}{2} \sum_{i,j} x^i M_{ij} x^j + i \sum_i \lambda_i x^i} = \frac{(2\pi \; e^{i\pi/2})^{\frac{n}{2}} \; e^{\frac{i}{2} \sum_{i,j} \lambda_i (M^{-1})^{ij} \lambda_j}}{\sqrt{\det M}}$$

(1.2)

where $M = [M_{ij}]$ is a non-singular symmetric $n \times n$ matrix. In this way, the Feynman propagator (1.1) can be evaluated formally for any field theory which is at most quadratic in the field variables. If this is not the case, then one can expand the argument of the exponential in (1.1), approximate it by a quadratic form as in (1.2), and then take the formula (1.2) as an approximation for the integral. For a finite-dimensional integral this is the well-known stationary phase (otherwise known as the saddle-point or steepest-descent) approximation [64]. In the framework of path integration, it is usually referred to as the Wentzel-Kramers-Brillouin (or WKB for short) approximation [101, 147]. Since the result (1.2) is determined by substituting into the exponential integrand the global minimum (i.e. classical value) of the quadratic form and multiplying it by a term involving the second variation of that form (i.e. the fluctuation determinant), this approach to functional integration is also called the semi-classical approximation. In this sense, (1.1) interprets quantum mechanics as a sum over paths fluctuating about the classical trajectories (those with minimal length $L[C_{qq'}]$) of a dynamical system. When the semi-classical approximation is exact, one can think of the Gaussian integration formula (1.2) as a 'localization' of the complicated looking integral on the left-hand side of (1.2) onto the global minimum of the quadratic form there.

For a long time, these were the only examples of exactly solvable path integrals. In 1968 Schulman [146] found that a path integral describing the precession of a spin vector was given exactly by its WKB approximation. This was subsequently generalized by Dowker [38] who proved the exactness of the semi-classical approximation for the path integral describing free geodesic motion on compact group manifolds. It was not until the late 1970's that more general methods, beyond the restrictive range of the standard WKB method, were developed. In these methods, the Feynman path integral is calculated rigorously in discretized form (i.e. over piecewise-linear paths) by a careful regularization prescription [93], and then exploiting information provided by functional analysis, the theory of special functions, and the theory of differential equations (see [31] and references therein). With these tricks the list of exactly solvable path integrals has significantly increased over the last 15 years, so that today one is able to essentially evaluate analytically the path integral for any quantum mechanical problem for which the Schrödinger equation can be solved exactly. We refer to [91] and [62] for an overview of these methods and a complete classification of the known examples of exactly solved quantum mechanical path integrals to present date.

The situation is somewhat better in quantum field theory, which represents the real functional integrals of interest from a physical standpoint. There are many non-trivial examples of classically integrable models (i.e. ones whose classical equations of motion are 'exactly solvable'), for example the sine-Gordon model, where the semi-classical approximation describes the exact spectrum of the quantum field theory [176]. Indeed, for any classically integrable dynamical system one can canonically transform the phase space variables so that, using Hamilton-Jacobi theory [55], the path integral can be formally manipulated to yield a result which if taken naively would imply the exactness of the WKB approximation for *any* classically integrable system [147]. This is not really the case, because the canonical transformations used in the phase space path integral do not respect the ordering prescription used for the properly discretized path integral and consequently the integration measure is not invariant under these transformations [30]. However, as these problems stem mainly from ordering ambiguities in the discretization of the path integral, in quantum field theory these ordering ambiguities could be removed by a suitable renormalization, for instance by an operator commutator ordering prescription. This has led to the conjecture that properly interpreted results of semi-classical approximations in integrable field theories reproduce features of the exact quantum spectrum [176]. One of the present motivations for us is to therefore develop a systematic way to implement realizations of this conjecture.

Another class of field theories where the path integral is exactly solvable in most cases is supersymmetric theories and topological quantum field theories (see [22] for a concise review). Topological field theories have lately been of much interest in both the mathematics and physics literature. A field theory is topological if it has only global degrees of freedom. This means, for example, that its classical equations of motion eliminate all propagating degrees of freedom from the problem (so that the effective quantum action vanishes). In particular, the theory cannot depend on any metric of the space on which the fields are defined. The observables of these quantum field theories therefore describe geometrical and topological invariants of the spacetime which are computable by the conventional techniques of quantum field theory and are of prime interest in mathematics. Physically, topological quantum field theories bear resemblances to many systems of longstanding physical interest and it is hoped that this special class of field theories might serve to provide insight into the structure of more complicated physical systems and a testing ground for new approaches to quantum field theory. There is also a conjecture that topological quantum field theories represent different (topological) phases of their more conventional counterparts (e.g. 4-dimensional Yang-Mills theory). Furthermore, from a mathematical point of view, these field theories provide novel representations of some global invariants whose properties are frequently transparent in the path integral approach.

Topological field theory essentially traces back to the work of Schwarz [148] in 1978 who showed that a particular topological invariant, the Ray-Singer analytic torsion, could be represented as the partition function of a certain quantum field theory. The most important historical work for us, however, is the observation made by Witten [166] in 1982 that the supersymmetry algebra of supersymmetric quantum mechanics describes exactly the DeRham complex of a manifold, where the supersymmetry charge is the exterior derivative. This gives a framework for understanding Morse theory in terms of supersymmetric quantum mechanics in which the quantum partition function computes exactly the Euler characteristic of the configuration manifold, i.e. the index of the DeRham complex.

Witten's partition function computed the so-called Witten index [167], the difference between the number of bosonic and fermionic zero energy states. In order for the supersymmetry to be broken in the ground state of a supersymmetric model, the Witten index must vanish. As supersymmetry, i.e. a boson-fermion symmetry, is not observed in nature, it is necessary to have some criterion for dynamical supersymmetry breaking if supersymmetric theories are to have any physical meaning. Witten's construction was subsequently generalized by Alvarez-Gaumé [5], and Friedan and Windey [48], to give supersymmetric field theory proofs of the Atiyah-Singer index theorem [41]. In this way, the partition function is reduced to an integral over the configuration manifold $\mathcal{M}$. This occurs because the supersymmetry of the action causes only zero modes of the fields, i.e. points on $\mathcal{M}$, to contribute to the path integral, and the integrals over the remaining fluctuation modes are Gaussian. The resulting integral encodes topological information about the manifold $\mathcal{M}$ and represents a huge reduction of the original infinite-dimensional path integration.

This field began to draw more attention around 1988 when Witten introduced topological field theories in a more general setting [169] (see also [170]). A particular supersymmetric non-abelian gauge theory was shown by Witten to describe a theory with only global degrees of freedom whose observables are the Donaldson invariants, certain differential invariants which are used for the study of differentiable structures on 4-manifolds. Subsequent work then put these ideas into a general framework so that today the formal field theoretic structures of Witten's actions are well-understood [22]. Furthermore, because of their topological nature, these field theories have become the focal point for the description of topological effects in quantum systems using quantum field theory, for instance for the description of holonomy effects in physical systems arising from the adiabatic transports of particles [152] and extended objects such as strings [18] (i.e. Aharonov-Bohm type effects). In this way the functional integral has in recent years become a very popular tool lying on the interface between string theory, conformal field theory and topological quantum field theory in physics, and between topology and algebraic geometry in mathematics. Because of the consistent reliability of results that path integrals of these theories can produce when handled with care, functional integration has even acquired a certain degree of respectability among mathematicians.

1.2 Equivariant Localization Theory

The common feature of topological field theories is that their path integrals are described exactly by the semi-classical approximation. It would be nice to put semi-classically exact features of functional integrals, as well as the features which reduce them to integrals over finite-dimensional manifolds as described above, into some sort of general framework. More generally, we would like to have certain criteria available for when we expect partition functions of quantum theories to reduce to such simple expressions, or 'localize'. This motivates an approach to quantum integrability in which one can systematically study the properties of integrable field theories and their conjectured semi-classical "exactness" that we mentioned before. In this approach we focus on the general features and properties that path integrals appearing in this context have in common. Foremost among these is the existence of a large number of (super-)symmetries in the underlying dynamical theory, so that these functional integrals reduce to Gaussian ones and essentially represent finite-dimensional integrals[1]. The transition between the functional and finite-dimensional integrals can then be regarded as a rather drastic localization of the original infinite-dimensional integral, thereby putting it into a form that is useable for extracting physical and mathematical information. The mathematical framework for describing these symmetries, which turn out to be of a topological nature, is equivariant cohomology and the approach discussed in this paragraph is usually called 'equivariant localization

---

[1] The exact solvability features of path integrals in this context is similar to the solvability features of the Schrödinger equation in quantum mechanics when there is a large symmetry group of the problem. For instance, the $O(4)$ symmetry of the 3-dimensional Coulomb problem is what makes the hydrogen atom an exactly solvable quantum system [101].

theory'. This approach introduces an equivariant cohomological framework as a tool for developing geometric techniques for manipulating path integrals and examining their localization properties.

Historically, this subject originated in the mathematics literature in 1982 with the Duistermaat-Heckman theorem [39], which established the exactness of the semi-classical approximation for finite-dimensional oscillatory integrals (i.e. finite-dimensional versions of (1.1)) over compact symplectic manifolds in certain instances. The Duistermaat-Heckman theorem applies to classical systems whose trajectories all have a common period, so that the symmetry responsible for the localization here is the existence of a global Hamiltonian torus action on the manifold. Atiyah and Bott [9] showed that the Duistermaat-Heckman localization was a special case of a more general localization property of equivariant cohomology (with respect to the torus group action in the case of the Duistermaat-Heckman theorem). This fact was used by Berline and Vergne [19, 20] at around the same time to derive a quite general integration formula valid for Killing vectors on general compact Riemannian manifolds.

The first infinite-dimensional generalization of the Duistermaat-Heckman theorem is due to Atiyah and Witten [8], in the setting of a supersymmetric path integral for the index (i.e. the dimension of the space of zero modes) of a Dirac operator. They showed that a formal application of the Duistermaat-Heckman theorem on the loop space $L\mathcal{M}$ of a manifold $\mathcal{M}$ to the partition function of $N = \frac{1}{2}$ supersymmetric quantum mechanics (i.e. a supersymmetric spinning particle in a gravitational background) reproduced the well-known Atiyah-Singer index theorem correctly. The crucial idea was the interpretation of the fermion bilinear in the supersymmetric action as a loop space symplectic 2-form. This approach was then generalized by Bismut [23, 24], within a mathematically rigorous framework, to twisted Dirac operators (i.e. the path integral for spinning particles in gauge field backgrounds), and to the computation of the Lefschetz number of a Killing vector field V (a measure of the number of zeroes of V) acting on the manifold. Another nice infinite-dimensional generalization of the Duistermaat-Heckman theorem was suggested by Picken [139] who formally applied the theorem to the space of loops over a group manifold to localize the path integral for geodesic motion on the group, thus establishing the well-known semi-classical properties of these systems.

It was the beautiful paper by Stone [156] in 1989 that brought the Duistermaat-Heckman theorem to the attention of a wider physics audience. Stone presented a supersymmetric derivation of the Weyl character formula for $SU(2)$ using the path integral for spin and interpreted the result as a Duistermaat-Heckman localization. This supersymmetric derivation was extended by Alvarez, Singer and Windey [4] to more general Lie groups using fiber bundle theory, and the supersymmetries in both of these approaches are very closely related to equivariant cohomology. At around this time, other important papers concerning the quantization of spin appeared. Most notably, Nielsen and Rohrlich [117] (see also [50]) viewed the path integral for spin from a more geometrical point of view, using as action functional the solid angle swept out by the closed orbit of the spin. This approach was related more closely to geometric quantization and group representation theory by Alekseev, Faddeev and Shatashvili [2, 3], who calculated the coadjoint orbit path integral for unitary and orthogonal groups, and also for cotangent bundles of compact groups, Kac-Moody groups and the Virasoro group. The common feature is always that the path integrals are given exactly by a semi-classical localization formula that resembles the Duistermaat-Heckman formula.

The connections between supersymmetry and equivariant cohomology in the quantum mechanics of spin were clarified by Blau in [26], who related the Weinstein action invariant [165] to Chern-Simons gauge theory using the Duistermaat-Heckman integration formula. Based on this interpretation, and the observation of Gozzi et al. [57]–[59] of the hidden supersymmetry underlying any classical dynamical system, Blau, Keski-Vakkuri and Niemi [30] introduced a general supersymmetric (or equivariant cohomological) framework for investigating Duistermaat-Heckman (or WKB) localization formulas for generic (non-supersymmetric) phase space path integrals, leading to the fair amount of activity in this field which is today the foundation of equivariant localization theory. They showed formally that the partition function for the quantum mechanics of circle actions of isometries on symplectic manifolds localizes. Their method of proof involves formal techniques of Becchi-Rouet-Stora-Tyupin (or BRST for short) quantization of constrained systems [16]. BRST-cohomology is the fundamental structure in topological field theories, and such BRST supersymmetries are always the symmetries that are responsible for localization in these models.

1.3 Outline

In this Book we shall primarily explore the geometric features of the localization formalism for phase space path integrals. In particular, we shall focus on how these models can be used to extract information about integrable and topological quantum field theories. In this sense, the path integrals we study for the most part can be thought of as "toy models" serving as a testing ground for ideas in some more sophisticated field theories. The main idea behind this reasoning is that the localization in topological field theories is determined by their kinematical properties. The path integrals we shall focus on allow us to study their kinematical (i.e. geometrical and topological) aspects in isolation from their dynamical properties. These models are typically dynamically linear (i.e. free field theories) in some sense and the entire non-triviality of their path integrals lie in the large kinematical non-linearity that these theories possess. The appropriate relationship between topological field theories and more conventional, physical interacting quantum field theories (which are kinematically linear but dynamically highly non-linear) would then, in principle at least, allow one to incorporate the approaches and techniques which follow to generic physical models. Indeed, one of the central themes in what follows will be the interplay between physical, integrable and topological field theories, and we shall see that the equivariant localization formalism implies connections between these 3 classes of models and thus a sort of unified description of functional integration which provides alternative approximation techniques to the usual perturbative expansion in quantum field theory.

We shall therefore approach the localization formalism for path integrals in the following manner. Focusing on the idea of localizing a quantum partition function by reducing it using the large symmetry of the dynamical system to a sum or finite-dimensional integral in analogy with the classical Gaussian integration formula (1.2), we shall first analyse the symmetries responsible for the localization of finite-dimensional integrals (where the symmetry of the dynamics is represented by an equivariant cohomology). The main focus of this Book will then be the formal generalizations of these ideas to phase space path integrals, where the symmetry becomes a "hidden" supersymmetry of the dynamics representing the infinite-dimensional analog of equivariant cohomology. The subsequent generalization will be then the extension of these notions to both Poincaré supersymmetric quantum field theories (where the symmetry is represented by the supersymmetry of that model) and topological quantum field theories (where the symmetry is represented by a gauge symmetry). The hope is that these serve as testing grounds for the more sophisticated quantum field theories of real physical interest, such as quantum chromodynamics (QCD). This gives a geometric framework for studying quantum integrability, as well as insights into the structure of topological and supersymmetric field theories, and integrable models. In particular, from this analysis we can hope to uncover systematically the reasons why some quantum problems are exactly solvable, and the reasons why others aren't.

Briefly, the structure of this Book is as follows. In Chapter 2 we go through the main mathematical background for localization theory, i.e. equivariant cohomology, with reviews of some other mathematical ideas that will be important for later Chapters as well. In Chapter 3 we present the Duistermaat-Heckman theorem and its generalizations and discuss the connections they imply between equivariant cohomology and the notion of classical integrability of a dynamical system. Chapter 4 then goes through the formal supersymmetry and loop space equivariant cohomology arguments establishing the localization of phase space path integrals when there is a Riemannian structure on the phase space which is invariant under the classical dynamics of the system. Depending on the choice of localization scheme, different sets of phase space trajectories are lifted to a preferred status in the integral. Then all contributions to the functional integral come from these preferred paths along with a term taking into account the quantum fluctuations about these selected loops. Chapters 5 and 6 contain the main physical and mathematical applications of equivariant localization. There we use the fundamental isometry condition to construct numerous examples of localizable path integrals. In each case we evaluate and discuss the localization formulas from both physical and mathematical standpoints. Here we shall encounter numerous examples and gain much insight into the range of applicability of the localization formalism in general. We will also see here many interesting features of the localizable partition functions when interpreted as topological field theories, and we discuss in detail various other issues (e.g. coherent state quantization and coadjoint orbit character formulas) which are common to all the localizable examples that we find within this setting and which have been of interest in the more modern approaches to the quantization of dynamical systems. Chapter 7 then takes a slightly different approach to analysing localizable systems, this time by some geometric constructs of the full loop-expansion on the phase space. Here we shall discuss how the standard localization symmetries could be extended to more general ones, and we shall also show how the localization ideas could be applied to the formulation of a geometric approach to obtaining corrections to the standard WKB approximation for non-localizable partition functions. The analysis of this Chapter is a first step towards a systematic, geometric understanding of the reasons why the localization formulas may not apply to a given dynamical system. In Chapter 8 we turn our attention to field theoretical applications of equivariant localization and discuss the relationships that are implied between topological field theories, physical quantum field theories, and the localization formalism for dynamical systems. For completeness, 2 Appendices at the end of the book are devoted to an overview of some ideas in the BRST quantization formalism and some more mathematical ideas of equivariant cohomology, all of which play an important role in the development of the ideas in the main body of this Book.

We close this introductory Chapter with some comments about the style of this Book. Although we have attempted to keep things self-contained and at places where topics aren't developed in full detail we have included ample references for further reading, we do assume that the reader has a relatively solid background in many of the mathematical techniques of modern theoretical physics such as topology, differential geometry and group theory. All of the group theory that is used extensively in this Book can be found in [162] (or see [53] for a more elementary introduction), while most of the material discussing differential geometry, homology and cohomology, and index theorems can be found in the books [61, 32, 111] and the review articles [22, 41]. For a more detailed introduction to algebraic topology, see [98]. The basic reference for quantum field theory is the classic text [75]. Finally, for a discussion of the issues in supersymmetry theory and BRST quantization, see [16, 22, 69, 118, 155] and references therein.

2. Equivariant Cohomology and the Localization Principle

2.1 Example: The Height Function of the Sphere

To help motivate some of the abstract and technical formalism which follows, we start by considering the evaluation of a rather simple integral. Consider the 2-sphere S^2 of unit radius viewed in Euclidean 3-space $\mathbb{R}^3$ as a sphere standing on end on the xy-plane and centered at $z = a$ symmetrically about the z-axis. We introduce the usual spherical polar coordinates $x = \sin\theta\cos\phi$, $y = \sin\theta\sin\phi$ and $z = a - \cos\theta$ for the embedding of the sphere in 3-space as $S^2 = \{(x,y,z) \in \mathbb{R}^3 : x^2 + y^2 + (z-a)^2 = 1\}$, where $0 \leq \theta \leq \pi$ and $0 \leq \phi \leq 2\pi$. The height of the sphere off of the xy-plane is given by the height function z in $\mathbb{R}^3$ restricted to S^2,

$$h_0(\theta,\phi) = a - \cos\theta \tag{2.1}$$

We want to evaluate the oscillatory integral

$$Z_0(T) = \int_0^\pi \int_0^{2\pi} d\theta\, d\phi\, \sin\theta\, e^{iTh_0(\theta,\phi)} \tag{2.2}$$

which represents a 'toy' version of (1.1). The integration measure in (2.2) is the standard volume form on S^2, i.e. that which is obtained by restriction of the measure $dx\, dy\, dz$ of $\mathbb{R}^3$ to the sphere, and T is some real-valued parameter. It is straightforward to carry out the integration in (2.2) to get

$$Z_0(T) = 2\pi \int_{-1}^{+1} d\cos\theta\, e^{iT(a-\cos\theta)}$$
$$= \frac{2\pi i}{T}\left(e^{-iT(1+a)} - e^{iT(1-a)}\right) = \frac{4\pi}{T} e^{-iTa} \sin T \tag{2.3}$$

Although this integral is simple to evaluate explicitly, it illustrates 2 important features that will be the common theme throughout our discussion. The first characteristic is the second equality in (2.3). This shows that $Z_0(T)$ can be expressed as a sum of 2 terms which correspond, respectively, to the 2 extrema of the height function (2.1) – one from the north pole $\theta = \pi$, which is the maximum of (2.1), and the other from the south pole $\theta = 0$, which is its minimum. The relative minus sign between these 2 terms arises from the fact that the signature of the Hessian matrix of h_0 at its maximum is negative while that at its minimum is positive, i.e. the maximum of h_0 is unstable in the 2 directions along the sphere, each of which, heuristically, constributes a factor of i to the sum in (2.3). Finally, the factor $2\pi i/T$ can be understood as the contribution from the 1-loop determinant when (2.1) is expanded to quadratic order in (θ, ϕ) and the standard WKB approximation to the integral is used. In other words, (2.3) coincides *exactly* with the Gaussian integral formula (1.2), except that it sums over both the minimum and maximum of the argument of the exponential in (2.2).

The second noteworthy feature here is that there is a symmetry responsible for the simple evaluation of (2.2). This symmetry is associated with the interplay between the globally-defined (i.e. single-valued) integration measure and the integrand function (2.1) (see the first equality in (2.3)). Both the height function (2.1) and the integration measure in (2.2) are independent of the polar coordinate ϕ of S^2. This is what led to the simple evaluation of (2.2), and it means, in particular, that the quantities integrated in (2.2) are invariant under the translations $\phi \to \phi + \phi_0$, $\phi_0 \in [0, 2\pi)$, which correspond to rigid rotations of the sphere about the z-axis. These translations generate the circle group $S^1 \sim U(1)$. The existence of a group acting on S^2 which serves as a mechanism for the 'localization' of $Z_0(T)$ onto the stationary points of h_0 gives us hope that we could understand this feature by exploiting non-trivial global features of the quotient space S^2/S^1 within a mathematically rigorous framework. Our hopes are immediately dashed because the 2-sphere with its points identified under this (continuous) circular symmetry globally has the same properties as the mathematically trivial interval $[0, 1]$ (the space where θ/π lives), i.e.

$$S^2/U(1) \simeq [0, 1] \tag{2.4}$$

We shall see that the reason we cannot examine this space in this way is because the circle action above leaves fixed the north pole, at $\theta = \pi$, and the south pole, at $\theta = 0$, of the sphere. The fixed points in this case are at the 2 extrema of the height function h_0. The correct mathematical framework that we should use to describe this situation should take into proper account of this group action on the space – this is 'equivariant cohomology'.

Equivariant cohomology has over the past few years become an increasingly popular tool in theoretical physics, primarily in studies of topological models such as topological gauge theories, topological string theory, and topological gravity. This theory, and its connection with the ideas of this Section, will be the topic of this Chapter. Beginning with a quick review of the DeRham theory, which has for quite a while now been at the forefront of many of the developments of modern theoretical physics (see [41] for a comprehensive review), we shall then develop the framework which describes the topology of a space when there is an action of some Lie group on it. This is reminiscent of how one changes ordinary derivatives to gauge-covariant ones in a gauge field theory to properly incorporate local gauge invariance of the model. We shall ultimately end up discussing the important localization property of integration in equivariant cohomology, and we will see later on that the localization theorems are then fairly immediate consequences of this general formalism.

We close this Section with a comment about the above example. Although it may seem to serve merely as a toy model for some ideas that we wish to pursue, we shall see that this example can be considered as the classical partition function of a spin system (i.e. a classical rotor). A quantum mechanical generalization of it will therefore be associated with the quantization of spin. If we think of the sphere as the Lie group quotient space $SU(2)/U(1)$, then, as we shall discuss extensively later on, this example has a nice generalization to the so-called 'homogeneous' spaces of the form G/T, where G is a Lie group and T is a maximal torus of G. These sets of examples, known as 'coadjoint orbits', will frequently occur as non-trivial verifications of the localization formalisms.

2.2 A Brief Review of DeRham Cohomology

To introduce some notation and to provide a basis for some of the more abstract concepts that will be used throughout this Book, we begin with an elementary 'lightning' review of DeRham cohomology theory and how it probes the topological features of a space. Throughout we shall be working on an abstract topological space (i.e. a set with a collection of open subsets which is closed under unions and finite intersections), and we always regard 2 topological spaces as the same if there is an invertible mapping between the 2 spaces which preserves their open sets, i.e. a bi-continuous function or 'homeomorphism'. To carry out calculus on these spaces, we have to introduce a smooth structure on them (i.e. one that is infinitely-continuously differentiable – or C^∞ for short) which is done in the usual way by turning to the notion of a differentiable manifold.

Let $\mathcal{M}$ be a C^∞ manifold of dimension n, i.e. $\mathcal{M}$ is a paracompact Hausdorff topological space which can be covered by open sets U_i, $\mathcal{M} = \bigcup_i U_i$, each of which is homeomorphic to n-dimensional Euclidean space $\mathbb{R}^n$ and the local homeomorphisms so used induce C^∞ coordinate transformations on the overlaps of patches in $\mathbb{R}^n$. This means that locally, in a neighbourhood of any point on $\mathcal{M}$, we can treat the manifold as a copy of the more familiar $\mathbb{R}^n$, but globally the space $\mathcal{M}$ may be very different from Euclidean space. One way to characterize the global properties of $\mathcal{M}$, i.e. its topology, which make it quite different from $\mathbb{R}^n$ is through the theory of homology and its dual theory, cohomology. Of particular importance to us will be the DeRham theory [32]. We shall always assume that $\mathcal{M}$ is orientable and path-connected (i.e. any 2 points in $\mathcal{M}$ can be joined by a continuous path in $\mathcal{M}$). We shall usually assume, unless otherwise stated, that $\mathcal{M}$ is compact. In the non-compact case, we shall assume certain regularity conditions at infinity so that results for the compact case hold there as well.

Around each point of the manifold we choose an open set U which is a copy of $\mathbb{R}^n$. In $\mathbb{R}^n$ we have the natural notion of tangent vectors to a point, and so we can use the locally defined homeomorphisms to define tangent vectors to a point $x \in \mathcal{M}$. Using the local coordinatization provided by the homeomorphism onto $\mathbb{R}^n$, a general linear combination of tangent vectors is denoted as

$$V = V^\mu(x) \frac{\partial}{\partial x^\mu} \tag{2.5}$$

where throughout we use the Einstein summation convention for repeated upper and lower indices. A linear combination such as (2.5) will be refered to here as a vector field. Its components $V^\mu(x)$ are C^∞ functions on $\mathcal{M}$ and are specified by the introduction of local coordinates from $\mathbb{R}^n$. Acting on a smooth function $f(x)$, the quantity $V(f) \equiv V^\mu \partial_\mu f$ is the directional derivative of f in the direction of the vector components $\{V^\mu\}$. The local derivatives $\{\frac{\partial}{\partial x^\mu}\}_{\mu=1}^n$ span an n-dimensional vector space over $\mathbb{R}$ which is called the tangent space to $\mathcal{M}$ at x and it is denoted by $T_x\mathcal{M}$. The disjoint union of all tangent spaces of the manifold,

$$T\mathcal{M} = \coprod_{x \in \mathcal{M}} T_x \mathcal{M} \tag{2.6}$$

is called the tangent bundle of $\mathcal{M}$.

Any vector space W has a dual vector space W^* which is the space of linear functionals $\text{Hom}_\mathbb{R}(W, \mathbb{R})$ on $W \to \mathbb{R}$. The dual of the tangent space $T_x\mathcal{M}$ is called the cotangent space $T_x^*\mathcal{M}$ and its basis elements dx^μ are defined by

$$dx^\mu \left(\frac{\partial}{\partial x^\nu} \right) = \delta^\mu_\nu \tag{2.7}$$

The disjoint union of all the cotangent spaces of $\mathcal{M}$,

$$T^*\mathcal{M} = \coprod_{x \in \mathcal{M}} T_x^* \mathcal{M} \tag{2.8}$$

is called the cotangent bundle of $\mathcal{M}$.

The space $(T_x^*\mathcal{M})^{\otimes k}$ is the space of n-multilinear functionals on $T_x\mathcal{M} \times \cdots \times T_x\mathcal{M}$ whose elements are the linear combinations

$$T = T_{\mu_1 \cdots \mu_k}(x) dx^{\mu_1} \otimes \cdots \otimes dx^{\mu_k} \tag{2.9}$$

The object (2.9) is called a rank-$(k,0)$ tensor and its components are C^∞ functions of $x \in \mathcal{M}$. Similarly, the associated dual space $(T_x\mathcal{M})^{\otimes \ell}$ consists of the linear combinations

$$\tilde{T} = \tilde{T}^{\mu_1 \cdots \mu_\ell}(x) \frac{\partial}{\partial x^{\mu_1}} \otimes \cdots \otimes \frac{\partial}{\partial x^{\mu_\ell}} \tag{2.10}$$

2.2 A Brief Review of DeRham Cohomology

which are called $(0, \ell)$ tensors. The elements of $(T_x^*\mathcal{M})^{\otimes k} \otimes (T_x\mathcal{M})^{\otimes \ell}$ are called (k, ℓ) tensors and one can define tensor bundles analogously to the tangent and cotangent bundles above. Under a local C^∞ change of coordinates on $\mathcal{M}$ represented by the diffeomorphism $x \to x'(x)$, (2.9) and (2.10) along with the usual chain rules

$$\frac{\partial}{\partial x^\mu} = \frac{\partial x'^\lambda}{\partial x^\mu}\frac{\partial}{\partial x'^\lambda} \quad , \quad dx^\mu = \frac{\partial x^\mu}{\partial x'^\lambda} dx'^\lambda \tag{2.11}$$

imply that the components of a generic rank (k, ℓ) tensor field $T^{\mu_1 \cdots \mu_\ell}_{\nu_1 \cdots \nu_k}(x)$ transform as

$$T'^{\lambda_1 \cdots \lambda_\ell}_{\rho_1 \cdots \rho_k}(x') = \frac{\partial x'^{\lambda_1}}{\partial x^{\mu_1}} \cdots \frac{\partial x'^{\lambda_\ell}}{\partial x^{\mu_\ell}} \frac{\partial x^{\nu_1}}{\partial x'^{\rho_1}} \cdots \frac{\partial x^{\nu_k}}{\partial x'^{\rho_k}} T^{\mu_1 \cdots \mu_\ell}_{\nu_1 \cdots \nu_k}(x) \tag{2.12}$$

Such local coordinate transformations can be thought of as changes of bases (2.11) on the tangent and cotangent spaces.

We are now ready to define the DeRham complex of a manifold $\mathcal{M}$. Given the tensor product of copies of the cotangent bundle as above, we define a multi-linear anti-symmetric multiplication of elements of the cotangent bundle, called the exterior or wedge product, by

$$dx^{\mu_1} \wedge \cdots \wedge dx^{\mu_k} = \sum_{P \in S_k} \text{sgn}(P) dx^{\mu_{P(1)}} \otimes \cdots \otimes dx^{\mu_{P(k)}} \tag{2.13}$$

where the sum is over all permutations P of $1, \ldots, k$ and $\text{sgn}(P)$ is the sign of P, defined as $(-1)^{t(P)}$ where $t(P)$ is the number of transpositions in P. For example, for 2 cotangent basis vector elements

$$dx \wedge dy = dx \otimes dy - dy \otimes dx \tag{2.14}$$

The space of all linear combinations of the basis elements (2.13),

$$\alpha = \frac{1}{k!}\alpha_{\mu_1 \cdots \mu_k}(x) dx^{\mu_1} \wedge \cdots \wedge dx^{\mu_k} \tag{2.15}$$

is the antisymmetrization $\mathcal{A}(T_x^*\mathcal{M})^{\otimes k}$ of the k-th tensor power of the cotangent bundle. The disjoint union, over all $x \in \mathcal{M}$, of these vector spaces is called the k-th exterior power $\Lambda^k \mathcal{M}$ of $\mathcal{M}$. Its elements (2.15) are called differential k-forms whose components are C^∞ functions on $\mathcal{M}$ which are completely antisymmetric in their indices $\mu_1, \ldots, \mu_k$. Notice that by the antisymmetry of the exterior product, if $\mathcal{M}$ is n-dimensional, then $\Lambda^k \mathcal{M} = 0$ for all $k > n$. Furthermore, $\Lambda^0 \mathcal{M} = C^\infty(\mathcal{M})$, the space of smooth functions on $\mathcal{M} \to \mathbb{R}$, and $\Lambda^1 \mathcal{M} = T^*\mathcal{M}$ is the cotangent bundle of $\mathcal{M}$.

The exterior product of a p-form α and a q-form β is the $(p+q)$-form $\alpha \wedge \beta = \frac{1}{(p+q)!}(\alpha \wedge \beta)_{\mu_1 \cdots \mu_{p+q}}(x) dx^{\mu_1} \wedge \cdots \wedge dx^{\mu_{p+q}}$ with local components

$$(\alpha \wedge \beta)_{\mu_1 \cdots \mu_{p+q}}(x) = \sum_{P \in S_{p+q}} \text{sgn}(P) \alpha_{\mu_{P(1)} \cdots \mu_{P(p)}}(x) \beta_{\mu_{P(p+1)} \cdots \mu_{P(p+q)}}(x)$$
$$\tag{2.16}$$

The exterior product of differential forms makes the direct sum of the exterior powers

$$\Lambda\mathcal{M} = \bigoplus_{k=0}^{n} \Lambda^k \mathcal{M} \qquad (2.17)$$

into a graded-commutative algebra called the exterior algebra of $\mathcal{M}$. In $\Lambda\mathcal{M}$, the exterior product of a p-form α and a q-form β obeys the graded-commutativity property

$$\alpha \wedge \beta = (-1)^{pq} \beta \wedge \alpha \quad , \quad \alpha \in \Lambda^p \mathcal{M}, \beta \in \Lambda^q \mathcal{M} \qquad (2.18)$$

On the exterior algebra (2.17), we define a linear operator

$$d : \Lambda^k \mathcal{M} \to \Lambda^{k+1} \mathcal{M} \qquad (2.19)$$

on k-forms (2.15) by

$$(d\alpha)_{\mu_1 \cdots \mu_{k+1}}(x) = \sum_{P \in S_{k+1}} \operatorname{sgn}(P) \partial_{\mu_{P(1)}} \alpha_{\mu_{P(2)} \cdots \mu_{P(k+1)}}(x) \qquad (2.20)$$

and $d\alpha = \frac{1}{(k+1)!} (d\alpha)_{\mu_1 \cdots \mu_{k+1}}(x) dx^{\mu_1} \wedge \cdots \wedge dx^{\mu_{k+1}}$. The operator d is called the exterior derivative and it generalizes the notion of the differential of a function

$$df = \frac{\partial f(x)}{\partial x^\mu} dx^\mu \quad , \quad f \in \Lambda^0 \mathcal{M} = C^\infty(\mathcal{M}) \qquad (2.21)$$

to generic differential forms. It is a graded derivation, i.e. it satisfies the graded Leibniz property

$$d(\alpha \wedge \beta) = d\alpha \wedge \beta + (-1)^p \alpha \wedge d\beta \quad , \quad \alpha \in \Lambda^p \mathcal{M}, \beta \in \Lambda^q \mathcal{M} \qquad (2.22)$$

and it is nilpotent,

$$d^2 = 0 \qquad (2.23)$$

which follows from the commutativity of multiple partial derivatives of C^∞ functions. Thus the exterior derivative allows one to generalize the common notion of vector calculus to more general spaces other than $\mathbb{R}^n$. The collection of vector spaces $\{\Lambda^k \mathcal{M}\}_{k=0}^n$ and nilpotent derivations d form what is called the DeRham complex $\Lambda^*(\mathcal{M})$ of the manifold $\mathcal{M}$.

There are 2 important subspaces of the exterior algebra (2.17) as far as the map d is concerned. One is the kernel of d,

$$\ker d = \{\alpha \in \Lambda\mathcal{M} : d\alpha = 0\} \qquad (2.24)$$

whose elements are called closed forms, and the other is the image of d,

$$\operatorname{im} d = \{\beta \in \Lambda\mathcal{M} : \beta = d\alpha \text{ for some } \alpha \in \Lambda\mathcal{M}\} \qquad (2.25)$$

whose elements are called exact forms. Since d is nilpotent, we have im $d \subset$ ker d. Thus we can consider the quotient of the kernel of d by its image.

2.2 A Brief Review of DeRham Cohomology

The vector space of closed k-forms modulo exact k-forms is called the k-th DeRham cohomology group (or vector space) of $\mathcal{M}$,

$$H^k(\mathcal{M};\mathbb{R}) = \ker d|_{\Lambda^k \mathcal{M}} / \operatorname{im} d|_{\Lambda^{k-1}\mathcal{M}} \tag{2.26}$$

The elements of the vector space (2.26) are the equivalence classes of differential k-forms where 2 differential forms are equivalent if and only if they differ only by an exact form, i.e. if the closed form $\alpha \in \Lambda^k \mathcal{M}$ is a representative of the cohomology class $[\alpha] \in H^k(\mathcal{M};\mathbb{R})$, then so is the closed form $\alpha + d\beta$ for any differential form $\beta \in \Lambda^{k-1}\mathcal{M}$.

One important theorem in DeRham cohomology is Poincaré's lemma. This states that if $d\omega = 0$ in a star-shaped region S of the manifold $\mathcal{M}$ (i.e. one in which the affine line segment joining any 2 points in S lies in S), then one can write $\omega = d\theta$ in that region for some other differential form θ. Thus each representative of a DeRham cohomology class can be *locally* written as an exact form, but globally there may be an obstruction to extending the form θ over the entire manifold in a smooth way depending on whether or not $[\omega] \neq 0$ in the DeRham cohomology group.

The DeRham cohomology groups are related to the topology of the manifold $\mathcal{M}$ as follows. Consider the following q-dimensional subspace of $\mathbb{R}^{q+1}$,

$$\Delta^q = \left\{ (x_0, x_1, \ldots, x_q) \in \mathbb{R}^{q+1} : x_i \geq 0, \sum_{i=0}^{q} x_i = 1 \right\} \tag{2.27}$$

which is called the standard q-simplex. Geometrically, Δ^q is the convex hull generated by the vertices placed at unit distance along the axes of $\mathbb{R}^{q+1}$. We define the geometric boundary of the standard q-simplex as

$$\partial \Delta^q = \sum_{i=0}^{q} (-1)^i \hat{\Delta}^q_{(i)} \tag{2.28}$$

where $\hat{\Delta}^q_{(i)}$ is the $(q-1)$-simplex generated by all the vertices of Δ^q except the i-th one, and the sum on the right-hand side is the formal algebraic sum of simplices (where a minus sign signifies a change of orientation). A singular q-simplex of the manifold $\mathcal{M}$ is defined to be a continuous map $\sigma : \Delta^q \to \mathcal{M}$. A formal algebraic sum of q-simplices with integer coefficients is called a q-chain, and the collection of all q-chains in a manifold $\mathcal{M}$ is called the q-th chain group $C_q(\mathcal{M})$ of $\mathcal{M}$. It defines an abelian group under the formal addition. The boundary of a q-chain is the $(q-1)$-chain

$$\partial \sigma = \sum_{i=0}^{q} (-1)^i \sigma \bigg|_{\hat{\Delta}^q_{(i)}} \tag{2.29}$$

which is easily verified to give a nilpotent homomorphism

$$\partial : C_q(\mathcal{M}) \to C_{q-1}(\mathcal{M}) \tag{2.30}$$

of abelian groups. The collection of abelian groups $\{C_q(\mathcal{M})\}_{q \in \mathbb{Z}^+}$ and nilpotent homomorphisms ∂ form the singular chain complex $C_*(\mathcal{M})$ of the manifold $\mathcal{M}$.

Nilpotency of the boundary map (2.30) means that every q-chain in the image of $\partial|_{C_{q+1}}$, the elements of which are called the q-boundaries of $\mathcal{M}$, lies as well in the kernel of $\partial|_{C_q}$, whose elements are called the q-cycles of $\mathcal{M}$. The abelian group defined as the quotient of the group of all q-cycles modulo the group of all q-boundaries is called the q-th (singular) homology group of $\mathcal{M}$,

$$H_q(\mathcal{M}; \mathbb{Z}) = \ker \partial|_{C_q} / \operatorname{im} \partial|_{C_{q+1}} \qquad (2.31)$$

These groups are homotopy invariants of the manifold $\mathcal{M}$ (i.e. invariant under continuous deformations of the space), and in particular they are topological invariants and diffeomorphism invariants (i.e. invariant under C^∞ invertible bi-continuous mappings of $\mathcal{M}$). As such, they are invariant under local deformations of the space and depend only on the global characteristics of $\mathcal{M}$. Intuitively, they measure whether or not a manifold has 'holes' in it or not. If $H_q(\mathcal{M}; \mathbb{Z}) = 0$, then every q-cycle (intuitively a closed q-dimensional curve or surface) encloses a $q+1$-dimensional chain and $\mathcal{M}$ has no 'q-holes'. For instance, if $\mathcal{M}$ is simply-connected (i.e. every loop in $\mathcal{M}$ can be contracted to a point) then $H_1(\mathcal{M}; \mathbb{Z}) = 0$. A star-shaped region, such as a simplex, is simply-connected.

Given the abelian groups (2.31), we can form their duals using the universal coefficient theorem

$$H^q(\mathcal{M}; \mathbb{Z}) \simeq \operatorname{Hom}_{\mathbb{Z}}(H_q(\mathcal{M}; \mathbb{Z}), \mathbb{Z}) \oplus \operatorname{Ext}_{\mathbb{Z}}(H_{q-1}(\mathcal{M}; \mathbb{Z}), \mathbb{Z}) \qquad (2.32)$$

which is called the q-th singular cohomology group of $\mathcal{M}$ with integer coefficients. Here $\operatorname{Hom}_{\mathbb{Z}}(H_q(\mathcal{M}; \mathbb{Z}), \mathbb{Z}) = H_q(\mathcal{M}; \mathbb{Z})^*$ is the free part of the cohomology group, and $\operatorname{Ext}_{\mathbb{Z}}$ is the torsion subgroup of $H^q(\mathcal{M}; \mathbb{Z})$. The DeRham theorem then states that the DeRham cohomology groups are naturally isomorphic to the singular cohomology groups with real coefficients,

$$H^q(\mathcal{M}; \mathbb{R}) = H^q(\mathcal{M}; \mathbb{Z}) \otimes \mathbb{R} = H_q(\mathcal{M}; \mathbb{R})^* \qquad (2.33)$$

where the tensor product with the reals means that H^q is considered as an abelian group with real instead of integer coefficients, i.e. a vector space over $\mathbb{R}$ (this eliminates the torsion subgroup in (2.32)).

The crux of the proof of DeRham's theorem is Stokes' theorem,

$$\int_c d\omega = \oint_{\partial c} \omega \quad , \quad \omega \in \Lambda^q \mathcal{M} \quad , \quad c \in C_{q+1}(\mathcal{M}) \qquad (2.34)$$

which relates the integral of an exact $(q+1)$-form over a smooth $(q+1)$-chain c in $\mathcal{M}$ to an integral over the closed q-dimensional boundary ∂c of c. The map $(\omega, c) \to \int_c \omega$ on $H^q(\mathcal{M}; \mathbb{R}) \otimes H_q(\mathcal{M}; \mathbb{R}) \to \mathbb{R}$ defines a natural duality pairing between $H^q(\mathcal{M};\mathbb{R})$ and $H_q(\mathcal{M};\mathbb{R})$ and is the basis of the DeRham isomorphism. In particular, (2.34) generalizes to the global version of Stokes' theorem,

$$\int_{\mathcal{M}} d\omega = \oint_{\partial \mathcal{M}} \omega \quad , \quad \omega \in \Lambda^{n-1}\mathcal{M} \tag{2.35}$$

which relates the integral of an exact form over $\mathcal{M}$ to an integral over the closed $(n-1)$-dimensional boundary $\partial \mathcal{M}$ of $\mathcal{M}$. Here integration over a manifold is defined by partitioning the manifold up into open sets homeomorphic to $\mathbb{R}^n$, integrating a top form (i.e. a differential form of highest degree n on $\mathcal{M}$) locally over $\mathbb{R}^n$ as usual[1], and then summing up all of these contributions. In this way, we see how the DeRham cohomology of a manifold measures its topological (or global) features in an analytic way suited for the differential calculus of C^∞ manifolds. We refer to [98] and [32] for a more complete and leisurely introduction to this subject.

2.3 The Cartan Model of Equivariant Cohomology

We shall now generalize the constructions of the last Section to the case where there is a Lie group (i.e. a continuous group with a smooth structure whose group multiplication is also smooth) acting on the space. Then the construction of topological invariants for these spaces (i.e. structures that are the same for homeomorphic spaces) will be the foundation for the derivation of general integration formulas in the subsequent Chapters.

Many situations in theoretical physics involve not only a differentiable manifold $\mathcal{M}$, but also the action of some Lie group G acting on $\mathcal{M}$, which we denote symbolically by

$$G \times \mathcal{M} \to \mathcal{M}$$
$$(g,x) \to g \cdot x \tag{2.36}$$

By a group action we mean that $g \cdot x = x, \forall x \in \mathcal{M}$ if g is the identity element of G, and the group action represents the multiplication law of the group, i.e. $g_1 \cdot (g_2 \cdot x) = (g_1 g_2) \cdot x, \forall g_1, g_2 \in G$. We shall throughout assume that G is connected and that its action on $\mathcal{M}$ is smooth, i.e. for fixed $g \in G$, the function $x \to g \cdot x$ is a diffeomorphism of $\mathcal{M}$. Usually G is taken to be the symmetry group of the given physical problem. The common (infinite-dimensional) example in topological field theory is where $\mathcal{M}$ is the space of gauge connections of a gauge theory and G is the group of gauge transformations. The space $\mathcal{M}$ modulo this group action is then the moduli space of gauge orbits. Another example is in string theory where $\mathcal{M}$ is the space of metrics on a Riemann surface (a connected orientable 2-manifold) and G is the semi-direct product of

---

[1] The integral over $\mathcal{M}$ of a p-form with $p < \dim \mathcal{M}$ is always understood here to be zero.

20   2. Equivariant Cohomology and the Localization Principle

the Weyl and diffeomorphism groups of that 2-surface. Then $\mathcal{M}$ modulo this group action is the moduli space of the Riemann surface. In such instances we are interested in knowing the cohomology of the manifold $\mathcal{M}$ given this action of the group G. This cohomology is known as the G-equivariant cohomology of $\mathcal{M}$. Given the G-action on $\mathcal{M}$, the space of orbits $\mathcal{M}/G$ is the set of equivalence classes where x and x' are equivalent if and only if $x' = g \cdot x$ for some $g \in G$ (the topology of $\mathcal{M}/G$ is the induced topology from $\mathcal{M}$). If the G-action on $\mathcal{M}$ is free, i.e. $g \cdot x = x$ if and only if g is the identity element of G, $\forall x \in \mathcal{M}$, then the space of orbits $\mathcal{M}/G$ is also a differentiable manifold of dimension $\dim \mathcal{M} - \dim G$ and the G-equivariant cohomology is defined simply as the cohomology of the coset space $\mathcal{M}/G$,

$$H_G^k(\mathcal{M}) = H^k(\mathcal{M}/G) \tag{2.37}$$

However, if the group action is not free and has fixed points on $\mathcal{M}$, the space $\mathcal{M}/G$ can become singular. The dimension of the orbit $G \cdot x = \{g \cdot x : g \in G\}$ of a point $x \in \mathcal{M}$ is $\dim G - \dim G_x$, where $G_x = \{g \in G : g \cdot x = x\}$ is the isotropy subgroup of x. Consequently, in a neighbourhood of a fixed point x, the dimension $\dim \mathcal{M} - \dim G \cdot x$ of $\mathcal{M}/G$ can be larger than the dimension $\dim \mathcal{M} - \dim G$ of other fixed-point free coordinate neighbourhoods (because then the isotropy subgroup G_x of that fixed point x is non-trivial), and there is no smooth notion of dimensionality for the coset $\mathcal{M}/G$. A singular quotient space $\mathcal{M}/G$ is called an orbifold. In such instances, one cannot define the equivariant cohomology of $\mathcal{M}$ in a smooth way using (2.37) and more elaborate methods are needed to define this cohomology. This is the "right" cohomology theory that properly accounts for the group action and it is always defined in a manner such that if the group action is trivial, the cohomology reduces to the usual cohomological ideas of the classical DeRham theory.

There are many approaches to defining the equivariant cohomology of $\mathcal{M}$, but there is only one that will be used extensively in this Book. This is the Cartan model of equivariant cohomology and it is defined in a manner similar to the analytic DeRham cohomology which was reviewed in the last Section. However, the other models of equivariant cohomology are equally as important – the Weil algebra formulation relates the algebraic models to the topological definition of equivariant cohomology using universal bundles of Lie groups [9, 33, 81, 99], while the BRST model relates the Cartan and Weil models and moreover is the basis for the superspace formulation of topological Yang-Mills theory in 4-dimensions and other cohomological field theories [34, 81, 129]. These other models are outlined in Appendix B.

We begin by generalizing the notion of a differential form to the case where there is a group action on $\mathcal{M}$ as above. We say that a map $f : \mathcal{M}_1 \to \mathcal{M}_2$ between 2 manifolds with G-actions on them is equivariant with respect to the group action if

$$f(g \cdot x) = g \cdot f(x) \quad \forall x \in \mathcal{M}_1 \ , \ \forall g \in G \tag{2.38}$$

2.3 The Cartan Model of Equivariant Cohomology

We want to extend this notion of equivariance to differential forms. Consider the symmetric polynomial functions from the Lie algebra $\mathbf{g}$ of $G \equiv \exp(\mathbf{g})$ into the exterior algebra $\Lambda\mathcal{M}$ of the manifold $\mathcal{M}$. These maps form the algebra $S(\mathbf{g}^*) \otimes \Lambda\mathcal{M}$, where $S(\mathbf{g}^*)$ is called the symmetric algebra over the dual vector space $\mathbf{g}^*$ of $\mathbf{g}$ and it corresponds to the algebra of polynomial functions on $\mathbf{g}$. The action of $g \in G$ on an element $\alpha \in S(\mathbf{g}^*) \otimes \Lambda\mathcal{M}$ is given by

$$(g \cdot \alpha)(X) = g \cdot \left(\alpha(g^{-1}Xg)\right) \tag{2.39}$$

where $X \in \mathbf{g}$. Here we have used the natural coadjoint action of G on $\mathbf{g}^*$ and the induced G-action on $\Lambda\mathcal{M}$ from that on $\mathcal{M}$ as dictated by the tensor transformation law (2.12) with $x'(x) = g \cdot x$. From this it follows immediately that the equivariance condition (2.38) is satisfied for the polynomial maps $\alpha : \mathbf{g} \to \Lambda\mathcal{M}$ in the G-invariant subalgebra

$$\Lambda_G \mathcal{M} = (S(\mathbf{g}^*) \otimes \Lambda\mathcal{M})^G \tag{2.40}$$

where the superscript G denotes the (infinitesimal) G-invariant part. The elements of (2.40) are called equivariant differential forms [19, 21].

Elements of G are represented in terms of elements of the Lie algebra $\mathbf{g}$ through the exponential map,

$$g = e^{c^a X^a} \tag{2.41}$$

where c^a are constants and X^a are the generators of $\mathbf{g}$ obeying the Lie bracket algebra

$$[X^a, X^b] = f^{abc} X^c \tag{2.42}$$

with f^{abc} the antisymmetric structure constants of $\mathbf{g}$. Here and in the following we shall assume an implicit sum over the Lie algebraic indices $a, b, c, \ldots$. The space where the c^a's in (2.41) lie defines the group manifold of G. The generators X^a can be written as $X^a = \frac{\partial}{\partial c^a} g|_{c=0}$ and so the Lie algebra $\mathbf{g}$ can be regarded as the tangent space to the identity on the group manifold of the Lie group G. The strucutre constants in (2.42) define a natural representation of G of dimension $\dim G$, called the adjoint representation, whose (Hermitian) generators have matrix elements $(\text{ad } X^a)_{bc} \equiv if^{abc}$.

The smooth G-action on $\mathcal{M}$ can be represented locally as the continuous flow

$$g_t \cdot x = x(t) \quad , \quad t \in \mathbb{R}^+ \tag{2.43}$$

where g_t is a path in G starting at the identity $g_{t=0}$. The induced action on differential forms is defined by pullback, i.e. as

$$(g_t \cdot \alpha)(x) = \alpha(x(t)) \tag{2.44}$$

For example, we can represent the group action on C^∞ functions by diffeomorphisms on $\mathcal{M}$ which are connected to the identity, i.e.

22   2. Equivariant Cohomology and the Localization Principle

$$(g_t \cdot f)(x) = f(x(t)) = e^{tV(x(t))} f(x) \quad , \quad f \in \Lambda^0 \mathcal{M} \tag{2.45}$$

The action (2.45) represents the flow of the group on C^∞ functions on $\mathcal{M}$, where $V(x) = V^\mu(x) \frac{\partial}{\partial x^\mu}$ is a vector field on $\mathcal{M}$ representing a Lie algebra element. It is related to the flows (2.43) on the manifold by

$$\dot{x}^\mu(t) = V^\mu(x(t)) \tag{2.46}$$

which defines a set of curves in $\mathcal{M}$ which we will refer to as the integral curves of the group action. If V^a is the vector field representing the generator X^a of **g**, then the Lie algebra (2.42) is represented on C^∞ functions by

$$\left[V^a, V^b\right](h) = f^{abc} V^c(h) \quad , \quad \forall h \in \Lambda^0 \mathcal{M} \tag{2.47}$$

with Lie bracket represented by the ordinary commutator bracket. This defines a representation of G by vector fields in the tangent bundle $T\mathcal{M}$. In this setting, the group G is represented as a subgroup of the (infinite-dimensional) connected diffeomorphism group of $\mathcal{M}$ whose Lie algebra is generated by all vector fields of $\mathcal{M}$ with the commutator bracket.

The infinitesimal ($t \to 0$) action of the group on $\Lambda^0 \mathcal{M}$ can be expressed as

$$V(f) = i_V df \tag{2.48}$$

where

$$i_V : \Lambda^k \mathcal{M} \to \Lambda^{k-1} \mathcal{M} \tag{2.49}$$

is the nilpotent contraction operator, or interior multiplication, with respect to V and it is defined locally on k-forms (2.15) by

$$i_V \alpha = \frac{1}{(k-1)!} V^{\mu_1}(x) \alpha_{\mu_1 \mu_2 \cdots \mu_k}(x) dx^{\mu_2} \wedge \cdots \wedge dx^{\mu_k} \tag{2.50}$$

The operator i_V is a graded derivation (c.f. (2.22)) and the quantity $i_V T$ represents the component of a tensor T along the vector field V. The infinitesimal G-action on the higher-degree differential forms is generated by the Lie derivative along V

$$\mathcal{L}_V : \Lambda^k \mathcal{M} \to \Lambda^k \mathcal{M} \tag{2.51}$$

where

$$\mathcal{L}_V = di_V + i_V d \tag{2.52}$$

generates the induced action of G on $\Lambda \mathcal{M}$, i.e.

$$\mathcal{L}_V \alpha(x(0)) = \frac{d}{dt} \alpha(x(t)) \bigg|_{t=0} \tag{2.53}$$

This can be verifed by direct computation from expanding (2.44) about $t = 0$ using (2.12) and (2.46), and by noting that

2.3 The Cartan Model of Equivariant Cohomology

$$[\mathcal{L}_{V^a}, \mathcal{L}_{V^b}](\alpha) = f^{abc}\mathcal{L}_{V^c}(\alpha) \quad , \quad \forall \alpha \in \Lambda\mathcal{M} \tag{2.54}$$

Thus the Lie derivative in general defines a representation of G on $\Lambda\mathcal{M}$. The local components of $\mathcal{L}_V T$ for a general (k, ℓ) tensor field T are found by substituting into the tensor transformation law (2.12) the infinitesimal coordinate change $x'^\mu(x) = x^\mu(t) = x^\mu + tV^\mu(x)$. For example, on a vector field $W = W^\mu(x)\frac{\partial}{\partial x^\mu}$ we have

$$(\mathcal{L}_V W)^\mu = W^\nu \partial_\nu V^\mu - V^\nu \partial_\nu W^\mu \equiv [W, V]^\mu \tag{2.55}$$

Furthermore, the Lie derivative $\mathcal{L}_V$ is an ungraded derivation and its action on contractions is

$$[i_{V^a}, \mathcal{L}_{V^b}](\alpha) = f^{abc} i_{V^c}(\alpha) \tag{2.56}$$

We are now ready to define the Cartan model for the G-equivariant cohomology of $\mathcal{M}$ [21, 33, 99]. We assign a $\mathbb{Z}$-grading[2] to the elements of (2.40) by defining the degree of an equivariant differential form to be the sum of its ordinary form degree and twice the polynomial degree from the $S(\mathbf{g}^*)$ part. Let $\{\phi^a\}_{a=1}^{\dim G}$ be a basis of $\mathbf{g}^*$ dual to the basis $\{X^a\}_{a=1}^{\dim G}$ of $\mathbf{g}$, so that

$$\phi^a(X^b) = \delta^{ab} \tag{2.57}$$

With the above grading, the basis elements ϕ^a have degree 2. We define a linear map

$$D_\mathbf{g} : \Lambda_G^k \mathcal{M} \to \Lambda_G^{k+1} \mathcal{M} \tag{2.58}$$

on the algebra (2.40) by

$$D_\mathbf{g}\phi^a = 0 \quad , \quad D_\mathbf{g}\alpha = (\mathbf{1} \otimes d - \phi^a \otimes i_{V^a})\alpha \quad ; \quad \alpha \in \Lambda\mathcal{M} \tag{2.59}$$

The operator $D_\mathbf{g}$ is called the equivariant exterior derivative and it is a graded derivation. Its definition (2.59) means that its action on forms $\alpha \in S(\mathbf{g}^*) \otimes \Lambda\mathcal{M}$ is

$$(D_\mathbf{g}\alpha)(X) = (d - i_V)(\alpha(X)) \tag{2.60}$$

where $V = c^a V^a$ is the vector field on $\mathcal{M}$ representing the Lie algebra element $X = c^a X^a \in \mathbf{g}$. However, unlike the operators d and i_V, $D_\mathbf{g}$ is not nilpotent in general, but its square is given by the Cartan-Weil identity

$$D_\mathbf{g}^2 = -\phi^a \otimes (di_{V^a} + i_{V^a}d) = -\phi^a \otimes \mathcal{L}_{V^a} \tag{2.61}$$

Thus the operator $D_\mathbf{g}$ is nilpotent on the algebra $\Lambda_G \mathcal{M}$ of equivariant differential forms. The set of G-invariant algebras $\{\Lambda_G^k \mathcal{M}\}_{k \in \mathbb{Z}^+}$ and nilpotent

---
[2] A $\mathbb{Z}$-grading is usually refered to as a 'ghost number' in the physics literature. The equivalence between the 2 notions will become clearer when we deal with path integrals in Chapter 4 – see Appendices A and B for this algebraic correspondence.

derivations $D_{\mathbf{g}}$ thereon defines the G-equivariant complex $\Lambda_G^*(\mathcal{M})$ of the manifold $\mathcal{M}$.

Thus, just as in the last Section, we can proceed to define the cohomology of the operator $D_{\mathbf{g}}$. The space of equivariantly closed forms, i.e. $D_{\mathbf{g}}\alpha = 0$, modulo the space of equivariantly exact forms, i.e. $\alpha = D_{\mathbf{g}}\beta$, is called the G-equivariant cohomology group of $\mathcal{M}$,

$$H_G^k(\mathcal{M}) = \ker D_{\mathbf{g}}|_{\Lambda_G^k \mathcal{M}}/\operatorname{im} D_{\mathbf{g}}|_{\Lambda_G^{k-1}\mathcal{M}} \tag{2.62}$$

With this definition, the cohomology of the operator $D_{\mathbf{g}}$ for a fixed-point free G-action on $\mathcal{M}$ reduces to the DeRham cohomology of the quotient space $\mathcal{M}/G$, as in (2.37). The definition (2.62) of equivariant cohomology is known as the Cartan model [21, 33, 99]. Note that the definition of $D_{\mathbf{g}}$ in (2.60) resembles a gauge-covariant derivative.

We close this Section with a few remarks concerning the above construction. First of all, it follows from these definitions that $H_G^k(\mathcal{M})$ coincides with the ordinary DeRham cohomology of $\mathcal{M}$ if G is the trivial group consisting of only the identity element (i.e. $V \equiv 0$ in the above), and that the G-equivariant cohomology of a point is the algebra of G-invariant polynomials on $\mathbf{g}$, $H_G(\text{pt}) = S(\mathbf{g}^*)^G$, of the given degree. Secondly, if a form $\alpha \in \Lambda_G \mathcal{M}$ is equivariantly exact, $\alpha = D_{\mathbf{g}}\beta$, then its top-form component $\alpha^{(n)} \in \Lambda^n \mathcal{M}$ is exact in the ordinary DeRham sense. This follows because the i_V part of $D_{\mathbf{g}}$ lowers the form-degree by 1 so there is no way to produce a top-form by acting with i_V. Finally, in what follows we shall have occasion to also consider the C^∞ extension $\Lambda_G^\infty \mathcal{M}$ of $\Lambda_G \mathcal{M}$ to include arbitrary G-invariant smooth functions from $\mathbf{g}$ to $\Lambda \mathcal{M}$. In this extension we lose the $\mathbb{Z}$-grading described above, but we are left with a $\mathbb{Z}_2$-grading corresponding to the differential form being of even or odd degree [21] ($\mathbb{Z}_2 = \mathbb{Z}/2\mathbb{Z}$ is the cyclic group of order 2).

2.4 Fiber Bundles and Equivariant Characteristic Classes

The 'bundles' we introduced in Section 2.2 (tangent, cotangent, etc.) are all examples of a more general geometric entity known as a fiber bundle. The geometry and topology of fiber bundles will play an important role in the development of equivariant localization theory, and in this Section we briefly review the essential features that we shall need (see [22, 41, 34] for more detailed discussions). A fiber bundle consists of a quadruple $(E, \mathcal{M}, F, \pi)$, where E is a topological space called the total space of the fiber bundle, $\mathcal{M}$ is a topological space called the base space of the fiber bundle (usually we take $\mathcal{M}$ to be a manifold), F is a topological space called the fiber, and $\pi : E \to \mathcal{M}$ is a surjective continuous map with $\pi^{-1}(x) = F$, $\forall x \in \mathcal{M}$, which is called the projection of the fiber bundle. A fiber bundle is also defined so

2.4 Fiber Bundles and Equivariant Characteristic Classes

that locally it is trivial, i.e. locally the bundle is a product $U \times F$ of an open neighbourhood $U \subset \mathcal{M}$ of the base and the fibers, and $\pi : U \times F \to U$ is the projection onto the first coordinate[3]. In the case of the tangent bundle, for instance, the fibers are $F = T_x \mathbb{R}^n \simeq \mathbb{R}^n$ and the projection map is defined on $T\mathcal{M} \to \mathcal{M}$ by $\pi : T_x \mathcal{M} \to x$. In fact, in this case the fibration spaces are vector spaces, so that the tangent bundle is an example of a vector bundle. If the fiber of a bundle is a Lie group G, then the fiber bundle is called a principal fiber bundle with structure group G. It has a right, smooth and free action of G on the total space E and the base $\mathcal{M}$ which gives a local representation of the group in the fibers. This action also embeds a copy of the group G inside E.

The vector and tensor fields introduced in Section 2.2 should more precisely be defined as 'sections' of the associated bundles, i.e. smooth maps $s : \mathcal{M} \to E$ which take a point $x \in \mathcal{M}$ into the fiber $\pi^{-1}(x)$ over x. Although we shall be a bit abusive in our discussion by considering these as genuine functions on $\mathcal{M}$, for simplicity and ease of notation, it should be kept in mind that it is only locally where these objects admit such a functional interpretation. Thus, for instance, the tangent bundle is $T\mathcal{M} = \{(x, V) : x \in \mathcal{M}, V \in T_x\mathcal{M}\}$ and locally a section of $T\mathcal{M}$ can be written as $(x, V^\mu(x)\frac{\partial}{\partial x^\mu})$.

The set of frames (i.e. bases) on the tangent bundle $T\mathcal{M}$ form a principal $GL(n, \mathbb{R})$-bundle over $\mathcal{M}$, called the frame bundle, whose points are $(x; (e_1, \ldots, e_n))$ where $x \in \mathcal{M}$ and $(e_1, \ldots, e_n)$ is a linear basis for $T_x\mathcal{M}$. If $\mathcal{M}$ has a metric (i.e. a globally defined inner product on each tangent space $T_x\mathcal{M}$), then we can restrict the basis to an orthonormal basis and obtain a principal $O(n-N, N)$-bundle, where $(n-N, N)$ is the signature of the metric. If $\mathcal{M}$ is furthermore orientable, then we can further restrict to an oriented orthonormal basis, defined by the equivalence classes with respect to the equivalence relation $e \equiv M \cdot f$ where $\det M > 0$, and get a principal $SO(n-N, N)$-bundle. When $\mathcal{M}$ is a space-time manifold, the Lie group $SO(n-N, N)$ is then referred to as the local Lorentz group of $\mathcal{M}$ (or of the tangent bundle $T\mathcal{M}$). The associated spin group $\mathrm{spin}(n-N, N)$ is defined as a double cover of the local Lorentz group, i.e. $SO(n-N, N) \simeq \mathrm{spin}(n-N, N)/\mathbb{Z}_2$ (for instance, $\mathrm{spin}(2) = U(1)$ and $\mathrm{spin}(3) = SU(2)$). A principal $\mathrm{spin}(n-N, N)$-bundle over $\mathcal{M}$ whose fibers form a double cover of those of the oriented orthonormal frame bundle is called a spin bundle and is said to define a spin structure on the manifold $\mathcal{M}$.

Conversely, to any principal G-bundle $P \to \mathcal{M}$ there is an associated vector bundle. Let W be the representation space for a representation ρ of G. Since G acts smoothly and freely on the right of P, locally on $P \times W$ there is the G-action $(p, v) \to (p \cdot g^{-1}, \rho(g) \cdot v)$ where $g \in G$. This defines the associated vector bundle $(P \times W)/G \to P/G$ for the representation ρ, which

---
[3] The topology on the total spaces $E = \coprod_{x \in \mathcal{M}} \pi^{-1}(x)$ of fiber bundles is usually taken as the induced topology from the erection of points from $\mathcal{M}$.

has fiber the vector space W. For instance, for the trivial representation ρ, $(P \times W)/G = P/G \times W$. In this way, we can naturally identify sections (e.g. differential forms) on $(P \times W)/G$ with equivariant functions $f : P \to W$, i.e. $f(p \cdot g) = \rho(g^{-1}) \cdot f(p)$. Notice that for a vector bundle $E \to \mathcal{M}$, the bundle of differential forms on $\mathcal{M}$ with values in E is defined as

$$\Lambda^k(\mathcal{M}, E) \equiv \Lambda^k \mathcal{M} \otimes E \qquad (2.63)$$

where tensor products and direct (Whitney) sums of bundles are defined locally by the corresponding algebraic combination of their fiber spaces.

Intuitively, a fiber bundle 'pins' some geometrical or topological object over each point of a manifold $\mathcal{M}$ (e.g. a vector space in the case of a vector bundle). For instance, if $\mathcal{M} = \mathbb{R}^n$, then the tangent bundle $T\mathbb{R}^n$ associates the vector space $W = \mathbb{R}^n$ to each point of $\mathbb{R}^n$. In fact in this case, the tangent bundle is globally given by $T\mathbb{R}^n = \mathbb{R}^n \times W$, the product of its base and fibers. We then say that the bundle is trivial, in that the erecting of points into vector spaces is done without any 'twistings' of the fibers. However, a general vector bundle is only *locally* trivial and globally the fibers can twist in a very complicated fashion. One way to characterize the non-triviality of fiber bundles is through special cohomology classes of the base manifold $\mathcal{M}$ called characteristic classes [104]. A non-trivial characteristic class in this sense signifies the non-triviality of the vector bundle.

As we shall see, all of these notions can be generalized to the case of the equivariant cohomology of a manifold which signifies the non-triviality of an equivariant bundle. First, we define what we mean by an equivariant bundle [19, 20]. We say that a fiber bundle $E \xrightarrow{\pi} \mathcal{M}$ is a G-equivariant bundle if there are G-actions on both E and $\mathcal{M}$ which are compatible with each other in the sense that

$$g \cdot \pi(x) = \pi(g \cdot x) \qquad \forall x \in E \; , \; \forall g \in G \qquad (2.64)$$

This means that the bundle projection π is a G-equivariant map. The action of the group G on differential forms with values in the bundle is generated by the Lie derivatives $\mathcal{L}_{V^a}$.

In the ordinary DeRham case, when there are 'twists' in the given bundle one needs to specify how to 'connect' different fibers. This is done using a connection Γ which is a geometrical object (such as a 1-form) defined over $\mathcal{M}$ with values in E whose action on sections of the bundle specifies their parallel transport along fibers, as required. The parallel transport is generated by the covariant derivative associated with Γ,

$$\nabla = d + \Gamma \qquad (2.65)$$

The derivative operator (2.65) is a linear derivation which associates to each section of the given vector bundle a 1-form in $\Lambda^1(\mathcal{M}, E)$. If V is a tangent vector on $\mathcal{M}$, its action on a section s is defined in local coordinates by

2.4 Fiber Bundles and Equivariant Characteristic Classes

$$(\nabla s)^\alpha(V) = V^\mu(\partial_\mu s^\alpha + (\Gamma_\mu)^\alpha_\beta s^\beta) \tag{2.66}$$

where $(\Gamma_\mu)^\alpha_\beta dx^\mu$ is a 1-form on $\mathcal{M}$ with values in $\text{End}(E)$ (α, β are the vector space indices in E). If $x(t)$ is a path in $\mathcal{M}$, then $\dot{x}(t)$ is a tangent vector along the path and the equation

$$(\nabla s)(\dot{x}(t)) = 0 \tag{2.67}$$

determines parallel transport along the path allowing us to connect different fibers of the bundle. The first order differential equation (2.67) can admit topologically non-trivial solutions if either the space $\mathcal{M}$ is multiply-connected ($H_1(\mathcal{M}; \mathbb{Z}) \neq 0$), or if the connection Γ has non-trivial curvature $F \neq 0$ (see below). The latter condition characterizes the non-triviality of the bundle, so that $F = 0$ on a trivial bundle and the solutions to (2.67) are straight lines. At each point p of the total space E, there is a natural vertical tangent space V_p in the tangent space $T_p E$ along the fiber of E. A choice of connection above is just equivalent to a choice of horizontal component H_p in the tangent space so that $T_p E = V_p \oplus H_p$.

When the bundle $P \to \mathcal{M}$ is a principal G-bundle, we require further that this splitting into horizontal and vertical components be G-equivariant (i.e. $H_p \to H_{p \cdot g}$ under the action of $g \in G$). In this case, $V_p \simeq \mathbf{g}$ so that the connection Γ is a globally-defined 1-form with values in the Lie algebra $\mathbf{g}$, i.e. $\Gamma \in \Lambda^1(P, \mathbf{g})$. The horizontal subspace can then be taken to be $H_p = \ker \Gamma$. Horizontality and G-equivariance mean, respectively, that Γ satisfies

$$i_V \Gamma = X \quad , \quad \mathcal{L}_V \Gamma = -\text{ad}(X)\Gamma \equiv -[X, \Gamma] \tag{2.68}$$

where V is the vector field on P representing $X \in \mathbf{g}$ and $\text{ad}(X)$ denotes the infinitesimal adjoint action of X. This infinitesimal action can be exponentiated to give the finite, total adjoint action of the group on Γ as

$$\Gamma_{p \cdot g} = \text{Ad}(g^{-1})\Gamma = g^{-1}\Gamma g \tag{2.69}$$

Let us briefly look at some examples. If G is a matrix group (e.g. $SU(N)$, $SO(N)$, etc.), then we can regard matrix elements of $g \in G$ as functions on G and $G \to$ pt as a principal G-bundle. Then the unique solution to the connection conditions (2.68) for a Lie group is called the Cartan-Maurer (matrix) 1-form

$$\Xi = g^{-1}dg \tag{2.70}$$

For a general G-bundle $P \xrightarrow{\pi} \mathcal{M}$, in a local trivialization $U \times G \to \pi^{-1}(U)$, the connection must look like

$$\Gamma_{(x,g)} = g^{-1}dg + g^{-1}A_\mu g dx^\mu \tag{2.71}$$

where $A = A_\mu dx^\mu = A^a_\mu X^a \otimes dx^\mu$ is a Lie algebra valued 1-form on $\mathcal{M}$, which is usually refered to in this context as a gauge connection or gauge field. The transformation laws across local patch boundaries on $\mathcal{M}$, labelled by a G-valued transition matrix g (the automorphisms of the bundle), act on gauge connections via pull-back,

$$A \to A^g \equiv g^{-1}Ag + g^{-1}dg \tag{2.72}$$

which is the familiar form of the gauge transformation law in a gauge field theory [22]. Another example is where the bundle is the tangent bundle $T\mathcal{M}$ equipped with a Riemannian metric g. Then Γ is the (affine) Levi-Civita-Christoffel connection $\Gamma^{\lambda}_{\mu\nu}(g)$ associated with g. Its transformation law under local changes of coordinates is determined by (2.72) when the diffeomorphisms of the tangent bundle are regarded as the automorphisms of the associated frame bundle. In this case, the parallel transport equation (2.67) determines the geodesics of the Riemannian manifold $(\mathcal{M}, g)$ (i.e. the "straight lines", or paths of minimal distance, with respect to the curved geometry g).

The failure of the covariant derivative ∇ on a principal G-bundle to define a complex is measured by its curvature

$$F \equiv \nabla^2 = dA + [A \stackrel{\wedge}{,} A]/2 \tag{2.73}$$

The curvature 2-form (2.73) is horizontal,

$$i_V F = 0 \tag{2.74}$$

and in a local trivialization of the bundle, F transforms in the adjoint representation of G,

$$F \to g^{-1}F_{\mu\nu}(x)g dx^{\mu} \wedge dx^{\nu} = (g^{-1}X^a g)F^a_{\mu\nu} \otimes dx^{\mu} \wedge dx^{\nu} \tag{2.75}$$

and so it can be regarded as an element of $\Lambda^2(\mathcal{M}, \text{Ad } P)$, where Ad P is the vector bundle associated to P by the adjoint representation of G. Furthermore, from its definition (2.73), the curvature F obeys the Bianchi identity

$$[\nabla, F] = dF + [A \stackrel{\wedge}{,} F] = 0 \tag{2.76}$$

When the bundle being considered is a G-equivariant bundle, we assume that the covariant derivative (2.65) is G-invariant,

$$[\nabla, \mathcal{L}_{V^a}] = 0 \tag{2.77}$$

Mimicking the equivariant exterior derivative (2.58), we define the equivariant covariant derivative

$$\nabla_{\mathbf{g}} = \mathbf{1} \otimes \nabla - \phi^a \otimes i_{V^a} \tag{2.78}$$

which is considered as an operator on the algebra $\Lambda_G(\mathcal{M}, E)$ of equivariant differential forms on $\mathcal{M}$ with values in E. In a local trivialization $E = U \times W$, $U \subset \mathcal{M}$, this algebra looks like

2.4 Fiber Bundles and Equivariant Characteristic Classes

$$\Lambda_G(U, E) = (S(\mathbf{g}^*) \otimes \Lambda U \otimes W)^G \tag{2.79}$$

Recalling the Cartan-Weil identity (2.61), we define the equivariant curvature of the connection (2.78)

$$F_\mathbf{g} = (\nabla_\mathbf{g})^2 + \phi^a \otimes \mathcal{L}_{V^a} \tag{2.80}$$

which, using (2.77), then satisfies the equivariant Bianchi identity

$$[\nabla_\mathbf{g}, F_\mathbf{g}] = 0 \tag{2.81}$$

Notice that if G is the trivial group, these identities reduce to the usual notions of curvature, etc. discussed above. Expanding out (2.80) explicitly using (2.77) gives

$$F_\mathbf{g} = \mathbf{1} \otimes F + \mu \tag{2.82}$$

where

$$\mu = \phi^a \otimes \mathcal{L}_{V^a} - [\phi^a \otimes i_{V^a}, \mathbf{1} \otimes \nabla] \tag{2.83}$$

is called the moment map of the G-action with respect to the connection ∇. The moment map μ is a G-equivariant extension of the ordinary curvature 2-form (2.73) from a covariantly-closed 2-form, in the sense of (2.76), to an equivariant one in the sense of (2.81).

When evaluated on an element $X \in \mathbf{g}$, represented by a vector field $V \in T\mathcal{M} \otimes W$, we write

$$F_\mathbf{g}(X) = F + \mu(X) \equiv F + \mu_V \equiv F_V \tag{2.84}$$

where

$$\mu_V = \mathcal{L}_V - [i_V, \nabla] \tag{2.85}$$

generates the induced G-action on the fibers of the bundle. The moment map in this way can be regarded locally as a function $\mu : \Lambda U \otimes W \to \mathbf{g}^*$. Furthermore, using the equivariant Bianchi identity (2.81) we see that it obeys the important property

$$\nabla \mu_V = i_V F \tag{2.86}$$

so that a non-trivial moment map produces a non-zero vertical component of the curvature of the connection ∇ (c.f. (2.74)). Later on, we shall encounter 2 important instances of equivariant bundles on $\mathcal{M}$, one associated with a Riemannian structure, and the other with a symplectic structure. In the latter case the moment map is associated with the Hamiltonian of a dynamical system.

Now we are ready to define the notion of an equivariant characteristic class. First, we recall how to construct conventional characteristic classes [104]. Given a Lie group H with Lie algebra $\mathbf{h}$, we say that a real- or complex-valued function P is an invariant polynomial on $\mathbf{h}$ if it is invariant under the natural adjoint action of H on $\mathbf{h}$,

$$P(h^{-1}Yh) = P(Y) \quad \forall h \in H, \forall Y \in \mathbf{h} \tag{2.87}$$

An invariant polynomial P can be used to define characteristic classes on principal fiber bundles with structure group H. If we consider the polynomial P in such a setting as a function on $\mathbf{h}$-valued 2-forms on $\mathcal{M}$, then the H-invariance (2.87) of P implies that

$$dP(\alpha) = rP(\nabla\alpha) \quad , \quad \alpha \in \Lambda^2 \mathcal{M} \otimes \mathbf{h} \tag{2.88}$$

where r is the degree of P. In particular, taking the argument α to be the curvature 2-form $\alpha = F = \nabla^2$ on the principal H-bundle $E \xrightarrow{\pi} \mathcal{M}$ (which is locally an $\mathbf{h}$-valued 2-form), we have

$$dP(F) = 0 \tag{2.89}$$

as a consequence of the Bianchi identity (2.76) for F. This means that $P(F)$ defines a (DeRham) cohomology class of $\mathcal{M}$.

What is particularly remarkable about this cohomology class is that it is independent of the particular connection ∇ used to define the curvature F. To see this, consider the simplest case where the invariant polynomial is just $P(\alpha) = \mathrm{tr}\,\alpha^n$ [4], with tr the invariant Cartan-Killing linear form of the Lie algebra $\mathbf{h}$ (usually the ordinary operator trace). Consider a continuous one-parameter family of connections ∇_t, $t \in \mathbb{R}$, with curvatures $F_t = \nabla_t^2$. Then

$$\frac{d}{dt}F_t = \left[\nabla_t, \frac{d}{dt}\nabla_t\right] \tag{2.90}$$

and applying this to the invariant polynomial $\mathrm{tr}\,F_t^n$ gives

$$\begin{aligned}
\frac{d}{dt}\,\mathrm{tr}\,F_t^n &= n\,\mathrm{tr}\left(\frac{d}{dt}F_t\right)F_t^{n-1} \\
&= n\,\mathrm{tr}\left[\nabla_t, \left(\frac{d}{dt}\nabla_t\right)F_t^{n-1}\right] \\
&= d\,\mathrm{tr}\left(\frac{d}{dt}\nabla_t\right)F_t^{n-1}
\end{aligned} \tag{2.91}$$

where d is the exterior derivative and in the last equality we have applied (2.88). This means that any continuous deformation of the $2n$-form $\mathrm{tr}\,F^n$ changes it by an exact form, so that the cohomology class determined by it is independent of the choice of connection. In general, the invariant polynomial $P(F) \in \Lambda\mathcal{M}$ is called a characteristic class of the given H-bundle.

This notion and construction of characteristic classes can be generalized almost verbatum to the equivariant case [21]. Taking instead the G-equivariant curvature (2.80) as the argument of the G-invariant polynomial P, (2.89) generalizes to

---

[4] Occasionally, for ease of notation, we shall denote exterior products of differential forms as ordinary multiplication. For instance, we define $\alpha^{\wedge n} \equiv \alpha^n$.

2.4 Fiber Bundles and Equivariant Characteristic Classes

$$D_{\mathbf{g}} P(F_{\mathbf{g}}) = r P(\nabla_{\mathbf{g}} F_{\mathbf{g}}) = 0 \tag{2.92}$$

and now the resulting equivariant characteristic classes $P(F_{\mathbf{g}})$ of the given G-equivariant bundle are elements of the algebra $\Lambda_G \mathcal{M}$. These are denoted by $P_{\mathbf{g}}(F)$, or when evaluated on an element $X \in \mathfrak{g}$ with associated vector field $V \in TU \otimes W$, we write

$$P_{\mathbf{g}}(F)(X) = P(F_V) \equiv P_V(F) \tag{2.93}$$

The equivariant cohomology class of $P_{\mathbf{g}}(F)$ is independent of the chosen connection on the bundle. Consequently, on a trivial vector bundle $\mathcal{M} \times W$ we can choose a flat connection[5], $F = 0$, and then

$$P^{\mathcal{M} \times W}(F_{\mathbf{g}})(X) = P^{\mathcal{M} \times W}(\mu_V) = P(\rho(X)) \tag{2.94}$$

where ρ is the representation of G defined by the G-action on the fibers W.

There are 4 equivariant characteristic classes that commonly appear in the localization formalism for topological field theories, all of which are to be understood as elements of the completion $\Lambda_G^\infty \mathcal{M}$. These can all be found and are extensively discussed in [21]. The first one is related to the invariant polynomial $\operatorname{tr} e^\alpha$ and is used for G-equivariant complex vector bundles (i.e. one in which the fibers are vector spaces over the complex numbers $\mathbb{C}$). It is called the G-equivariant Chern character

$$\operatorname{ch}_{\mathbf{g}}(F) = \operatorname{tr} e^{F_{\mathbf{g}}} \tag{2.95}$$

The other 3 are given by determinants of specific polynomials. On a G-equivariant real vector bundle we define the equivariant Dirac $\hat{A}$-genus

$$\hat{A}_{\mathbf{g}}(F) = \sqrt{\det\left[\frac{\frac{1}{2} F_{\mathbf{g}}}{\sinh(\frac{1}{2} F_{\mathbf{g}})}\right]} \tag{2.96}$$

where the inverse of an inhomogeneous polynomial of differential forms is always to be understood in terms of the power series

$$(1+x)^{-1} = \sum_{k=0}^\infty (-1)^k x^k \tag{2.97}$$

On a complex fiber bundle, the complex version of the equivariant $\hat{A}$-genus is the equivariant Todd class

$$\operatorname{td}_{\mathbf{g}}(F) = \det\left[\frac{F_{\mathbf{g}}}{e^{F_{\mathbf{g}}} - 1}\right] \tag{2.98}$$

---

[5] A non-trivial vector bundle can always be considered as a trivial one endowed with a non-trivial curvature $F \neq 0$. This point of view is quite useful in certain applications to topological gauge theories.

When G is the trivial group, these all reduce to the conventional characteristic classes [104] defined by replacing $F_{\mathbf{g}} \to F$ in the above. Just as for the ordinary $\hat{A}$-genus and Todd classes, their equivariant generalizations inherit the multiplicativity property under Whitney sums of bundles,

$$\hat{A}_{\mathbf{g}}^{E \oplus F} = \hat{A}_{\mathbf{g}}^{E} \hat{A}_{\mathbf{g}}^{F} \quad , \quad \mathrm{td}_{\mathbf{g}}^{E \oplus F} = \mathrm{td}_{\mathbf{g}}^{E} \, \mathrm{td}_{\mathbf{g}}^{F} \tag{2.99}$$

Finally, on an orientable real bundle we can define the equivariant generalization of the Euler class,

$$E_{\mathbf{g}}(F) = \mathrm{Pfaff}(F_{\mathbf{g}}) \tag{2.100}$$

where the Pfaffian (or Salam-Mathiews determinant) of a $2N \times 2N$ antisymmetric matrix $M = [M_{ij}]$ is defined as

$$\begin{aligned}\mathrm{Pfaff}\ M &= \epsilon^{i_1 \cdots i_{2N}} M_{i_1 i_2} \cdots M_{i_{2N-1} i_{2N}} \\ &= \frac{1}{2^N N!} \sum_{P \in S_{2N}} \mathrm{sgn}(P) \prod_{k=1}^{N} M_{P(2k-1), P(2k)}\end{aligned} \tag{2.101}$$

with the property that

$$\det M = (\mathrm{Pfaff}\ M)^2 \tag{2.102}$$

The sign of the Pfaffian when written as the square root of the determinant as in (2.102) is chosen so that it is the product of the upper skew-diagonal eigenvalues in a skew-diagonalization of the antisymmetric matrix M. In (2.101), $\epsilon^{i_1 \cdots i_N}$ is the antisymmetric tensor with the convention $\epsilon^{123 \cdots N} = +1$. Pfaffians arise naturally, as we will see, as fermionic determinants from the integration of fermion bilinears in supersymmetric and topological field theories. Transformations which change the orientation of the bundle change the sign of the Pfaffian.

When $F = R$ is the Riemann curvature 2-form associated with the tangent bundle TM (which can be regarded as a principal $SO(2n)$-bundle) of a closed manifold $\mathcal{M}$ of even dimension $2n$, the integral over $\mathcal{M}$ of the ordinary Euler class is the integer

$$\chi(\mathcal{M}) = \frac{(-1)^n}{(4\pi)^n n!} \int_{\mathcal{M}} E(R) \tag{2.103}$$

where

$$\chi(\mathcal{M}) = \sum_{k=0}^{2n} (-1)^k \dim_{\mathbb{R}} H^k(\mathcal{M}; \mathbb{R}) \tag{2.104}$$

is the famous topological invariant called the Euler characteristic of the manifold $\mathcal{M}$. That (2.104) can be written as an integral of a density in (2.103) is a celebrated result of differential topology known as the Gauss-Bonnet theorem. The generalization of (2.103) to the case of an arbitrary vector bundle with Euler class $E(F)$ of a curvature 2-form $F \in \Lambda^2(\mathcal{M}, E)$ is called the Gauss-Bonnet-Chern theorem. In that case, the cohomology groups appearing in the alternating sum (2.104) get replaced by the cohomology groups $H^k(\mathcal{M}; E)$ of the twisted derivative operator $\nabla : \Lambda^k(\mathcal{M}, E) \to \Lambda^{k+1}(\mathcal{M}, E)$.

Similarly, with $F = F_A$ the curvature of a gauge connection A on a principle H-bundle over a $2k$-dimensional manifold $\mathcal{M}$, the integral over $\mathcal{M}$ of the k-th term in the expansion of the conventional version of the Chern class (2.95) (which defines the k-th Chern class) is the number

$$c_k(\mathcal{M}) = \left(-\frac{1}{2\pi i}\right)^k \int_{\mathcal{M}} \operatorname{tr} F_A^k \qquad (2.105)$$

which is a topological invariant of $\mathcal{M}$ called the k-th Chern number of $\mathcal{M}$ (or, more precisely, of the complex vector bundle $(E, \mathcal{M}, W, \pi)$). The Chern number is always an integer for closed orientable manifolds. Thus the equivariant characteristic classes defined above lead to interesting equivariant generalizations of some classical topological invariants. In the next Chapter we will see that their topological invariance in both the ordinary DeRham and the equivariant cases are a consequence of the topological invariance of the integrations there. We shall see later on that they appear in most interesting ways within the formalism of localization formulas and topological field theory functional integration.

2.5 The Equivariant Localization Principle

We now discuss a very interesting property of equivariant cohomology which is the fundamental feature of all localization theorems. It also introduces the fundamental geometric constraint that will be one of the issues of focus in what follows. In most of our applications we will be concerned with the following situation. Let $\mathcal{M}$ be a compact orientable manifold without boundary and let V be a vector field over $\mathcal{M}$ corresponding to some action of the circle group $G = U(1) \sim S^1$ on $\mathcal{M}$. In this case the role of the multiplier $\phi \in S(\mathbf{u}(1)^*)$, which is a linear functional on the 1-dimensional Lie algebra of $U(1)$, will not be important for the discussion that follows. Indeed, we can regard ϕ as just some external parameter in this case and 'localize' algebraically by setting $\phi = -1$. As shown in [9] (see also Appendix B), the operations of evaluating ϕ on Lie algebra elements and the formation of equivariant cohomology commute for abelian group actions, so that all results below will coincide independently of the interpretation of ϕ. In particular, for a free $U(1)$-action on $\mathcal{M}$ we have

$$(S(\mathbf{u}(1)^*) \otimes \Lambda \mathcal{M})^{U(1)} = S(\mathbf{u}(1)^*) \otimes \Lambda(\mathcal{M}/U(1)) \qquad (2.106)$$

so that in this case the multipliers ϕ play no cohomological role and the equivariant cohomology just restricts to the cohomology of the quotient space

$\mathcal{M}/U(1)$. The corresponding equivariant exterior derivative is now denoted as

$$D_{\mathbf{u}(1)} \equiv D_V = d + i_V \tag{2.107}$$

and it is now considered as an operator on the algebra

$$\Lambda_V \mathcal{M} = \{\alpha \in \Lambda \mathcal{M} : \mathcal{L}_V \alpha = 0\} \tag{2.108}$$

It was Atiyah and Bott [9] and Berline and Vergne [19, 20] who first noticed that equivariant cohomology is determined by the fixed point locus of the G-action. In our simplified case here, this is the set

$$\mathcal{M}_V = \{x \in \mathcal{M} : V(x) = 0\} \tag{2.109}$$

This fact is at the very heart of the localization theorems in both the finite dimensional case and in topological field theory, and it is known as the equivariant localization principle. In this Section we shall establish this property in 2 analytic ways. For a more algebraic description of this principle using the Weil algebra and the topological definition of equivariant cohomology, see [9].

Our first argument for localization involves an explicit proof at the level of differential forms. Given an integral $\int_\mathcal{M} \alpha$ over $\mathcal{M}$ of an equivariantly closed differential form $\alpha \in \Lambda_V \mathcal{M}$, $D_V \alpha = 0$, we wish to show that this integral depends only on the fixed-point set (2.109) of the $U(1)$-action on $\mathcal{M}$. To show this, we shall explicitly construct a differential form λ on $\mathcal{M} - \mathcal{M}_V$ satisfying $D_V \lambda = \alpha$. This is just the equivariant version of the Poincaré lemma. Thus the form α is equivariantly exact away from the zero locus $\mathcal{M}_V$, and we recall that this implies that the top-form component of α is exact. Since integration over $\mathcal{M}$ picks up the top-form component of any differential form, and since $\partial \mathcal{M} = \emptyset$ by hypothesis here, it follows from Stokes' theorem (2.35) that the integral $\int_\mathcal{M} \alpha$ only receives contributions from an arbitrarily small neighbourhood of $\mathcal{M}_V$ in $\mathcal{M}$, i.e. the integral 'localizes' onto the smaller subspace $\mathcal{M}_V$ of $\mathcal{M}$.

To construct λ, we need to impose the following geometric restriction on the manifold $\mathcal{M}$. We assume that $\mathcal{M}$ has a globally-defined $U(1)$-invariant Riemannian structure on it, which means that it admits a globally-defined metric tensor

$$g = \frac{1}{2} g_{\mu\nu}(x) dx^\mu \otimes dx^\nu \tag{2.110}$$

which is invariant under the $U(1)$-action generated by V, i.e. for which

$$\mathcal{L}_V g = 0 \tag{2.111}$$

or in local coordinates on $\mathcal{M}$,

$$g_{\mu\lambda} \partial_\nu V^\lambda + g_{\nu\lambda} \partial_\mu V^\lambda + V^\lambda \partial_\lambda g_{\mu\nu} = 0 \tag{2.112}$$

2.5 The Equivariant Localization Principle

Alternatively, this Lie derivative constraint can be written as

$$g_{\nu\lambda}\nabla_\mu V^\lambda + g_{\mu\lambda}\nabla_\nu V^\lambda = 0 \tag{2.113}$$

where ∇ is the covariant derivative (2.65) constructed from the Levi-Civita-Christoffel connection

$$\Gamma^\lambda_{\mu\nu} = \frac{1}{2}g^{\lambda\rho}\left(\partial_\mu g_{\nu\rho} + \partial_\nu g_{\mu\rho} - \partial_\rho g_{\mu\nu}\right) \tag{2.114}$$

associated with g on the tangent bundle $T\mathcal{M}$. Here $g^{\mu\nu}$ is the matrix inverse of $g_{\mu\nu}$ and the covariant derivative acts on the vector field V in the usual way as

$$\nabla_\mu V^\nu = \partial_\mu V^\nu + \Gamma^\nu_{\lambda\mu}V^\lambda \tag{2.115}$$

with a plus sign for $(0,k)$-tensors and a minus sign for $(k,0)$-tensors in front of Γ, as in (2.115). Notice that by construction the Levi-Civita-Christoffel connection is torsion-free, $\Gamma^\lambda_{\mu\nu} = \Gamma^\lambda_{\nu\mu}$, and it is compatible with the the metric g, $\nabla_\lambda g_{\mu\nu} = 0$, which together mean that ∇ preserves the inner product in the fibers of the tangent bundle.

The equivalent equations (2.111)–(2.113) are called the Killing equations and in this case we say that V is a Killing vector field of the metric g. Since the map $V \to \mathcal{L}_V$ is linear, the space of Killing vectors of a Riemannian manifold $(\mathcal{M}, g)$ generate the Lie algebra of a Lie group acting on $\mathcal{M}$ by diffeomorphisms which is called the isometry group of $(\mathcal{M}, g)$. We shall describe this group in detail in Chapters 5 and 6. The Killing equations here are assumed to hold globally over the entire manifold $\mathcal{M}$. If both $\mathcal{M}$ and G are compact, then such a metric can always be obtained from an arbitrary Riemannian metric h on $\mathcal{M}$ by averaging h over the group manifold of G in its (G-invariant) Haar measure, i.e. $g = \int_G D\tilde{g}\,(\tilde{g}\cdot h)$. However, we shall have occasion to also consider more general vector field flows which aren't necessarily closed or when the manifold $\mathcal{M}$ isn't compact, as are the cases in many physical applications. In such cases the Lie derivative constraint (2.111) is a very stringent one on the manifold. This feature of the localization formalism, that the manifold admit a globally defined metric with the property (2.111) whose components $g_{\mu\nu}(x) = g_{\nu\mu}(x)$ are globally-defined C^∞ functions on $\mathcal{M}$, is the crux of all finite- and infinite-dimensional localization formulas and will be analysed in detail later on in this Book. For now, we content ourselves with assuming that such a metric tensor has been constructed.

Any metric tensor defines a duality between vector fields and differential 1-forms, i.e. we can consider the metric tensor (2.110) as a map

$$g : T\mathcal{M} \to T^*\mathcal{M} \tag{2.116}$$

which takes a vector field V into its metric dual 1-form

$$\beta \equiv g(V, \cdot) = g_{\mu\nu}(x)V^\nu(x)dx^\mu \tag{2.117}$$

Non-degeneracy, $\det g(x) \neq 0$, $\forall x \in \mathcal{M}$, of the metric tensor implies that this defines an isomorphism between the tangent and cotangent bundles of $\mathcal{M}$. The 1-form β satisfies

$$D_V^2 \beta = \mathcal{L}_V \beta = 0 \tag{2.118}$$

since $\mathcal{L}_V V = 0$ and V is a Killing vector of g. This means that β is an equivariant differential 1-form. Furthermore, we have

$$D_V \beta = K_V + \Omega_V \tag{2.119}$$

where K_V is the globally-defined C^∞-function

$$K_V = g(V, V) = g_{\mu\nu}(x) V^\mu(x) V^\nu(x) \tag{2.120}$$

and

$$\Omega_V = d\beta = dg(V, \cdot) \tag{2.121}$$

is the 2-form with local components

$$(\Omega_V)_{\mu\nu} = g_{\mu\lambda} \nabla_\nu V^\lambda - g_{\nu\lambda} \nabla_\mu V^\lambda \tag{2.122}$$

Consequently, away from zero locus $\mathcal{M}_V$ of the vector field V, the 0-form part K_V of $D_V \beta$ is non-zero and hence $D_V \beta$ is invertible on $\mathcal{M} - \mathcal{M}_V$. Again we understand here the inverse of an inhomogeneous differential form with non-zero scalar term in analogy with the formula (2.97).

We can now define an inhomogenous differential form by

$$\xi = \beta (D_V \beta)^{-1} \tag{2.123}$$

on $\mathcal{M} - \mathcal{M}_V$, which satisfies $D_V \xi = 1$ and $\mathcal{L}_V \xi = 0$ owing to the equivariance (2.118) of β. Thus we can define an equivariant differential form $\lambda = \xi \alpha$, and since α is equivariantly closed it follows that

$$\alpha = 1 \cdot \alpha = (D_V \xi) \alpha = D_V(\xi \alpha) \tag{2.124}$$

Thus, as claimed above, any equivariantly closed form is equivariantly exact away from $\mathcal{M}_V$, and in particular the top-form component of an equivariantly closed form is exact away from $\mathcal{M}_V$. This establishes the equivariant localization property mentioned above.

The other argument we wish to present here for equivariant localization is less explicit and involves cohomological arguments. First, consider an ordinary closed form ω, $d\omega = 0$. For any other differential form λ, we have

$$\int_\mathcal{M} (\omega + d\lambda) = \int_\mathcal{M} \omega \tag{2.125}$$

by Stokes' theorem (2.35) since $\partial \mathcal{M} = \emptyset$. This means that the integral $\int_\mathcal{M} \omega$ of a closed form ω depends only on the cohomology class defined by ω, not

2.5 The Equivariant Localization Principle

on the particular representative. Since the map $\omega \to \int_{\mathcal{M}} \omega$ in general defines a linear map on $\Lambda^k \mathcal{M} \to \Lambda^{n-k}(\text{pt}) = \delta^{nk}\mathbb{R}$, it follows that this map descends to a map on $H^n(\mathcal{M}; \mathbb{R}) \to H^0(\text{pt}; \mathbb{R}) = \mathbb{R}$. The same is true for equivariant integration. Since, for a general G-action on $\mathcal{M}$, integration of a differential form picks up the top-form component which for an equivariantly exact form is exact, for any equivariantly-closed differential form α we can again invoke Stokes' theorem to deduce

$$\int_{\mathcal{M}} (\alpha + D_\mathbf{g} \lambda) = \int_{\mathcal{M}} \alpha \qquad (2.126)$$

so that the integral of an equivariantly closed form depends only on the equivariant cohomology class defined by it, and not on the particular representative. Note, however, that equivariant integration for general Lie groups G takes a far richer form. In analogy with the DeRham case above, the integration of equivariant differential forms defines a map on $H_G(\mathcal{M}) \to H_G(\text{pt}) = S(\mathbf{g}^*)^G$. This we define by

$$\left(\int_{\mathcal{M}} \alpha\right)(X) = \int_{\mathcal{M}} \alpha(X) \quad , \quad X \in \mathbf{g} \qquad (2.127)$$

with integration over the $\Lambda\mathcal{M}$ part of α in the ordinary DeRham sense. Later on, we shall also consider the dual Lie algebra elements ϕ^a in a more 'dynamical' situation where they are a more integral part of the cohomological description above. We shall see then how this definition of integration should be accordingly modified. In any case, the arguments below which lead to the equivariant localization principle generalize immediately to the non-abelian case as well.

Given that the integral $\int_{\mathcal{M}} \alpha$ depends only on the equivariant cohomology class defined by α, we can choose a particular representative of the cohomology class making the localization manifest. Taking the equivariant differential form β defined in (2.117), we consider the integral

$$\mathcal{Z}(s) = \int_{\mathcal{M}} \alpha\, e^{-sD_V \beta} \qquad (2.128)$$

viewed as a function of $s \in \mathbb{R}^+$. We assume that (2.128) is a regular function of $s \in \mathbb{R}^+$ and that its $s \to 0$ and $s \to \infty$ limits exist. Its $s \to 0$ limit is the integral of interest, $\int_{\mathcal{M}} \alpha$, while from the identities (2.119) and (2.120) we see that the integrand of (2.128) is an increasingly sharply Gaussian peaked form around $\mathcal{M}_V \subset \mathcal{M}$ as $s \to \infty$. The crucial point here is that the equivariant differential form which is the integrand of (2.128) is equivariantly cohomologous to α for all $s \in \mathbb{R}^+$. This can be seen by applying Stokes' theorem to get

38   2. Equivariant Cohomology and the Localization Principle

$$\begin{aligned}\frac{d}{ds}\mathcal{Z}(s) &= -\int_{\mathcal{M}} \alpha(D_V\beta)\, e^{-sD_V\beta} \\ &= -\int_{\mathcal{M}} \{D_V(\alpha\beta\, e^{-sD_V\beta}) + \beta D_V(\alpha\, e^{-sD_V\beta})\} \\ &= s\int_{\mathcal{M}} \alpha\beta(\mathcal{L}_V\beta)\, e^{-sD_V\beta} = 0\end{aligned}$$ (2.129)

where we have used the fact that α is equivariantly closed and the equivariance property (2.118) of β. Therefore the integral (2.128) is independent of the parameter $s \in \mathbb{R}^+$, and so its $s \to 0$ and $s \to \infty$ limits coincide. Hence, we may evaluate the integral of interest as

$$\int_{\mathcal{M}} \alpha = \lim_{s \to \infty} \int_{\mathcal{M}} \alpha\, e^{-sD_V\beta}$$ (2.130)

which establishes the localization of $\int_{\mathcal{M}} \alpha$ to $\mathcal{M}_V$.

It should be pointed out though that there is nothing particularly unique about the choice of β in (2.130). Indeed, the same steps leading to (2.130) can be carried out for an *arbitrary* equivariant differential form β, i.e. any one with the property (2.118). In this general case, the localization of $\int_{\mathcal{M}} \alpha$ is onto the subspace of $\mathcal{M}$ which is the support for the non-trivial equivariant cohomology of α, i.e. $\int_{\mathcal{M}} \alpha$ localizes to the points where $D_V\beta = 0$. Different choices of representatives β for the equivariant cohomology classes then lead to potentially different localizations other than the one onto $\mathcal{M}_V$. This would lead to seemingly different expressions for the integral in (2.130), but of course these must all coincide in some way. In principle this argument for localization could also therefore work without the assumption that V is a Killing vector for some metric on $\mathcal{M}$, but it appears difficult to make general statements in that case. Nonetheless, as everything at the end will be equivariantly closed by our general arguments above, it is possible to reduce the resulting expressions further to $\mathcal{M}_V$ by applying the above localization arguments once more, now to the localized expression. We shall examine situations in which V isn't necessarily a Killing vector field in Chapter 7.

2.6 The Berline-Vergne Theorem

The first general localization formula using only the general equivariant cohomological arguments presented in the last Section was derived by Berline and Vergne [19, 20]. This formula, as well as some of the arguments leading to the equivariant localization principle, have since been established in many different contexts suitable to other finite dimensional applications and also to path integrals [9, 11, 21, 23, 24]. The proof presented here introduces a

2.6 The Berline-Vergne Theorem

method that will generalize to functional integrals. For now, we assume the fixed-point set $\mathcal{M}_V$ of the $U(1)$-action on $\mathcal{M}$ consists of discrete isolated points, i.e. $\mathcal{M}_V$ is a submanifold of $\mathcal{M}$ of codimension $n = \dim \mathcal{M}$ [6]. We shall discuss the generalization to the case where $\mathcal{M}_V$ has non-zero dimension later on. If we assume that $\mathcal{M}$ is compact, then $\mathcal{M}_V$ is a finite set of points.

We wish to evaluate explicitly the right-hand side of the localization formula (2.130). To do this, we introduce an alternative way of evaluating integrals over differential forms which is based on a more algebraic description of the exterior bundle of $\mathcal{M}$. We introduce a set of nilpotent anticommuting (fermionic) variables η^μ, $\mu = 1, \ldots, n$,

$$\eta^\mu \eta^\nu = -\eta^\nu \eta^\mu \tag{2.131}$$

which generate the exterior algebra $\Lambda \mathcal{M}$. The variables η^μ are to be identified with the local basis vectors dx^μ of $\Lambda^1 \mathcal{M} = T^* \mathcal{M}$ with the exterior product of differential forms replaced by the ordinary product of the η^μ variables with the algebra (2.131). The k-th exterior power $\Lambda^k \mathcal{M}$ is then generated by the products $\eta^{\mu_1} \cdots \eta^{\mu_k}$ and this definition turns $\Lambda \mathcal{M}$ into a graded Grassmann algebra with the generators η^μ having grading 1. For instance, suppose the differential form α is the sum

$$\alpha = \alpha^{(0)} + \alpha^{(1)} + \ldots + \alpha^{(n)} \quad , \quad \alpha^{(k)} \in \Lambda^k \mathcal{M} \tag{2.132}$$

with $\alpha^{(k)}$ the k-form component of α and $\alpha^{(0)}(x)$ its 0-form component which is a C^∞-function on $\mathcal{M}$. The k-form component of α for $k > 0$ then has the form

$$\alpha^{(k)}(x, \eta) = \alpha^{(k)}_{\mu_1 \cdots \mu_k}(x) \eta^{\mu_1} \cdots \eta^{\mu_k} \quad , \quad k > 0 \tag{2.133}$$

and from this point of view differential forms are functions $\alpha(x, \eta)$ on the exterior bundle which is now the $2n$-dimensional supermanifold $\mathcal{M} \otimes \Lambda \mathcal{M}$ with local coordinates (x, η).

The integration of a differential form is now defined by introducing the Berezin rules for integrating Grassmann variables [17],

$$\int d\eta^\mu \, \eta^\mu = 1 \quad , \quad \int d\eta^\mu \, 1 = 0 \tag{2.134}$$

Since the η^μ's are nilpotent, any function of them is a polynomial in η^μ and consequently the rules (2.134) unambiguously define the integral of any function of the anticommuting variables η^μ. For instance, it is easily verified that with this definition of integration we have[7]

---

[6] We shall assume here that n is even. This restriction is by no means necessary but it will allow us to shorten some of the arguments in this Section.

[7] If we introduce a second independent set $\{\bar\eta^\mu\}$ of Grassmann variables, then the formula (2.135) generlizes to arbitrary (not necessarily even) dimensions n as

$$\int d^n\eta \; e^{\frac{1}{2}\eta^\mu M_{\mu\nu}\eta^\nu} = \text{Pfaff } M \qquad (2.135)$$

where $d^n\eta \equiv d\eta^n \, d\eta^{n-1} \cdots d\eta^1$. Note that under a local change of basis $\eta^\mu \to A^\mu_\nu \eta^\nu$ of the Grassmann algebra the antisymmetry property (2.131) and the Berezin rules (2.134) imply that $\int d^n\eta \to \det A \int d^n\eta$. It follows from this that the Berezin integral in (2.135) is invariant under similarity transformations. (2.135) is the fermionic analog of the Gaussian integration formula (1.2). The differentiation of Grassmann variables, for which the integration in (2.134) is the antiderivative thereof, is defined by the anticommutator

$$\left[\frac{\partial}{\partial \eta^\mu}, \eta^\nu\right]_+ = \delta^\nu_\mu \qquad (2.136)$$

With these definitions, the integration by parts formula $\int d\eta^\mu \, \frac{d}{d\eta^\mu} f(\eta^\mu) = 0$ always holds, since $\int d\eta^\mu \, f(\eta^\mu) = \frac{d}{d\eta^\mu} f(\eta^\mu)$.

Given these definitions, we can now alternatively write the integral of any differential form over $\mathcal{M}$ as an integral over the cotangent bundle $\mathcal{M} \otimes \Lambda^1 \mathcal{M}$. Thus given the localization formula (2.130) with the 1-form β in (2.117) and the identities (2.119)–(2.122), we have

$$\int_{\mathcal{M}} \alpha = \lim_{s \to \infty} \int_{\mathcal{M} \otimes \Lambda^1 \mathcal{M}} d^n x \, d^n \eta \, \alpha(x, \eta) \qquad (2.137)$$
$$\times \exp\left(-s g_{\mu\nu}(x) V^\mu(x) V^\nu(x) - \frac{s}{2}(\Omega_V)_{\mu\nu}(x)\eta^\mu\eta^\nu\right)$$

where the measure $d^n x \, d^n \eta$ on $\mathcal{M} \otimes \Lambda^1 \mathcal{M}$ is coordinate-independent because the measures $d^n x \equiv dx^1 \wedge \cdots \wedge dx^n$ and $d^n \eta$ transform inversely to each other. To evaluate the large-s limit of (2.137), we use the delta-function representations

$$\delta(V) = \lim_{s \to \infty} \left(\frac{s}{\pi}\right)^{n/2} \sqrt{\det g} \; e^{-s g_{\mu\nu} V^\mu V^\nu} \qquad (2.138)$$

$$\delta(\eta) = \lim_{s \to \infty} (-s)^{-n/2} \frac{1}{\text{Pfaff } \Omega_V} \; e^{-\frac{s}{2}(\Omega_V)_{\mu\nu}\eta^\mu\eta^\nu} \qquad (2.139)$$

as can be seen directly from the respective integrations in local coordinates on $\mathcal{M}$ and $\Lambda^1 \mathcal{M}$. Notice that from the Killing equations (2.113), the matrix $(\Omega_V)_{\mu\nu}$ is given by

$$(\Omega_V)_{\mu\nu} = 2g_{\mu\lambda} \nabla_\nu V^\lambda \qquad (2.140)$$

Thus using (2.138) and (2.139) we can write (2.137) as

$$\int \prod_{\mu=1}^n d\bar{\eta}^\mu \, d\eta^\mu \; e^{-\bar{\eta}^\mu M_{\mu\nu}\eta^\nu} = \det M$$

$$\int_{\mathcal{M}} \alpha = (-\pi)^{n/2} \int_{\mathcal{M} \otimes \Lambda^1 \mathcal{M}} d^n x \, d^n \eta \, \alpha(x, \eta) \frac{\text{Pfaff } \Omega_V(x)}{\sqrt{\det g(x)}} \delta(V(x)) \delta(\eta) \quad (2.141)$$

where we note the cancellation of the factors of $s^{n/2}$ between (2.138) and (2.139). The integration over $\Lambda^1 \mathcal{M}$ in (2.141) kills off all k-form components of the form α except its C^∞-function part $\alpha^{(0)}(x) \equiv \alpha(x, 0)$, while the integration over $\mathcal{M}$ localizes it onto a sum over the points in $\mathcal{M}_V$. This yields

$$\int_{\mathcal{M}} \alpha = (-\pi)^{n/2} \sum_{p \in \mathcal{M}_V} \frac{\alpha^{(0)}(p)}{|\det dV(p)|} \frac{\text{Pfaff } \Omega_V(p)}{\sqrt{\det g(p)}} \quad (2.142)$$

where the factor $|\det dV(p)|$ comes from the Jacobian of the coordinate transformation $x \to V(x)$ used to transform $\delta(V(x))$ to a sum of delta-functions $\sum_{p \in \mathcal{M}_V} \delta(x - p)$ localizing onto the zero locus $\mathcal{M}_V$. Substituting in the identity (2.140) and noting that at a point $p \in \mathcal{M}_V$ we have $\nabla V(p) = dV(p)$, the expression (2.142) reduces to

$$\int_{\mathcal{M}} \alpha = (-2\pi)^{n/2} \sum_{p \in \mathcal{M}_V} \frac{\alpha^{(0)}(p)}{\text{Pfaff } dV(p)} \quad (2.143)$$

where we emphasize the manner in which the dependence of orientation in the Pfaffian has been transfered from the numerator to the denominator in going from (2.142) to (2.143). This is the (non-degenerate form of the) Berline-Vergne integration formula, and it is our first example of what we shall call a localization formula. It reduces the original integral over the n-dimensional space $\mathcal{M}$ to a sum over a discrete set of points in $\mathcal{M}$ and it is valid for any equivariantly-closed differential form α on a manifold with a globally-defined circle action (and Riemannian metric for which the associated diffeomorphism generator is a Killing vector). In general, the localization formulas we shall encounter will always at least reduce the dimensionality of the integration of interest. This will be particularly important for path integrals, where we shall see that localization theory can be used to reduce complicated infinite-dimensional integrals to finite sums or finite-dimensional integrals.

We close this Chapter by noting the appearance of the operator in the denominator of the expression (2.143). For each $p \in \mathcal{M}_V$, it is readily seen that the operator $dV(p)$ appearing in the argument of the Pfaffian in (2.143) is just the invertible linear transformation $L_V(p)$ induced by the Lie derivative acting on the tangent spaces $T_p \mathcal{M}$, i.e. by the induced infinitesimal group action on the tangent bundle (see (2.55)). Explicitly, this operator is defined on vector fields $W = W^\mu(x) \frac{\partial}{\partial x^\mu}|_{x=p} \in T_p \mathcal{M}$ by

$$L_V(p) W = \partial_\nu V^\mu(p) W^\nu(p) \frac{\partial}{\partial x^\mu} \bigg|_{x=p} \quad (2.144)$$

Note however that $dV(p)$ is not covariant in general and so this is only true right on the tangent space $T_p \mathcal{M}$ and not in general on the entire tangent bundle TM. A linear transformation on the whole of TM can only be induced from the Lie derivative by introducing a (metric or non-metric) connection $\Gamma^\lambda_{\mu\nu}$ of TM and inducing an operator from ∇V, as in the matrix (2.140). We shall return to this point later on in a more specific setting.

3. Finite-Dimensional Localization Theory for Dynamical Systems

We shall now proceed to study a certain class of integrals that can be considered to be toy models for the functional integrals that we are ultimately interested in. The advantage of these models is that they are finite-dimensional and therefore rigorous mathematical theorems concerning their behaviour can be formulated. In the infinite-dimensional cases, although the techniques used will be standard methods of supersymmetry and topological field theory, a lot of rigor is lost due the ill-definedness of infinite-dimensional manifolds and functional integrals. A lot can therefore be learned by looking closely at some finite-dimensional cases.

We shall be interested in certain oscillatory integrals $\int_{\mathcal{M}} d\mu\ e^{iTH}$ representing the Fourier-Laplace transform of some smooth measure $d\mu$ on a manifold $\mathcal{M}$ in terms of a smooth function H. The common method of evaluating such integrals is the stationary phase approximation which expresses the fact that for large-T the main contributions to the integral come from the critical points of H. The main result of this Chapter is the Duistermaat-Heckman theorem [39] which provides a criterion for the stationary phase approximation to an oscillatory integral to be exact. Although this theorem was originally discovered within the context of symplectic geometry, it turns out to have its most natural explanation in the setting of equivariant cohomology and equivariant characteristic classes [9],[19]–[21]. The Duistermaat-Heckman theorem, and its various extensions that we shall discuss towards the end of this Chapter, are precisely those which originally motivated the localization theory of path integrals.

For physical applications, we shall be primarily interested in a special class of differentiable manifolds known as 'symplectic' manifolds. As we shall see in this Chapter, the application of the equivariant cohomological ideas to these manifolds leads quite nicely to the notion of a Hamiltonian from a mathematical perspective, as well as some standard ideas in the geometrical theory of classical integrability. Furthermore, the configuration space of a topological field theory is typically an (infinite-dimensional) symplectic manifold (or phase space) [22] and we shall therefore restrict our attention for the remainder of this Book to the localization theory for oscillatory integrals over symplectic manifolds.

3.1 Symplectic Geometry

Symplectic geometry is the natural mathematical setting for the geometrical formulation of classical mechanics and the study of classical integrability [1, 6]. It also has applications in other branches of physics, such as geometrical optics [65]. In elementary classical mechanics [55], one is introduced to the Hamiltonian formalism of classical dynamics as follows. For a dynamical system defined on some manifold $\mathcal{M}$ (usually $\mathbb{R}^n$) with coordinates $(q^1, \ldots, q^n)$, we introduce the canonical momenta p_μ conjugate to each variable q^μ from the Lagrangian of the system and then the Hamiltonian $H(p,q)$ is obtained by a Legendre transformation of the Lagrangian. In this way one has a description of the dynamics on the $2n$-dimensional space of the (p,q) variables which is called the phase space of the dynamical system. With this construction the phase space is the cotangent bundle $\mathcal{M} \otimes \Lambda^1 \mathcal{M}$ of the configuration manifold $\mathcal{M}$. The equations of motion can be represented through the time evolution of the phase space coordinates by Hamilton's equations. For most elementary dynamical systems, this description is sufficient. However, there are relatively few examples of mechanical systems whose equations of motion can be solved by quadratures and it is desirable to seek other more general formulations of this elementary situation in the hopes of being able to formulate rigorous theorems about when a classical mechanical system has solvable equations of motion, or is 'integrable'. Furthermore, the above notion of a 'phase space' is very local and is strictly speaking only globally valid when the phase space is $\mathbb{R}^{2n}$, a rather restrictive class of systems. Motivated by the search for more non-trivial integrable models in both classical and quantum physics, theoretical physicists have turned to the general theory of symplectic geometry which encompasses the above local description in a coordinate-free way suitable to the methods of modern differential geometry. In this Section we shall review the basic ideas of symplectic geometry and how these descriptions tie in with the more familiar ones of elementary classical mechanics.

A symplectic manifold is a differentiable manifold $\mathcal{M}$ of even dimension $2n$ together with a globally-defined non-degenerate closed 2-form

$$\omega = \frac{1}{2}\omega_{\mu\nu}(x)dx^\mu \wedge dx^\nu \tag{3.1}$$

called the symplectic form of $\mathcal{M}$. By closed we mean as usual that

$$d\omega = 0 \tag{3.2}$$

or in local coordinates

$$\partial_\mu \omega_{\nu\lambda} + \partial_\nu \omega_{\lambda\mu} + \partial_\lambda \omega_{\mu\nu} = 0 \tag{3.3}$$

Thus ω defines a DeRham cohomology class in $H^2(\mathcal{M};\mathbb{R})$. By non-degenerate we mean that the components $\omega_{\mu\nu}(x)$ of ω define an invertible $2n \times 2n$ antisymmetric matrix globally on the manifold $\mathcal{M}$, i.e.

$$\det \omega(x) \neq 0 \quad \forall x \in \mathcal{M} \tag{3.4}$$

The manifold $\mathcal{M}$ together with its symplectic form ω defines the phase space of a dynamical system, as we shall see below.

Since ω is closed, it follows from the Poincaré lemma that locally there exists a 1-form

$$\theta = \theta_\mu(x) dx^\mu \tag{3.5}$$

such that

$$\omega = d\theta \tag{3.6}$$

or in local coordinates

$$\omega_{\mu\nu} = \partial_\mu \theta_\nu - \partial_\nu \theta_\mu \tag{3.7}$$

The locally-defined 1-form θ is called the symplectic potential or canonical 1-form of $\mathcal{M}$. When ω is generated globally as above by a symplectic potential θ it is said to be integrable. Diffeomorphisms of $\mathcal{M}$ that leave the symplectic 2-form invariant are called canonical or symplectic transformations. These are determined by C^∞-maps that act on the symplectic potential as

$$\theta \xrightarrow{F} \theta_F = \theta + dF \tag{3.8}$$

or in local coordinates

$$\theta_\mu(x) \xrightarrow{F} \theta_{F,\mu}(x) = \theta_\mu(x) + \partial_\mu F(x) \tag{3.9}$$

so that by nilpotency of the exterior derivative it follows that ω is invariant under such transformations,

$$\omega = d\theta \xrightarrow{F} \omega_F = d\theta_F \equiv \omega \tag{3.10}$$

The function $F(x)$ is called the generating function of the canonical transformation.

The symplectic 2-form determines a bilinear function $\{\cdot,\cdot\}_\omega : \Lambda^0 \mathcal{M} \otimes \Lambda^0 \mathcal{M} \to \Lambda^0 \mathcal{M}$ called the Poisson bracket. It is defined by

$$\{f,g\}_\omega = \omega^{-1}(df, dg) \quad , \quad f,g \in \Lambda^0 \mathcal{M} \tag{3.11}$$

or in local coordinates

$$\{f,g\}_\omega = \omega^{\mu\nu}(x) \partial_\mu f(x) \partial_\nu g(x) \tag{3.12}$$

where $\omega^{\mu\nu}$ is the matrix inverse of $\omega_{\mu\nu}$. Note that the local coordinate functions themselves have Poisson bracket

$$\{x^\mu, x^\nu\}_\omega = \omega^{\mu\nu}(x) \tag{3.13}$$

The Poisson bracket is anti-symmetric,

$$\{f,g\}_\omega = -\{g,f\}_\omega \quad , \tag{3.14}$$

it obeys the Leibniz property

$$\{f,gh\}_\omega = g\{f,h\}_\omega + h\{f,g\}_\omega \tag{3.15}$$

and it satisfies the Jacobi identity

$$\{f,\{g,h\}_\omega\}_\omega + \{g,\{h,f\}_\omega\}_\omega + \{h,\{f,g\}_\omega\}_\omega = 0 \tag{3.16}$$

This latter property follows from the fact (3.3) that ω is closed. These 3 properties of the Poisson bracket mean that it defines a Lie bracket. Thus the Poisson bracket makes the space of C^∞-functions on $\mathcal{M}$ into a Lie algebra which we call the Poisson algebra of $(\mathcal{M},\omega)$.

The connection with the elementary formulation of classical mechanics discussed above is given by a result known as Darboux's theorem [65], which states that this connection is always possible locally. More precisely, Darboux's theorem states that locally there exists a system of coordinates $(p_\mu, q^\mu)_{\mu=1}^n$ on $\mathcal{M}$ in which the symplectic 2-form looks like

$$\omega = dp_\mu \wedge dq^\mu \tag{3.17}$$

so that they have Poisson brackets

$$\{p_\mu, p_\nu\}_\omega = \{q^\mu, q^\nu\}_\omega = 0 \quad , \quad \{p_\mu, q^\nu\}_\omega = \delta_\mu^\nu \tag{3.18}$$

These coordinates are called canonical or Darboux coordinates on $\mathcal{M}$ and from (3.18) we see that they can be identified with the usual canonical momentum and position variables on the phase space $\mathcal{M}$ [55]. In these coordinates the symplectic potential is

$$\theta = p_\mu dq^\mu \tag{3.19}$$

and the transformation (3.8) becomes

$$\theta = p_\mu dq^\mu \xrightarrow{F} \theta + dF = \theta_F = P_\mu dQ^\mu \tag{3.20}$$

where $(P_\mu, Q^\mu)_{\mu=1}^n$ are also canonical coordinates according to (3.10). It follows that

$$p_\mu dq^\mu - P_\mu dQ^\mu = dF \tag{3.21}$$

where both (p_μ, q^μ) and (P_μ, Q^μ) are canonical momentum and position variables on $\mathcal{M}$. (3.21) is the usual form of a canonical transformation determined by the generating function F [55].

Smooth real-valued functions H on $\mathcal{M}$ (i.e. elements of $\Lambda^0\mathcal{M}$) will be called classical observables. Exterior products of ω with itself determine non-trivial closed $2k$-forms on $\mathcal{M}$ (i.e. non-zero cohomology classes $[\omega^k] \in H^{2k}(\mathcal{M};\mathbb{R}))$. In particular, the $2n$-form

$$d\mu_L = \omega^n/n! = \sqrt{\det \omega(x)}\, d^{2n}x \tag{3.22}$$

defines a natural volume element on $\mathcal{M}$ which is invariant under canonical transformations. It is called the Liouville measure, and in the local Darboux coordinates (3.17) it becomes the familiar phase space measure [55]

$$(-1)^{n(n-1)/2}\omega^n/n! = dp_1 \wedge \cdots \wedge dp_n \wedge dq^1 \wedge \cdots \wedge dq^n \tag{3.23}$$

3.2 Equivariant Cohomology on Symplectic Manifolds

In this Section we shall specialize the discussion of Chapter 2 to the case where the differentiable manifold $\mathcal{M}$ is a symplectic manifold of dimension $2n$. Consider the action of some connected Lie group G on $\mathcal{M}$ generated by the vector fields V^a with the commutator algebra (2.47). We assume that the action of G on $\mathcal{M}$ is symplectic so that it preserves the symplectic structure,

$$\mathcal{L}_{V^a}\omega = 0 \tag{3.24}$$

or in other words G acts on $\mathcal{M}$ by symplectic transformations. Since ω is closed this means that
$$di_{V^a}\omega = 0 \tag{3.25}$$

Let $L \to \mathcal{M}$ be a complex line bundle with connection 1-form the symplectic potential θ. If θ also satisfies
$$\mathcal{L}_{V^a}\theta = 0 \tag{3.26}$$

then the associated covariant derivative $\nabla = d + \theta$ is G-invariant, and according to the general discussion of Section 2.4 this defines a G-equivariant bundle. By definition (see Section 2.4) the structure group of this symplectic line bundle acts by canonical transformations. As such, ω represents the first Chern class of this $U(1)$-bundle, and, if $\mathcal{M}$ is closed, it defines an integer cohomology class in $H^2(\mathcal{M}; \mathbb{Z})$ (as the Chern numbers generated by ω are then integers).

The associated moment map $H : \mathcal{M} \to \mathbf{g}^*$ evaluated on a Lie algebra element $X \in \mathbf{g}$ with associated vector field V is called the Hamiltonian corresponding to V,

$$H_V = \mathcal{L}_V - [i_V, \nabla] = i_V\theta = V^\mu\theta_\mu \tag{3.27}$$

From (3.6) and (3.26) it then follows that

$$dH_V = -i_V\omega \tag{3.28}$$

or equivalently this follows from the general property (2.86) of the moment map since ω is the curvature of the connection θ. In local coordinates, this last equation reads

… 3. Finite-Dimensional Localization Theory for Dynamical Systems

$$\partial_\mu H_V(x) = V^\nu(x)\omega_{\mu\nu}(x) \tag{3.29}$$

In particular, the components H^a of the moment map

$$H = \phi^a \otimes H^a \tag{3.30}$$

satisfy

$$dH^a = -i_{V^a}\omega \tag{3.31}$$

Comparing with the symplecticity condition (3.25) on the group action, we see that this is equivalent to the statement that the closed 1-forms $i_{V^a}\omega$ are exact. If $H^1(\mathcal{M};\mathbb{R}) = 0$ this is certainly true, but in the following we will want to consider multiply connected phase spaces as well. We therefore impose this exactness requirement from the onset on the action of G on $\mathcal{M}$, i.e. the equivariance requirement (3.26) on the symplectic potential θ. When such a Hamiltonian function exists as a globally-defined C^∞-map on $\mathcal{M}$, we shall say that the group action is Hamiltonian. A vector field V which satisfies (3.28) is said to be the Hamiltonian vector field associated with H_V, and we shall call the triple $(\mathcal{M}, \omega, H_V)$, i.e. a symplectic manifold with a Hamiltonian G-action on it, a Hamiltonian system or a dynamical system.

The integral curves (2.46) defined by the flows (or time-evolution) of a Hamiltonian vector field V as in (3.29) define the Hamilton equations of motion

$$\dot{x}^\mu(t) = \omega^{\mu\nu}(x(t))\partial_\nu H_V(x(t)) = \{x^\mu, H_V\}_\omega \tag{3.32}$$

The Poisson bracket of the Hamiltonian with any other function f determines the (infinitesimal) variation (or time-evolution) of f along the classical trajectories of the dynamical system (compare with (2.55)),

$$\{f, H_V\}_\omega = \mathcal{L}_V f = \frac{d}{dt}f(x(t))\Big|_{t=0} \tag{3.33}$$

In the canonical coordinates defined by (3.17) the equations (3.32) read

$$\dot{q}^\mu = \frac{\partial H}{\partial p_\mu} \quad, \quad \dot{p}_\mu = -\frac{\partial H}{\partial q^\mu} \tag{3.34}$$

which are the usual form of the Hamilton equations of motion encountered in elementary classical mechanics [55]. Thus we see that the above formalisms for symplectic geometry encompass all of the usual ideas of classical Hamiltonian mechanics in a general, coordinate-independent setting.

The equivariant curvature of the above defined equivariant bundle is given by the equivariant extension of the symplectic 2-form,

$$\omega_\mathbf{g} = \mathbf{1} \otimes \omega + \phi^a \otimes H^a \tag{3.35}$$

and evaluated on $X \in \mathbf{g}$ we have

$$(D_\mathbf{g}\omega_\mathbf{g})(X) = (d - i_V)(\omega + H_V) = 0 \tag{3.36}$$

3.2 Equivariant Cohomology on Symplectic Manifolds

which is equivalent to the definition (3.28) of the Hamiltonian vector field V. In fact, the extension (3.35) is the *unique* equivariant extension of the symplectic 2-form ω [129], i.e. the unique extension of ω from a closed 2-form to an equivariantly-closed one. Thus, we see that finding an equivariantly-closed extension of ω is equivalent to finding a moment map for the G-action. If ω defines an integer cohomology class $[\omega] \in H^2(\mathcal{M}; \mathbb{Z})$, then the line bundle $L \to \mathcal{M}$ introduced above can be thought of as the prequantum line bundle of geometric quantization [172], the natural geometric framework (in terms of symplectic geometry) for the coordinate independent formulation of quantum mechanics. Within this framework, the equivariant curvature 2-form $\omega_V = \omega_{\mathbf{g}}(X)$ above is refered to as the prequantum operator. We shall say more about some of the general ideas of geometric quantization later on. Notice that if (3.26) (or (3.27)) does hold, then θ is also the equivariant symplectic potential for the equivariant extension (3.35), i.e. $\omega_{\mathbf{g}} = D_{\mathbf{g}}\theta$.

From (3.31) it follows that the Poisson algebra of the Hamiltonians H^a is given by

$$\{H^a, H^b\}_\omega = \omega(V^a, V^b) = \omega_{\mu\nu} V^{a,\mu} V^{b,\nu} = V^{a,\mu} \partial_\mu H^b = \mathcal{L}_{V^a} H^b = -\mathcal{L}_{V^b} H^a \tag{3.37}$$

From the Jacobi identity (3.16) it follows that the map $H^a \to V^a$ is a homomorphism of the Lie algebras $(\Lambda^0 \mathcal{M}, \{\cdot, \cdot\}_\omega) \to (T\mathcal{M}, [\cdot, \cdot])$ since

$$V^{\{H^a, H^b\}_\omega} = [V^a, V^b] \tag{3.38}$$

However, the inverse of this map does not necessarily define a homomorphism. The Hamiltonian function which corresponds to the commutator of 2 group generators may differ from the Poisson bracket of the pertinent Hamiltonian functions as

$$\{H^a, H^b\}_\omega = f^{abc} H^c + c^{ab} \tag{3.39}$$

where $c^{ab} \equiv c(X^a, X^b) = -c^{ba}$ is a 2-cocycle in the Lie algebra cohomology of G [77] (see Appendix A), i.e.

$$c([X_1, X_2], X_3) + c([X_2, X_3], X_1) + c([X_3, X_1], X_2) = 0 \quad \forall X_1, X_2, X_3 \in \mathbf{g} \tag{3.40}$$

If $H^2(G) = 0$ then we can set $c^{ab} = 0$ and the map $X^a \to H^a$ determines a homomorphism between the Lie algebra $\mathbf{g}$ and the Poisson algebra of C^∞-functions on $\mathcal{M}$.

The appearance of the 2-cocycle c^{ab} in (3.39) is in fact related to the possible non-invariance of the symplectic potential under G (c.f. eq. (3.26)). From the symplecticity (3.24) of the group action and (3.31) it follows that

$$\mathcal{L}_{V^a} \theta = (i_{V^a} d + d i_{V^a}) \theta = dg^a \tag{3.41}$$

locally in a neighbourhood $\mathcal{N}$ in $\mathcal{M}$ wherein $\omega = d\theta$ and $V^a \neq 0$. Here the locally-defined linear functions $g^a \equiv g(X^a) = -H^a + i_{V^a} \theta$ obey the consistency condition

50   3. Finite-Dimensional Localization Theory for Dynamical Systems

$$\{H(X_1), g(X_2)\}_\omega - \{H(X_2), g(X_1)\}_\omega = g([X_1, X_2]) \qquad \forall X_1, X_2 \in \mathbf{g} \quad (3.42)$$

which follows from (3.39). However, if there exists a locally-defined function f such that

$$g^a = \{H^a, f\}_\omega \quad , \quad a = 1, \ldots, \dim G \quad (3.43)$$

then we can remove the functions g^a by the canonical transformation $\theta \to \theta_f = \theta + df$ so that the symplectic potential θ_f is G-invariant. Indeed, the 1-form θ_f obeys

$$\mathcal{L}_{V^a} \theta_f = 0 \quad (3.44)$$

which implies that in the neighbourhood $\mathcal{N}$,

$$i_{V^a} \theta_f = H^a + C \quad (3.45)$$

where C is a constant. This constant is irrelevant here because we can introduce a function K in $\mathcal{N}$ such that

$$\{H^a, K\}_\omega = V^{a,\mu} \partial_\mu K = 1 \quad (3.46)$$

and defining $F = f + CK$ we find

$$i_{V^a} \theta_F = H^a \quad (3.47)$$

However, notice that the G-invariance (3.47) of the symplectic potential in general holds only locally in $\mathcal{M}$, and furthermore the canonical transformation $\theta \to \theta_f$ above does not remove the functions g^a for the entire Lie algebra $\mathbf{g}$, but only for a closed subalgebra of $\mathbf{g}$ which depends on the function f and on the phase space $\mathcal{M}$ where G acts [65, 123]. In this subspace, the symplectic potential is G-invariant and the identity (3.27) relating the Hamiltonians to the symplectic potential by $H^a = i_{V^a}\theta$ holds (so that θ is a local solution to the equivariant Poincaré lemma). In general though, on the entire Lie algebra $\mathbf{g}$, defining $h^a = -i_{V^a} dF$ in the above we have

$$i_{V^a} \theta = H^a + h^a \quad (3.48)$$

and then the Poisson bracket (3.37) implies that the 2-cocycle appearing in (3.39) is given by

$$c^{ab} = f^{abc} h^c - \mathcal{L}_{V^a} h^b + \mathcal{L}_{V^b} h^a \quad (3.49)$$

Thus it is only when $c^{ab} = 0$ for all a, b that the G-action of the vector fields V^a lifts isomorphically to the Poisson action of the corresponding Hamiltonians H^a on $\mathcal{M}$. Notice that this is certainly true on the Cartan subalgebra of the Lie algebra $\mathbf{g}$ (i.e. its maximal commuting subalgebra), since $H^2(U(1)) = H^2(S^1) = 0$. We shall see in Chapter 4 that the dynamical systems for which the equivariance condition (3.27) holds determine a very special class of quantum theories.

3.3 Stationary-Phase Approximation and the Duistermaat-Heckman Theorem

We now start examining localization formulas for a specific class of phase space integrals. We shall concentrate for the time being on the case of an abelian circle action on the manifold $\mathcal{M}$, as we did in Section 2.6. We shall also assume that the Hamiltonian H defined as in the last Section is a Morse function. This means that the critical points p of the Hamiltonian, defined by $dH(p) = 0$, are isolated and the Hessian matrix of H,

$$\mathcal{H}(x) = \left[\frac{\partial^2 H(x)}{\partial x^\mu \partial x^\nu} \right] \tag{3.50}$$

at each critical point p is a non-degenerate matrix, i.e.

$$\det \mathcal{H}(p) \neq 0 \tag{3.51}$$

The Hamiltonian vector field V is defined by (3.29) and it represents the action of some 1-parameter group on the phase space $\mathcal{M}$. We shall assume here that the orbits (2.46) of V generate the circle group $U(1) \sim S^1$. Later on we shall consider more general cases. Notice that the critical points of H coincide with zero locus $\mathcal{M}_V$ of the vector field V.

There is an important quantity of physical interest for the statistical mechanics of a classical dynamical system called the partition function. It is constructed as follows. Each point x of the phase space $\mathcal{M}$ represents a classical state of the dynamical system which in canonical coordinates is specified by its configuration q and its momentum p. The energy of this state is determined by the Hamiltonian H of the dynamical system which as usual is its energy function. According to the general principles of classical statistical mechanics [144] the partition function is built by attaching to each point $x \in \mathcal{M}$ the Boltzmann weight $e^{iTH(x)}$ and 'summing' them over all states of the system. Here the parameter iT is 'physically' to be identified with $-\beta/k_B$ where k_B is Boltzmann's constant and β is the inverse temperature. However, for mathematical ease in the following, we shall assume that the parameter T is real. In the canonical position and momentum coordinates we would just simply integrate up the Boltzmann weights. However, we would like to obtain a quantity which is invariant under transformations which preserve the (symplectic) volume of the phase space $\mathcal{M}$ (i.e. those which preserve the classical equations of motion (3.32) and hence the density of classical states), and so we integrate using the Liouville measure (3.22) to obtain a canonically invariant quantity. This defines the classical partition function of the dynamical system,

$$Z(T) = \int_\mathcal{M} \frac{\omega^n}{n!} e^{iTH} = \int_\mathcal{M} d^{2n}x \, \sqrt{\det \omega(x)} \, e^{iTH(x)} \tag{3.52}$$

The partition function determines all the usual thermodynamic quantities of the dynamical system [144], such as its free energies and specific heats, as well as all statistical averages in the canonical ensemble of the classical system.

However, it is very seldom that one can actually obtain an exact closed form for the partition function (3.52) as the integrals involved are usually rather complicated. But there is a method of approximating the integral (3.52), which is very familiar to both physicists and mathematicians, called the stationary-phase approximation [65, 72, 172]. This method is often employed when one encounters oscillatory integrals such as (3.52) to obtain an idea of its behaviour, at least for large T. It works as follows. Notice that for $T \to \infty$ the integrand of $Z(T)$ oscillates very rapidly and begins to damp to 0. The integral therefore has an asymptotic expansion in powers of $1/T$. The larger T gets the more the integrand tends to localize around its stationary values wherever the function $H(x)$ has extrema (equivalently where $dH(p) = 0$)[1]. To evaluate these contributions, we expand both H and the Liouville density in (3.52) in a neighbourhood U_p about each critical point $p \in \mathcal{M}_V$ in a Taylor series, where as usual integration in U_p can be thought of as integration in the more familiar $\mathbb{R}^{2n}$. We expand the exponential of all derivative terms in H of order higher than 2 in the exponential power series, and in this way we are left with an infinite series of Gaussian moment integrals with Gaussian weight determined by the bilinear form defined by the Hessian matrix (3.50) of H at p. The lowest order contribution is just the normalization of the Gaussian (see (1.2)), while the k-th order moments are down by powers of $1/T^k$ compared to the leading term. Carrying out these Gaussian integrations, taking into careful account the signature of the Hessian at each point, and summing over all points $p \in \mathcal{M}_V$, in this way we obtain the standard lowest-order stationary-phase approximation to the integral (3.52),

$$Z(T) = \left(\frac{2\pi i}{T}\right)^n \sum_{p \in \mathcal{M}_V} (-i)^{\lambda(p)} \, e^{iTH(p)} \sqrt{\frac{\det \omega(p)}{\det \mathcal{H}(p)}} + \mathcal{O}(1/T^{n+1}) \qquad (3.53)$$

where $\lambda(p)$ is the Morse index of the critical point p, defined as the number of negative eigenvalues in a diagonalization of the symmetric Hessian matrix of H at p. We shall always ignore a possible regular function of T in the large-T expansion (3.53). The higher-order terms in (3.53) are found from the higher-moment Gaussian integrals [157] and they will be analysed in Chapter 7. For now, we concern ourselves only with the lowest-order term in the stationary-phase series of (3.52).

---

[1] Usually one argues that the phase will concentrate around the points where H is minimized (the ground state) since this should be the dominant contribution for $T \to \infty$. However, the localization is properly determined by *all* points where $dH(p) = 0$ since the contribution from other extrema turn out to be of the same order of magnitude as those from the minima [72].

The field of equivariant localization theory was essentially born in 1982 when Duistermaat and Heckman [39] found a general class of Hamiltonian systems for which the leading-order of the stationary-phase approximation gives the *exact* result for the partition function (3.52) (i.e. for which the $\mathcal{O}(1/T^{n+1})$ correction terms in (3.53) all vanish). Roughly speaking, the Duistermaat-Heckman theorem goes as follows. Let $\mathcal{M}$ be a compact symplectic manifold. Suppose that the vector field V defined by (3.29) generates the global Hamiltonian action of a torus group $\mathcal{T} = (S^1)^m$ on $\mathcal{M}$ (where we shall usually assume that $m = 1$ for simplicity). Since the critical point set of the Hamiltonian H coincides with the fixed-point set $\mathcal{M}_V$ of the $\mathcal{T}$-action on $\mathcal{M}$ we can apply the equivariant Darboux theorem to the Hamiltonian system at hand [65]. This generalization of Darboux's theorem tells us that not only can we find a local canonical system of coordinates in a neighbourhood of each critical point in which the symplectic 2-form looks like (3.17), but these coordinates can further be chosen so that the origin $p_\mu = q^\mu = 0$ of the coordinate neighbourhood represents the fixed point p of the given compact group action on $\mathcal{M}$. This means that in these canonical coordinates the torus action is (locally) linear and has the form of n canonical rotation generators (rotations in each (p_μ, q^μ) plane) [39]

$$V = \sum_{\mu=1}^{n} \frac{\lambda_\mu(p)}{i} \left(p_\mu \frac{\partial}{\partial q^\mu} - q^\mu \frac{\partial}{\partial p_\mu} \right) \quad , \quad p \in \mathcal{M}_V \qquad (3.54)$$

where $\lambda_\mu(p)$ are weights that will be specified shortly. From the Hamilton equations (3.29) it follows that the Hamiltonian near each critical point p can be written in the quadratic form

$$H(x) = H(p) + \sum_{\mu=1}^{n} \frac{i\lambda_\mu(p)}{2} \left(p_\mu^2 + q_\mu^2 \right) \qquad (3.55)$$

In these coordinates the flows determined by the Hamilton equations of motion (3.34) are the circles $p_\mu(t), q^\mu(t) \sim e^{i\lambda_\mu t}$ about the critical points, which gives an explicit representation of the Hamiltonian $\mathcal{T}$-action locally on $\mathcal{M}$ and the group action preserves the Darboux coordinate neighbourhood. Thus each neighbourhood integration above is purely Gaussian and so all higher-order terms in the stationary-phase evaluation of (3.52) vanish and the partition function is given exactly by the leading term in (3.53) of its stationary-phase series[2]. This theorem therefore has the potential of supplying a large class of dynamical systems whose partition function (and hence all thermodynamic and statistical observables) can be evaluated exactly.

Atiyah and Bott [9] pointed out that the basic principle underlying the Duistermaat-Heckman theorem is not that of stationary-phase, but rather of

---

[2] Of course the proof is completed by showing that there is no regular function of T contributing in this case to (3.53) – for details we refer to [39].

the more general localization properties of equivariant cohomology that we discussed in the last Chapter. Suppose that the Hamiltonian vector field V generates a global, symplectic circle action on the phase space $\mathcal{M}$. Suppose further that $\mathcal{M}$ admits a globally defined Riemannian structure for which V is Killing vector, as in Section 2.5. Recall from the last Section that the symplecticity of the circle action implies that $\omega + H$ is the equivariant extension of the symplectic 2-form ω, i.e. $D_V(\omega + H) = 0$. Since integration over the $2n$-dimensional manifold $\mathcal{M}$ picks up the $2n$-degree component of any differential form, it follows that the partition function (3.52) can be written as

$$Z(T) = \int_{\mathcal{M}} \alpha \qquad (3.56)$$

where α is the inhomogeneous differential form

$$\alpha = \frac{1}{(iT)^n} e^{iT(H+\omega)} = \frac{1}{(iT)^n} e^{iTH} \sum_{k=0}^{n} \frac{(iT)^k}{k!} \omega^k \qquad (3.57)$$

whose $2k$-form component is $\alpha^{(2k)} = e^{iTH} \omega^k / (iT)^{n-k} k!$. Since $H + \omega$ is equivariantly closed, it follows that $D_V \alpha = 0$. Thus we can apply the Berline-Vergne localization formula (2.143) to the integral (3.56) to get

$$Z(T) = \left(\frac{2\pi i}{T}\right)^n \sum_{p \in \mathcal{M}_V} \frac{e^{iTH(p)}}{\text{Pfaff } dV(p)} \qquad (3.58)$$

In the case at hand the denominator of (3.58) at a critical point p is found from the Hamilton equations (3.29) which give

$$dV(p) = \omega^{-1}(p) \mathcal{H}(p) \qquad (3.59)$$

and so we see how the determinant factors appear in the formula (3.53). However, we have to remember that the Pfaffian also encodes a specific choice of sign when taking the square root determinant. The sign of the Pfaffian Pfaff $dV(p)$ can be determined by examining it in the equivariant Darboux coordinates above in which the matrix $\omega(p)$ is skew-diagonal with skew-eigenvalues 1 and the Hessian $\mathcal{H}(p)$ which comes from (3.55) is diagonal with eigenvalues $i\lambda_\mu(p)$ each of multiplicity 2. It follows that in these coordinates the matrix $dV(p)$ is skew-diagonal with skew-eigenvalues $i\lambda_\mu(p)$. Introducing the eta-invariant $\eta(\mathcal{H}(p))$ of $\mathcal{H}(p)$, defined as the difference between the number of positive and negative eigenvalues of the Hessian of H at p, i.e. its spectral asymmetry, we find

$$\eta(\mathcal{H}(p)) = 2 \sum_{\mu=1}^{n} \text{sgn } i\lambda_\mu(p) \qquad (3.60)$$

which is related to the Morse index of H at p by

$$\eta(\mathcal{H}(p)) = 2n - 2\lambda(p) \tag{3.61}$$

Using the identity $\pm 1 = e^{i\frac{\pi}{2}(\pm 1 - 1)}$ it follows that

$$\operatorname{sgn} \operatorname{Pfaff} dV(p) = \prod_{\mu=1}^{n} \operatorname{sgn} i\lambda_\mu(p) = e^{i\frac{\pi}{2}(\frac{1}{2}\eta(\mathcal{H}(p))-n)} = e^{-i\frac{\pi}{2}\lambda(p)} = (-i)^{\lambda(p)} \tag{3.62}$$

and so substituting (3.59) and (3.62) into (3.58) we arrive finally at the Duistermaat-Heckman integration formula

$$Z(T) = \left(\frac{2\pi i}{T}\right)^n \sum_{p \in \mathcal{M}_V} (-i)^{\lambda(p)} e^{iTH(p)} \sqrt{\frac{\det \omega(p)}{\det \mathcal{H}(p)}} \tag{3.63}$$

Recall from Section 2.6 that $dV(p)$ is associated with the anti-self-adjoint linear operator $L_V(p)$ which generates the infinitesimal circle (or torus) action on the tangent space $T_p\mathcal{M}$. From the above it then follows that the complex numbers $\lambda_\mu(p)$ introduced in (3.54) are just the weights (i.e. eigenvalues of the Cartan generators) of the complex linear representation of the circle (or torus) action in the tangent space at p and the determinant factors from (3.58) appear in terms of them as the products

$$e(p) = (-1)^{\lambda(p)/2} \prod_{\mu=1}^{n} \lambda_\mu(p) \tag{3.64}$$

as if each unstable mode contributes a factor of i to the integral for $Z(T)$ above. In fact, the Pfaffian Pfaff $dV(p)$ which appears in (3.58) is none other than the equivariant Euler characteristic class $E_V(\mathcal{N}_p) = \operatorname{Pfaff} dV(p)$ (see (2.100)) of the normal bundle $\mathcal{N}_p$ in $\mathcal{M}$ of each critical point $p \in \mathcal{M}_V$. The normal bundle is defined as the bundle of points normal to the directions of the critical point set $\mathcal{M}_V$, so that in a neighbourhood near $\mathcal{M}_V$ we can write the local coordinates as $x = p + p_\perp$ with $p \in \mathcal{M}_V$ and $p_\perp \in \mathcal{N}_p$. By its construction, Pfaff $dV(p)$ is taken over $\mathcal{N}_p$ (see Section 2.6). Thus the terms in (3.58) define an equivariant cohomology class in $H^{2n}_{U(1)}(\mathcal{M})$. From (3.54) it follows that the induced circle action on $\mathcal{N}_p$ is through non-trivial irreducible representations and we can therefore decompose the normal bundle at $p \in \mathcal{M}_V$ into a direct (Whitney) sum of 2-plane bundles with respect to this group action,

$$\mathcal{N}_p = \bigoplus_{\mu=1}^{n} N_p^{(\mu)} \tag{3.65}$$

(3.54) then implies that the equivariant Euler class of $N_p^{(\mu)}$ is simply $E_V(N_p^{(\mu)}) = i\lambda_\mu(p)/2$. Taking into account the proper orientation of $\mathcal{N}_p$ induced by the Hamiltonian vector field near $x = p$ and the Liouville measure, and using the

56   3. Finite-Dimensional Localization Theory for Dynamical Systems

multiplicativity of the Euler class under Whitney sums of bundles [21], we find that the equivariant Euler class of the normal bundle at p is

$$E_V(\mathcal{N}_p) = \prod_{\mu=1}^{n} E_V(N_p^{(\mu)}) \equiv e(p) \tag{3.66}$$

which is just the weight product (3.64). Thus, for Hamiltonians that generate circle actions, the 1-loop contribution to the classical partition function (i.e. the Duistermaat-Heckman formula in the form (3.58)) describes the equivariant cohomology of the phase space with respect to the Hamiltonian circle action on $\mathcal{M}$. The particular value of the Duistermaat-Heckman formula depends on the equivariant cohomology group $H_{U(1)}^{2n}(\mathcal{M})$ of the manifold $\mathcal{M}$. All the localization formulas we shall derive in this Book will be represented by equivariant characteristic classes, so that the partition functions of the physical systems we consider provide representations for the equivariant cohomology of the phase space $\mathcal{M}$. This is a consequence of the cohomological localization principle of Section 2.5.

The remarkable cohomological derivation of the Duistermaat-Heckman formula above, which followed from the quite general principles of equivariant cohomology of the last Chapter, suggests that one could try to develop more general types of localization formulas from these general principles in the hopes of being able to generate more general types of integration formulas for the classical partition function. Moreover, given the localization criteria of the last Chapter this has the possibility of expanding the set of dynamical systems whose partition functions are exactly solvable. We stress again that the crucial step in this cohomological derivation is the assumption that the Hamiltonian flows of the dynamical system globally generate isometries of a metric g on $\mathcal{M}$, i.e. the Hamiltonian vector field V is a global Killing vector of g (equivalently, as we will see, for $\mathcal{M}$ compact, the classical flows $x(t)$ trace out a torus $\mathcal{T}$ in $\mathcal{M}$). This geometric condition and a classification of the dynamical systems for which these localization constraints do hold true will be one of our main topics in what follows. The extensions and applications of the Duistermaat-Heckman localization formula and the general formalism of equivariant cohomology for dynamical systems will be the focus of the remainder of this Chapter.

3.4 Morse Theory and Kirwan's Theorem

There is a very interesting and useful connection between the Duistermaat-Heckman theorem and the Morse theory determined by the non-degenerate Hamiltonian H. Morse theory relates the structure of the critical points of a Morse function H to the topology of the manifold $\mathcal{M}$ on which it is defined. We very briefly now review some of the basic ideas in Morse theory (see [111] for a comprehensive introduction). Given a Morse function H as above, we define its Morse series

$$M_H(t) = \sum_{p \in \mathcal{M}_V} t^{\lambda(p)} \qquad (3.67)$$

which is a finite sum because the non-degeneracy of H implies that its critical points are all discrete and the compactness of $\mathcal{M}$ implies that the critical point set $\mathcal{M}_V$ is finite. The topology of the manifold $\mathcal{M}$ now enters the problem through the Poincaré series of $\mathcal{M}$, which is defined by

$$P_\mathcal{M}(t; \mathbb{F}) = \sum_{k=0}^{2n} \dim_\mathbb{F} H^k(\mathcal{M}; \mathbb{F}) t^k \qquad (3.68)$$

where $\mathbb{F}$ is some algebraic field (usually $\mathbb{R}$ or $\mathbb{C}$). The fundamental result of Morse theory is the inequality

$$M_H(t) \geq P_\mathcal{M}(t; \mathbb{F}) \qquad (3.69)$$

for all fields $\mathbb{F}$. If equality holds in (3.69) for all fields $\mathbb{F}$, then we say that H is a perfect Morse function. The inequality (3.69) leads to various relations between the critical points of H and the topology of $\mathcal{M}$. These are called the Morse inequalities, and the only feature of them that we shall really need in the following is the fact that the number of critical points of H of a given Morse index $k \geq 0$ is always at least the number $\dim_\mathbb{R} H^k(\mathcal{M}; \mathbb{R})$. This puts a severe restriction on the types of non-degenerate functions that can exist as C^∞-maps on a manifold of a given topology.

Another interesting relation is obtained when we set $t = -1$ in the Morse and Poincaré series. In the former series we get

$$M_H(-1) = \sum_{p \in \mathcal{M}_V} \text{sgn} \det \mathcal{H}(p) \qquad (3.70)$$

while (2.104) shows that in the latter series the result is the Euler characteristic $\chi(\mathcal{M})$ of $\mathcal{M}$. That these 2 quantities are equal is known as the Poincaré-Hopf theorem, and employing further the Gauss-Bonnet-Chern theorem (2.103) we find

$$\sum_{p \in \mathcal{M}_V} \text{sgn} \det \mathcal{H}(p) = \frac{(-1)^n}{(4\pi)^n n!} \int_\mathcal{M} E(R) \qquad (3.71)$$

with $E(R)$ the Euler class constructed from a Riemann curvature 2-form R on $\mathcal{M}$. This relation gives a very interesting connection between the structure of the critical point set of a non-degenerate function and the topology and geometry of the phase space $\mathcal{M}$. We remark that one can also define equivariant versions of the Morse and Poincaré series using the topological definition of equivariant cohomology [111] which is suitable to the equivariant cohomological ideas that we formulated earlier on. These equivariant generalizations which localize topological integrals such as (3.71) onto the zero locus of a vector field is the basis of the Mathai-Quillen formalism and its application to the construction of topological field theories [27, 29, 34, 81, 99, 121]. We shall discuss some of these ideas in Chapter 8.

In regards to the Duistermaat-Heckman theorem, there is a very interesting Morse theoretical result due to Kirwan [88]. Kirwan showed that the only Morse functions for which the stationary phase approximation can be exact are those which have only even Morse indices $\lambda(p)$. This theorem includes the cases where the Duistermaat-Heckman integration formula is exact, and under the assumptions of the Duistermaat-Heckman theorem it is a consequence of the circle action (see the previous Section). However, this result is even stronger – it means that when one constructs the full stationary-phase series as described in the last Section [134], if that series converges uniformly in $1/T$ to the exact partition function $Z(T)$, then the Morse index of every critical point of H must be even. From the Morse inequalities mentioned above this furthermore gives a relation between equivariant localization and the topology of the phase space of interest – if the manifold $\mathcal{M}$ has non-trivial cohomology groups of odd dimension, then the stationary phase series diverges for any Morse function defined on $\mathcal{M}$ and in particular the Duistermaat-Heckman localization formula for such phase spaces can never give the exact result for $Z(T)$. In this way, Kirwan's theorem rules out a large number of dynamical systems for which the stationary phase approximation could be exact in terms of the topology of the underlying phase space where the dynamical system lives. Moreover, an application of the Morse lacunary principle [111] shows that, when the stationary-phase approximation is exact so that H has only even Morse indices, H is in fact a perfect Morse function and its Morse inequalities become equalities. We shall not go into the rather straightforward proof of Kirwan's theorem here, but refer to [88] for the details. In the following we can therefore use Kirwan's theorem as an initial test using the topology of the phase space to determine which dynamical systems will localize in the sense of the Duistermaat-Heckman theorem. In Chapter 7 we shall see the direct connection between the higher order terms in the saddle-point series for the partition function and Kirwan's theorem, and more generally the geometry and topology of the manifold $\mathcal{M}$.

3.5 Examples: The Height Function of a Riemann Surface

We now present some concrete examples of the equivariant localization formalism presented above. One of the most common examples in both Morse theory and localization theory is the dynamical system whose phase space is a compact Riemann surface Σ^g of genus g (i.e. a closed surface with g 'handles') and whose Hamiltonian h_{Σ^g} is the height function on Σ^g

3.5 Examples: The Height Function of a Riemann Surface

[29, 85, 111, 153, 157]. For instance, we have already encountered the case of the Riemann sphere $\Sigma^0 = S^2$ in Section 2.1 with the height function h_{Σ^0} given by (2.1). The symplectic 2-form is the usual volume form

$$\omega_{\Sigma^0} = d\cos\theta \wedge d\phi \qquad (3.72)$$

induced by the Euclidean metric $g_{\mu\nu} = \delta_{\mu\nu}$ of $\mathbb{R}^3$ from the embedding of S^2 in 3-dimensional space. The partition function

$$Z_{\Sigma^0}(T) = \int_{\Sigma^0} \omega_{\Sigma^0} \, e^{iTh_{\Sigma^0}} \qquad (3.73)$$

is given by the expression (2.3) which is precisely the value anticipated from the Duistermaat-Heckman theorem. The relative minus sign in the last line of (2.3) comes from the fact that the Morse index of the maximum $\theta = \pi$ is 2 while that of $\theta = 0$ is 0. The vector field generating the compact group action on S^2 associated with rigid rotations of the sphere is $V = \frac{\partial}{\partial\phi}$, and the corresponding moment map is just h_{Σ^0}.

The Poincaré series of the 2-sphere is[3]

$$P_{S^2}(t; \mathbb{F}) = \sum_{k=0}^{2} \dim_{\mathbb{F}} H^k(S^2; \mathbb{F}) t^k = 1 + t^2 \qquad (3.74)$$

which coincides with the Morse series (3.67) for the height function h_{Σ^0}. Thus, consistent with Kirwan's theorem, we see that h_{Σ^0} is a perfect Morse function with even Morse indices. Notice that the Hamiltonian vector field $V = \frac{\partial}{\partial\phi}$ here generates an isometry of the standard round metric $d\theta \otimes d\theta + \sin^2\theta \, d\phi \otimes d\phi$ induced by the flat Euclidean metric of $\mathbb{R}^3$. The differential form (2.123) with this metric is $\xi = d\phi$, which as expected is ill-defined at the 2 poles of S^2. Now the partition function can be written as

$$Z_{\Sigma^0}(T) = -\frac{1}{iT} \int_{\Sigma^0} d\left(e^{iTh_{\Sigma^0}} d\phi \right) \qquad (3.75)$$

thus receiving contributions from only the critical points $\theta = 0, \pi$, the endpoints of the integration range for $\cos\theta$, in agreement with the explicit evaluation in Section 2.1. The partition function (2.3) represents the equivariant cohomology classes in[4]

---

[3] In general, if $\mathcal{M}$ is path-connected, as we always assume here, then $H_0(\mathcal{M}; \mathbb{Z}) = \mathbb{Z}$ and if $\mathcal{M}$ is closed, then $H_{2n}(\mathcal{M}; \mathbb{Z}) = \mathbb{Z}$. The intermediate homology groups depend on whether or not $\mathcal{M}$ has 'holes' in it or not.

[4] Equivariant cohomology groups are usually computed using so-called classifying bundles of Lie groups (the topological definition of equivariant cohomology) – see [111], for example.

60   3. Finite-Dimensional Localization Theory for Dynamical Systems

$$H^2_{U(1)}(S^2) = \mathbb{Z} \oplus \mathbb{Z} \tag{3.76}$$

Intuitively, (3.76) follows from the fact that the single Lie algebra generator $\Phi \in \mathbb{R}$ and the invariant volume form (3.72) of S^2 are linearly independent, i.e. for any 2 functions $f_1, f_2 \in S(\mathbf{u(1)}^*)$, the equivariant cohomology classes are spanned by the linearly independent generators $f_1(\Phi)(\omega_{\Sigma^0} + \Phi h_{\Sigma^0})$ and $f_2(\Phi)$.

As we shall see later on, the above example for the Riemann sphere is essentially the only Hamiltonian system to which the geometric equivariant localization constraints apply on a simply-connected phase space (i.e. $H_1(\mathcal{M}; \mathbb{Z}) = 0$). The situation is much different on a multiply-connected phase space, which as we shall see is due to the fact that the non-trivial first homology group of the phase space severely restricts the allowed $U(1)$ group actions on it and hence the Morse functions thereon. For example, consider the case of a genus 1 Riemann surface [85, 153, 157], i.e. Σ^1 is the 2-torus $T^2 = S^1 \times S^1$. The torus can be viewed as a parallelogram in the complex plane with its opposite edges identified. We take as horizontal edge the line segment from 0 to 1 along the real axis and the other slanted edge the line segment from 0 to some complex number τ in the complex plane. The number τ is called the modular parameter of the torus and we can take it to lie in the upper complex half-plane

$$\mathbb{C}^+ = \{z \in \mathbb{C} : \text{Im } z > 0\} \tag{3.77}$$

Geometrically, τ determines the inner and outer radii of the 2 circles of the torus, and it labels the inequivalent complex structures of Σ^1 [5].

We view the torus embedded in 3-space as a doughnut standing on end on the xy-plane and centered symmetrically about the z-axis. If (ϕ_1, ϕ_2) are the angle coordinates on $S^1 \times S^1$, then the height function on Σ^1 can be written as

$$h_{\Sigma^1}(\phi_1, \phi_2) = r_2 - (r_1 + \text{Im } \tau \cos \phi_1) \cos \phi_2 \tag{3.78}$$

where $r_1 = |\text{Re } \tau| + \text{Im } \tau$ and $r_2 = |\text{Re } \tau| + 2 \text{ Im } \tau$ label the inner and outer radii of the torus. The symplectic volume form on T^2 is just that induced by the identification of Σ^1 as a parallelogram in the plane with its opposite edges identified, i.e. the Darboux 2-form

$$\omega_D = d\phi_1 \wedge d\phi_2 \tag{3.79}$$

---

[5] In algebraic geometry one would therefore say that $\mathbb{C}^+$ is the Teichmüller space of the torus. The Teichmüller space of a simply-connected Riemann surface is a point, so that there is a unique complex structure (i.e. a unique way of defining complex coordinates) in genus 0. This is a consequence of the celebrated Riemann uniformization theorem. We refer to [111] and [145] for an elementary introduction to Teichmüller spaces in algebraic geometry, while a more extensive treatment can be found in [74].

3.5 Examples: The Height Function of a Riemann Surface

The associated Hamiltonian vector field for this dynamical system has components

$$V^1_{\Sigma^1} = -(r_1 + \operatorname{Im} \tau \cos \phi_1) \sin \phi_2 \quad , \quad V^2_{\Sigma^1} = \operatorname{Im} \tau \sin \phi_1 \cos \phi_2 \quad (3.80)$$

The Hamiltonian (3.78) has 4 isolated non-degenerate critical points on $S^1 \times S^1$ – a maximum at $(\phi_1, \phi_2) = (0, \pi)$ (top of the outer circle), a minimum at $(0,0)$ (bottom of the outer circle), and 2 saddle points at $(\pi, 0)$ and (π, π) (corresponding to the bottom and top of the inner circle, respectively). The Morse index of the maximum is 2, that of the minimum is 0, and those of the 2 saddle points are both 1. According to Kirwan's theorem, the appearence of odd Morse indices, or equivalently the fact that

$$H_1(\Sigma^1; \mathbb{Z}) = \mathbb{Z} \oplus \mathbb{Z} \quad (3.81)$$

with each $\mathbb{Z}$ labelling the windings around the 2 independent non-contractable loops associated with each S^1-factor, implies that the Duistermaat-Heckman integration formula should fail in this case. Indeed, evaluating the right-hand side of the Duistermaat-Heckman formula (3.63) gives

$$\frac{2\pi i}{T} \sum_{p \in \mathcal{M}_{V_{\Sigma^1}}} \frac{e^{iTh_{\Sigma^1}(p)}}{e(p)}$$

$$= \frac{2\pi i}{T \sqrt{\operatorname{Im} \tau}} \left[r_2^{-1/2} \left(1 + e^{2iTr_2} \right) + |\operatorname{Re} \tau|^{-1/2} e^{2iT \operatorname{Im}\tau} \left(1 - e^{2iT|\operatorname{Re}\tau|} \right) \right]$$
(3.82)

which for the parameter values $iT = 1$ and $\tau = 1 + i$ gives the numerical value

$$2\pi \, e^3 \left(\frac{2}{\sqrt{3}} \sinh 3 + 2 \cosh 1 \right) \sim 1849.33 \quad (3.83)$$

On the other hand, an explicit evaluation of the partition function gives

$$Z_{\Sigma^1}(T) = \int_0^{2\pi} \int_0^{2\pi} d\phi_1 \, d\phi_2 \; e^{iTh_{\Sigma^1}(\phi_1, \phi_2)}$$

$$= 2\pi \, e^{r_2} \int_0^{2\pi} d\phi_1 \, J_0 \left(iT(r_1 + \operatorname{Im} \tau \cos \phi_1) \right)$$
(3.84)

with J_0 the regular Bessel function of order 0 [60]. For the parameter values above, a numerical integration in (3.84) gives $Z_{\Sigma^1} \sim 2117.13$ [6], contradicting the result (3.83). Thus even though in this case the Hamiltonian h_{Σ^1} is a

---

[6] All numerical integrations in this Book were performed using the mathematical software package *MATHEMATICA*.

perfect Morse function, it doesn't generate any torus action on the phase space here.

This argument can be extended to the case where the phase space is a hyperbolic Riemann surface Σ^g, $g > 1$ [153]. For $g > 1$, $\Sigma^g = \Sigma^1 \# \cdots \# \Sigma^1$ is the g-fold connected sum of 2-tori and therefore its first homology group is

$$H_1(\Sigma^g; \mathbb{Z}) = \bigoplus_{i=1}^{2g} \mathbb{Z} \qquad (3.85)$$

It can be viewed in $\mathbb{R}^3$ as g doughnuts stuck together on end and standing on the xy-plane. The height function on Σ^g now has $2g + 2$ critical points consisting of 1 maximum, 1 minimum and $2g$ saddle points. Again the maximum and minimum have Morse indices 2 and 0, respectively, while those of the $2g$ saddle points are all 1. As a consequence the perfect Morse function h_{Σ^g} generates no torus action on Σ^g.

The above non-exactness of the stationary-phase approximation (and even worse the divergence of the stationary-phase series for (3.78)) is a consequence of the fact that the orbits of the vector field (3.80) do not generate a global, compact group action on Σ^1. Here the orbits of the Hamiltonian vector field bifurcate at the saddle points (like the classical trajectories of the simple pendulum which cross each other in figure eights), and we shall see explicitly in Chapter 7 why its flows cannot generate isometries of any metric on Σ^1 as well as how this makes the stationary phase series diverge. The extensions of the equivariant localization principle to non-compact group actions and to non-compact phase spaces are not always immediate [29]. A version of the Duistermaat-Heckman theorem appropriate to both abelian and non-abelian group actions on non-compact manifolds has been presented by Prato and Wu in [141]. This non-compact version of the Duistermaat-Heckman theorem assumes that there is a component of the moment map which is regular and bounded from below (so that the Fourier-Laplace transform $Z(T)$ exists). The above examples illustrate the strong topological dependence of the dynamical systems to which equivariant localization is applicable. The height function restricted to a compact Riemann surface can only be used for Duistermaat-Heckman localization in genus 0, and the introduction of more complicated topologies restricts even further the class of Hamiltonian systems to which the localization constraints apply. We shall investigate this phenomenon in a more detailed geometric setting later on when we consider quantum localization techniques.

3.6 Equivariant Localization and Classical Integrability

In this Section we discuss an interesting connection between the equivariant localization formalism and integrable Hamiltonian systems [84, 85]. By an integrable dynamical system we mean this in the sense of the Liouville-Arnold

3.6 Equivariant Localization and Classical Integrability

theorem which is a generalized, coordinate independent version of the classical Liouville theorem that dictates when a given Hamiltonian system will have equations of motion whose solutions can be explicitly found by integrating by quadratures [35, 55]. The Liouville-Arnold theorem is essentially a global version of Darboux's theorem and it states that a Hamiltonian is integrable if one can find canonically conjugated action-angle variables $(I_\mu, \phi^\mu)_{\mu=1}^n$,

$$\{I_\mu, \phi^\nu\}_\omega = \delta_\mu^\nu \tag{3.86}$$

defined almost everywhere on the phase space $\mathcal{M}$, such that the Hamiltonian $H = H(I)$ is a functional of only the action variables [6]. The action variables themselves are supposed to be functionally-independent and in involution,

$$\{I_\mu, I_\nu\}_\omega = 0 \tag{3.87}$$

and from the Hamilton equations of motion (3.32) it follows that

$$\dot{I}_\mu(t) = \{I_\mu, H(I)\}_\omega = 0 \tag{3.88}$$

so that the time-evolution of the action variables is constant. Consequently, (3.87) implies that the action variables generate a Cartan subalgebra $(S^1)^n$ of the Poisson algebra of the phase space, and the I_μ therefore label a set of canonically invariant tori on the phase space which are called Liouville tori. The motion of $H(I)$ is constrained to the Liouville tori, and the system is therefore integrable in the sense that we have found n independent degrees of freedom for the classical motion. The I_μ's in simple problems are conserved quantities such as the total energy or angular momentum which generate a particular symmetry of the dynamics, such as time-independence or radial symmetry. The symplectic 2-form in the action-angle variables is

$$\omega = dI_\mu \wedge d\phi^\mu \tag{3.89}$$

and the corresponding symplectic potential which generates the Hamiltonian as the moment map of a global $U(1)$ group action on $\mathcal{M}$ as in (3.47) is

$$\theta_F \equiv \theta + dF = I_\mu d\phi^\mu \tag{3.90}$$

The connection between integrability and equivariant localization now becomes rather transparent. The above integrability requirement that H be a functional of some torus action generators is precisely the requirement of the Duistermaat-Heckman theorem. The global solutions to the Hamilton equations of motion in this case are

$$I_\mu(t) = I_\mu(0) \quad , \quad \phi^\mu(t) = \phi^\mu(0) + \omega^\mu(I)t \tag{3.91}$$

where $\omega^\mu(I) = \partial_{I_\mu} H(I)$. The classical trajectories of the dynamical system therefore move along the Liouville tori with constant angular velocity

$\dot{\phi}^\mu = \omega^\mu$ (equivalently they share a common period) which represents a large symmetry of the classical mechanics. The Hamiltonian vector field $\frac{\partial}{\partial \phi^\mu}$ associated with the action variable I_μ generates the μ-th circle action component of the full torus action on $\mathcal{M}$, and consequently any Hamiltonian which is a *linear* combination of the action variables will generate a torus action on $\mathcal{M}$ and meet the criteria of the Duistermaat-Heckman theorem. For quadratic and higher-order functionals of the action variables the associated Hamiltonian vector field in general generates a circle action which does not have a constant angular velocity on the phase space and the Duistermaat-Heckman formula will not hold. We shall see, however, that modified versions of the Duistermaat-Heckman localization formula can still be derived, so that *any* integrable model will provide an example of a partition function that localizes. For the height function of S^2 above, the action-angle variables are $I_1 = h_{\Sigma^0} = a - \cos\theta$, $\phi^1 = \phi$. We shall see some more general (higher-dimensional) examples in Chapter 5.

Recall that one of the primary assumptions in the localization framework above was that the phase space admit a Riemannian metric g which is *globally* invariant under the $U(1)$ action on it. The existence of a Riemannian geometry which is globally invariant under the classical dynamics of a given Hamiltonian system is a very strong requirement. A $U(1)$-invariant metric tensor always exists *locally* in the regions where H has no critical points. To see this, introduce local equivariant Darboux coordinates $(p_1, \ldots, p_n, q^1, \ldots, q^n)$ in that region in which the Hamiltonian vector field generates translations in q^1. This means that $H = p_1$ is taken as the radius of this equivariant Darboux coordinate system. The $U(1)$-invariant metric tensor can then be taken to be any metric tensor whose components are independent of the coordinate q^1 (e.g. $g_{\mu\nu} = \delta_{\mu\nu}$), which follows from the Killing equations (2.112). However, there may be global obstructions to extending these local metrics to metrics defined globally on the entire phase space in a smooth way. This feature is just equivalent to the well-known fact that any Hamiltonian system is locally integrable. This is easily seen from the local representation (3.54),(3.55) where we can define $p_\mu = I_\mu \cos\phi^\mu, q^\mu = I_\mu \sin\phi^\mu$. Then $H \sim \sum_{\mu=1}^n I_\mu^2$ and $V \sim \sum_{\mu=1}^n \frac{\partial}{\partial \phi^\mu}$ generates translations in the local angle variables ϕ^μ (rigid rotations of the local coordinate neighbourhood). Then locally the metric tensor components $g_{\mu\nu}$ should be taken to depend only on the action variables I_μ (i.e. g is radially symmetric in the coordinate neighbourhood).

However, local integrability does not necessarily ensure global integrability. For the latter to follow, it is necessary that the neighbourhoods containing the conserved charges I_μ be patched together in such a way as to yield a complete set of conserved charges defined almost everywhere on the phase space $\mathcal{M}$. Furthermore, global integrability also implements strong requirements on the behaviour of H in the vicinity of its critical points. As we shall see later on, the isometry group of a compact Riemannian manifold is also compact, so that the *global* existence of an invariant metric tensor in the above for a

3.6 Equivariant Localization and Classical Integrability

compact phase space is equivalent to the requirement that H generates the global action of a circle (or more generally a torus). This means that the Hamiltonian vector field V is a Cartan element of the algebra of isometries of the metric g (or equivalently H is a Cartan element of the corresponding Poisson algebra). In other words, H is a globally-defined action variable (or a functional thereof), so that the applicable Hamiltonians within the framework of equivariant localization determine integrable dynamical systems. Thus it is the isometry condition that puts a rather severe restriction on the Hamiltonian functions which generate the circle action through the relation (3.28). These features also appear in the infinite-dimensional generalizations of the localization formalism above and they will be discussed at greater length in Chapters 5 and 6.

We note also that for an integrable Hamiltonian H we can construct an explicit representation of the function F which appears in (3.47) and (3.90) above. Indeed, the function K in (3.46) can be constructed locally outside of the critical point set of H by assuming that a given action variable I_μ is such that

$$\frac{\partial H(I)}{\partial I_\mu} \neq 0 \tag{3.92}$$

In this case, the function K can be realized explicitly by

$$K(I,\phi) = \phi^\mu \cdot \left(\frac{\partial H}{\partial I_\mu}\right)^{-1} \tag{3.93}$$

and the condition (3.47) becomes

$$i_V \theta_F = I_\mu \frac{\partial H}{\partial I_\mu} + \{H, F\}_\omega = H \tag{3.94}$$

which is satisfied by

$$F = K \cdot \left(H - I_\mu \frac{\partial H}{\partial I_\mu}\right) + G(I) \tag{3.95}$$

where $G(I)$ is an arbitrary function of the action variables. Consequently, in a neighbourhood where action-angle variables can be introduced and where H does not admit critial points, we get an explicit realization of the function F in (3.47) and thus a locally invariant symplectic potential θ_F.

In fact, given the equivariantly closed 2-form $K_V + \Omega_V$ introduced in (2.119), we note that Ω_V is a closed 2-form (but not necessarily non-degenerate) and that the function K_V satisfies

$$dK_V = -i_V \Omega_V \tag{3.96}$$

as a consequence of (2.121) and (2.118), respectively. It follows that

$$V^\mu = \Omega_V^{\mu\nu} \partial_\nu K_V = \omega^{\mu\nu} \partial_\nu H \tag{3.97}$$

66   3. Finite-Dimensional Localization Theory for Dynamical Systems

and so the classical equations of motion for the 2 Hamiltonian systems $(\mathcal{M}, \omega, H)$ and $(\mathcal{M}, \Omega_V, K_V)$ coincide[7],

$$\dot{x}^\mu(t) = \{x^\mu, H\}_\omega = \{x^\mu, K_V\}_{\Omega_V} \tag{3.98}$$

This means that these 2 dynamical systems determine a bi-Hamiltonian structure. There are 2 interesting consequences of this structure. The first follows from the fact that if $H = H(I)$ as above is integrable, then these action-angle variables can be chosen so that in addition $K_V = K_V(I)$ is an integrable Hamiltonian. We can therefore replace H everywhere in (3.92)–(3.95) by the function K_V and ω by Ω_V, and after a bit of algebra we find that the 1-form $\theta_F^{(V)}$ above which generates Ω_V satisfies

$$K_V + \Omega_V = D_V \theta_F^{(V)} \tag{3.99}$$

and likewise

$$H + \omega = D_V \theta_F \tag{3.100}$$

Since both $H + \omega$ and $K_V + \Omega_V$ are equivariantly closed, we see that for an integrable bi-Hamiltonian system we can solve explicitly the equivariant version of the Poincaré lemma. The global existence of the 1-forms θ_F and $\theta_F^{(V)}$ is therefore connected not only to the non-triviality of the DeRham cohomology of $\mathcal{M}$, but also to the non-triviality of the equivariant cohomology associated with the equivariant exterior derivative D_V. Note that this derivation could also have been carried out for an arbitrary equivariant differential 1-form β with the definition (2.119) (c.f. eq. (2.124)). This suggests an intimate relationship between the localization formalism, and more generally equivariant cohomology, and the existence of bi-Hamiltonian structures for a given phase space.

Furthermore, it is well-known that the existence alone of a bi-Hamiltonian system is directly connected to integrability [6, 35]. If the symplectic 2-forms ω and Ω_V are such that the rank (1,1) tensor

$$L = \Omega_V \cdot \omega^{-1} \tag{3.101}$$

is non-trivial, then one can straightforwardly show [84] that

$$\dot{L} = V^\mu \partial_\mu L = [L, dV] \tag{3.102}$$

which is just the Lax equation, so that (L, dV) determines a Lax pair [35]. Under a certain additional assumption on the tensor L it can then be shown [84] that the quantities

---

[7] Here we assume that Ω_V is non-degenerate on $\mathcal{M}$ except possibly on submanifolds of $\mathcal{M}$ of codimension at least 2, since when it is degenerate some of the equations in (3.96) should be considered as constraints. On these submanifolds, the Hamiltonian K_V must then vanish in order to keep the equations of motion non-singular [6].

$$I_\mu = \frac{1}{\mu} \text{ tr } L^\mu \tag{3.103}$$

give variables which are in involution and which are conserved, i.e. which commute with the Hamiltonian H. If these quantities are in addition complete, i.e. the number of functionally independent variables (3.103) is half the phase space dimension, then the Hamiltonian system $(\mathcal{M}, \omega, H)$ is integrable in the sense of the Liouville-Arnold theorem. We refer to [84] for more details of how this construction works. Therefore the equivariant localization formalism for classical dynamical systems presents an alternative, geometric approach to the problem of integrability.

3.7 Degenerate Version of the Duistermaat-Heckman Theorem

In these last 3 Sections of this Chapter we shall quickly run through some of the generalizations of the Duistermaat-Heckman theorem which can be applied to more general dynamical systems. The first generalization we consider is to the case where H isn't necessarily non-degenerate and its critical point set $\mathcal{M}_V$ is now a submanifold of $\mathcal{M}$ of co-dimension $r = \dim \mathcal{M} - \dim \mathcal{M}_V$ [9, 19, 21, 23, 24, 39, 121]. In this case some modifications are required in the evaluation of the canonical localization integral (2.137) which was used in the derivation of the Berline-Vergne theorem with the differential form α given in (3.57). The Hessian of H now vanishes everywhere on $\mathcal{M}_V$ (because $dH = 0$ everywhere on $\mathcal{M}_V$), but we assume that it is non-vanishing in the directions normal to the critical submanifold $\mathcal{M}_V$ [111]. This defines the normal bundle $\mathcal{N}_V$ of $\mathcal{M}_V$ in $\mathcal{M}$, and the phase space is now locally the disjoint union

$$\mathcal{M} = \mathcal{M}_V \amalg \mathcal{N}_V \tag{3.104}$$

so that in a neighbourhood near $\mathcal{M}_V$ we can decompose the local coordinates on $\mathcal{M}$ as

$$x^\mu = x_0^\mu + x_\perp^\mu \tag{3.105}$$

where x_0 are local coordinates on $\mathcal{M}_V$, i.e. $V(x_0) = 0$, and $x_\perp$ are local coordinates on $\mathcal{N}_V$. Similarly, the tangent space at any point x near $\mathcal{M}_V$ can be decomposed as

$$T_x\mathcal{M} = T_x\mathcal{M}_V \oplus T_x\mathcal{N}_V \tag{3.106}$$

where $T_x\mathcal{N}_V$ is the space of vectors orthogonal to those in $T_x\mathcal{M}_V$. We can therefore decompose the Grassmann variables η^μ which generate the exterior algebra of $\mathcal{M}$ as

$$\eta^\mu = \eta_0^\mu + \eta_\perp^\mu \tag{3.107}$$

where η_0^μ generate the exterior algebra $\Lambda \mathcal{M}_V$ and $\eta_\perp^\mu$ generate $\Lambda \mathcal{N}_V$.

Under the usual assumptions used in deriving the equivariant localization principle, it follows that the tangent bundle, equipped with a Levi-Civita-Christoffel connection Γ associated with a $U(1)$-invariant metric tensor g as in (2.114), is an equivariant vector bundle. Recall that the Lie derivative $\mathcal{L}_V$ induces a non-trivial action of the group on the fibers of the tangent bundle which is mediated by the matrix dV. More precisely, this action is given by

$$\mathcal{L}_V = V^\mu \partial_\mu + dV^\mu i_{V^\mu} - dV \qquad (3.108)$$

and so the moment map associated with this equivariant bundle is the Riemann moment map [21]

$$\mu_V = \nabla V \qquad (3.109)$$

which as always is regarded as a matrix acting on the fiber spaces. Given the Killing equations for V, this moment map is related to the 2-form Ω_V by

$$(\Omega_V)_{\mu\nu} = 2g_{\mu\lambda}(\mu_V)^\lambda_\nu \qquad (3.110)$$

and the equivariant curvature of the bundle is

$$R_V = R + \mu_V \qquad (3.111)$$

where the Riemann curvature 2-form of the tangent bundle is

$$R^\mu_\nu = \frac{1}{2} R^\mu_{\nu\lambda\rho}(x) \eta^\lambda \eta^\rho \qquad (3.112)$$

and

$$R^\rho_{\sigma\mu\nu} = \partial_\mu \Gamma^\rho_{\nu\sigma} - \partial_\nu \Gamma^\rho_{\mu\sigma} + \Gamma^\rho_{\mu\lambda} \Gamma^\lambda_{\nu\sigma} - \Gamma^\rho_{\nu\lambda} \Gamma^\lambda_{\mu\sigma} \qquad (3.113)$$

are the components of the associated Riemann curvature tensor $R = d\Gamma + \Gamma \wedge \Gamma$. Note that, from the decomposition (3.106), the normal bundle inherits a $U(1)$-invariant connection from $T\mathcal{M}$, and the curvature and moment map on $T\mathcal{N}_V$ are just the restrictions of the corresponding objects defined on $T\mathcal{M}$.

Given these features of the 2-form Ω_V, it follows that the generators η^μ_0 of $\Lambda\mathcal{M}_V$ satisfy

$$(\Omega_V)_{\mu\nu}(x_0)\eta^\nu_0 = 2(g_{\mu\lambda} \partial_\nu V^\lambda)(x_0)\eta^\nu_0 = 0 \qquad (3.114)$$

since $\eta^\mu_0 \sim dx^\mu_0$ lie in a direction cotangent to $\mathcal{M}_V$. For large $s \in \mathbb{R}^+$ in (2.137) the integral will localize exponentially to a neighbourhood of $\mathcal{M}_V$, and so, in the linearization (3.105) of the coordinates perpendicular to $\mathcal{M}_V$ wherein we approximate this neighbourhood with a neighbourhood of the normal bundle $\mathcal{N}_V$, we can extend the integration over all values of $x_\perp$ there. We now introduce the scaled change of integration variables

$$x^\mu = x^\mu_0 + x^\mu_\perp \to x^\mu_0 + x^\mu_\perp/\sqrt{s} \quad , \quad \eta^\mu = \eta^\mu_0 + \eta^\mu_\perp \to \eta^\mu_0 + \eta^\mu_\perp/\sqrt{s} \qquad (3.115)$$

and expand the argument of the large-s exponential in (2.137) using the decompositions (3.115). The Jacobian determinants from the anticommuting

3.7 Degenerate Version of the Duistermaat-Heckman Theorem

$\eta_\perp^\mu$ variables and the commuting $x_\perp^\mu$ variables cancel each other, and so the integral (2.137) remains unchanged under this coordinate rescaling. A tedious but straightforward calculation using observations such as (3.114) shows that the large-s expansion of the argument of the exponential in (2.137) is given by [121]

$$\frac{s}{2}\Omega_V \overset{s\to\infty}{\longrightarrow} \frac{1}{2}(\Omega_V)_{\mu\nu}(x_0)\eta_\perp^\mu\eta_\perp^\nu + \frac{1}{2}(\Omega_V)_{\mu\sigma}(x_0)R^\sigma_{\nu\lambda\rho}(x_0)x_\perp^\mu x_\perp^\nu \eta_0^\lambda \eta_0^\rho$$
$$+\mathcal{O}(1/\sqrt{s}) \qquad (3.116)$$

$$sK_V \overset{s\to\infty}{\longrightarrow} \frac{1}{2}(\mu_V)^\rho_\mu(x_0)(\Omega_V)_{\rho\nu}(x_0)x_\perp^\mu x_\perp^\nu + \mathcal{O}(1/\sqrt{s})$$

where we have expanded the C^∞-functions in (3.116) in their respective Taylor series about the critical points.

Thus with the coordinate change (3.115), the integration over the normal part of the full integration domain

$$\mathcal{M} \otimes \Lambda^1\mathcal{M} = (\mathcal{M}_V \otimes \Lambda^1\mathcal{M}_V) \amalg (\mathcal{N}_V \otimes \Lambda^1\mathcal{N}_V) \qquad , \qquad (3.117)$$

i.e. over $(x_\perp, \eta_\perp)$ in (2.137), is Gaussian and can be carried out explicitly. The result is an integral over the critical submanifold

$$Z(T) = \left(\frac{2\pi i}{T}\right)^{\frac{r}{2}} \int_{\mathcal{M}_V \otimes \Lambda^1\mathcal{M}_V} d^r x_0 \, d^r \eta_0 \, e^{iT(H(x_0) + \omega(x_0,\eta_0))}$$
$$\times \frac{\text{Pfaff } \Omega_V(x_0)}{\sqrt{\det \Omega_V(x_0)(\mu_V(x_0) + R(x_0,\eta_0))}} \qquad (3.118)$$
$$= \left(\frac{2\pi i}{T}\right)^{\frac{r}{2}} \int_{\mathcal{M}_V} \frac{\text{ch}_V(iT\omega)|_{\mathcal{M}_V}}{E_V(R)|_{\mathcal{N}_V}}$$

where we have identified the equivariant Chern and Euler characters (2.95) and (2.100) of the respective fiber bundles. In (3.118) the equivariant Chern and Euler characters are restricted to the critical submanifold $\mathcal{M}_V$, and the determinant and Pfaffian there are taken over the normal bundle $\mathcal{N}_V$. Note that the above derivation has assumed that the critical submanifold $\mathcal{M}_V$ is connected. If $\mathcal{M}_V$ consists of several connected components, then the formula (3.118) means a sum over the contributions from each of these components. Notice also the role that the large equivariant cohomological symmetry of the dynamical system has played here – it renders the Jacobian for the rescaling transformation (3.115) trivial and reduces the required integrations to Gaussian ones. This symmetry appears as a sort of supersymmetry here (i.e. a symmetry between the scalar x^μ and Grassmann η^μ coordinates).

There are several comments in order here. First of all, if $\mathcal{M}_V$ consists of discrete isolated points, so that $r = 2n$, then, since the curvature of the normal bundle of a point vanishes and so the Riemann moment map μ_V coincides with the usual moment map dV on TM calculated at that point, the formula (3.118) reduces to the non-degenerate localization formula (3.58) and hence to the Duistermaat-Heckman theorem. Secondly, we recall that the equivariant characteristic classes in (3.118) provide representatives of the equivariant cohomology of M and the integration formula (3.118) is formally independent of the chosen metric on M. Thus the localization formulas are topological invariants of M, as they should be, and they represent types of 'index theorems'. This fact will have important implications later on in the formal applications to topological field theory functional integration. Finally, we point out that Kirwan's theorem generalizes to the degenerate case above [88]. In this case, since the Hessian is a non-singular symmetric matrix along the directions normal to M_V, we can orthogonally decompose the normal bundle, with the aid of some locally-defined Riemannian metric on M_V, into a direct sum of the positive- and negative-eigenvalue eigenspaces of $\mathcal{H}$. The dimension of the latter subspace is now defined as the index of M_V and Kirwan's theorem now states that the index of every connected component of M_V must be even when the localization formula (3.118) holds. The Morse inequalities for this degenerate case [111] then relate the exactness or failure of (3.118) as before to the homology of the underlying phase space M. One dynamical system to which the formula (3.118) could be applied is the height function of the torus when the torus is now viewed in 3-space as a doughnut sitting on a dinner plate (the xy-plane). This function has 2 extrema but they are now circles, instead of points, which are parallel to each other and one is a minimum and the other is a maximum. The critical submanifold of T^2 in this case consists of 2 connected components, $T_V^2 = S^1 \amalg S^1$.

3.8 The Witten Localization Formula

We have thus far only applied the localization formalism to abelian group actions on M. The first generalization of the Duistermaat-Heckman theorem to non-abelian group actions was presented by Guilleman and Prato [63] in the case where the induced action of the Cartan subgroup (or maximal torus) of G has only a finite number of isolated fixed points p_i and the stabilizer $\{g \in G : g \cdot p_i = p_i\}$ of all these fixed points coincides with the Cartan subgroup. The Guilleman-Prato localization formula reduces the integrals over the dual Lie algebra $\mathbf{g}^*$ to integrals over the dual of the Cartan subalgebra using the so-called Weyl integral formula [29]. With this reduction one can apply the standard abelian localization formalism above. This procedure of abelianization thus reduces the problem to the consideration of localization theory for functions of Cartan elements of the Lie group G, i.e. integrable Hamiltonian systems. Witten [171] proposed a more general non-abelian localization formalism and used it to study 2-dimensional Yang-Mills theory.

3.8 The Witten Localization Formula

In this Section we shall outline the basic features of Witten's localization theory.

Given a Lie group G acting on the phase space $\mathcal{M}$, we wish to evaluate the partition function with the general equivariant extension (3.35),

$$Z_G = \int_{\mathcal{M}} \frac{\omega^n}{n!} e^{-\phi^a \otimes H^a} \tag{3.119}$$

where as usual the Boltzmann weights are given by the symplectic moment map of the G-action on $\mathcal{M}$. There are 2 ways to regard the dual algebra functions ϕ^a in (3.119). We can give the ϕ^a fixed values, regarding them as the values of elements of $S(\mathbf{g}^*)$ acting on algebra elements, i.e. the ϕ^a are complex-valued parameters, as is unambiguously the case if G is abelian [9] (in which case we set $\phi = -iT$ in (3.119)). In this case we are integrating with a fixed element of the Lie group G, i.e. we are essentially in the abelian case. We shall see that various localization schemes reproduce features of character formulas for the action of the Lie group G on $\mathcal{M}$ at the quantum level. The other possibility is to regard the ϕ^a as dynamical variables and integrate over them. This case allows a richer intepretation and is the basis of non-abelian localization formulas and the localization formalism in topological field theory.

To employ this latter interpretation for the symmetric algebra elements, we need a definition for equivariant integration. The definition (2.127) gives a map on $\Lambda_G \mathcal{M} \to S(\mathbf{g}^*)^G$, but in analogy with ordinary DeRham integration we wish to obtain a map on $\Lambda_G \mathcal{M} \to \mathbb{C}$. The group G has a natural G-invariant measure on it, namely its Haar measure. Since $\mathbf{g}$ is naturally isomorphic to the tangent space of G at the identity, it inherits from the Haar measure a natural translation-invariant measure. Given this measure, the definition we take for equivariant integration is [171]

$$\oint_{\mathcal{M} \otimes \mathbf{g}^*} \alpha \equiv \lim_{s \to \infty} \frac{1}{\mathrm{vol}(G)} \int_{\mathbf{g}^*} \prod_{a=1}^{\dim G} \frac{d\phi^a}{2\pi} e^{-\frac{1}{2s}(\phi^a)^2} \int_{\mathcal{M}} \alpha \tag{3.120}$$

for $\alpha \in \Lambda_G \mathcal{M}$, where $\mathrm{vol}(G) = \int_G Dg = \int_{\mathbf{g}^*} \prod_a d\phi^a / 2\pi$ is the volume of the group G in its Haar measure. The parameter $s \in \mathbb{R}^+$ in (3.120) is used to regulate the possible divergence on the completion $\Lambda_G^\infty \mathcal{M}$. The definition (3.120) indeed gives a map on $\Lambda_G \mathcal{M} \to \mathbb{C}$, and the ϕ^a's in it can be regarded as local Euclidean coordinates on $\mathbf{g}^*$ such that the measure there coincides with the chosen Haar measure at the identity of G. Setting $\alpha = e^{\omega_\mathbf{g}}$ in (3.120), with $\omega_\mathbf{g}$ the equivariant extension (3.35) of the symplectic 2-form of $\mathcal{M}$, and performing the Gaussian ϕ^a-integrals, we arrive at Witten's localization formula for the partition function (3.119),

$$Z_G = \lim_{s \to \infty} \left(\frac{s}{2\pi}\right)^{\dim G/2} \frac{1}{\mathrm{vol}(G)} \int_{\mathcal{M}} \frac{\omega^n}{n!} e^{-\frac{s}{2}\sum_a (H^a)^2} \tag{3.121}$$

3. Finite-Dimensional Localization Theory for Dynamical Systems

The right-hand side of (3.121) localizes onto the extrema of the square of the moment map $\sum_a (H^a)^2$. The absolute minima of this function are the solutions to $H = \phi^a \otimes H^a = 0$. The contribution of the absolute minimum to Z_G (the dominant contribution for $s \to \infty$) is given by a simple cohomological formula [171]

$$Z_G^{\min} = \lim_{s \to \infty} \frac{(2\pi)^{\dim G/2}}{\text{vol}(G)} \int_{\mathcal{M}_0} e^{\omega + \frac{1}{2s}\Theta} \bigg|_{\mathcal{M}_0} \qquad (3.122)$$

where $\mathcal{M}_0 = H^{-1}(0)/G$ is the Marsden-Weinstein reduced phase space [97] (or symplectic quotient) and the localization of the global minima onto $\mathcal{M}_0$ is a consequence of the G-equivariance of the integration in (3.120) (as one can then integrate over the fibers of the bundle $H^{-1}(0) \to H^{-1}(0)/G$, with bundle projection π that takes $x \in H^{-1}(0)$ into its equivalence class $[x] \in \mathcal{M}_0$). Here we have assumed that G acts freely on $H^{-1}(0)$, and Θ is a certain element of the cohomology group $H^4(\mathcal{M}_0; \mathbb{R})$ that is defined as follows. In integrating over the ϕ^a's in (3.120) we note that $\sum_a (\phi^a)^2/2 \in H^4_G(\mathcal{M})$, so that when restricted to $H^{-1}(0)$ it is the pullback of some $\Theta \in H^4(\mathcal{M}_0; \mathbb{R})$. Therefore the equivariant cohomology class of $\sum_a (\phi^a)^2/2 \in H^4_G(\mathcal{M})$ is determined by this form Θ which then serves as a characteristic class of the principal G-bundle $H^{-1}(0) \to \mathcal{M}_0$. The Witten localization formula can in this way be used to describe the cohomology of the reduced phase space $\mathcal{M}_0$ of the given symplectic G-action on $\mathcal{M}$. We refer to [171] for further details of this construction.

However, the contributions from the other local extrema of $\sum_a (H^a)^2$, which correspond to the critical points of H as in the Duistermaat-Heckman integration formula, are in general very complicated functions of the limiting parameter $s \in \mathbb{R}^+$. For instance, in the simple abelian example of Sections 2.1 and 3.5 above where $G = U(1)$, $\mathcal{M} = S^2$ and $H = h_{\Sigma^0}$ is the height function (2.1) of the sphere, the Witten localization formula (3.121) above becomes

$$Z_{\Sigma^0} = \lim_{s \to \infty} \left(\frac{s}{2\pi}\right)^{1/2} \int_{-1}^{+1} d\cos\theta \; e^{-s(a-\cos\theta)^2/2} = \lim_{s \to \infty} (1 - I_+(s) - I_-(s)) \qquad (3.123)$$

where we have assumed that $|a| < 1$, and $I_\pm(s)$ are the transcendental error functions [60]

$$I_\pm(s) = \pm \left(\frac{s}{2\pi}\right)^{1/2} \int_{\pm 1}^{\pm\infty} dx \; e^{-s(a-x)^2/2} \qquad (3.124)$$

The 3 final terms in (3.123) are the anticipated contributions from the 3 critical points of $h_{\Sigma^0}^2 = (\cos\theta - a)^2$ – the absolute minimum at $\cos\theta = a$ contributes $+1$, while the local maxima at $\cos\theta = \pm 1$ contribute negative terms $-I_\pm$ to the localization formula. The complicated error functions arise

3.8 The Witten Localization Formula   73

because here the critical point at $\cos\theta = a$ is a degenerate critical point of the canonical localization integral in (2.137). The appearence of these error functions is in marked contrast with the elementary functions that appear as the contributions from the critical points in the usual Duistermaat-Heckman formula.

Another interesting application of the Witten localization formalism is that it can be used to derive integration formulas when the argument of the Boltzmann weight in the partition function is instead the square of the moment map. This can be done by reversing the arguments which led to the localization formula (3.121), and further localizing the Duistermaat-Heckman type integral (3.119) using the localization principle of Section 2.5. The result (for finite s) is then a sum of local contributions $\sum_m Z_m(s)$, where the functions $Z_m(s)$ can only be determined explicitly in appropriate instances [78, 171, 173] (see the simple abelian example above). Combining these ideas together, we arrive at the localization formula

$$\frac{1}{\operatorname{vol}(G)} \left(-\frac{iT}{\pi}\right)^{\dim G/2} \int_{\mathcal{M}} \frac{\omega^n}{n!} e^{iT \sum_a (H^a)^2}$$

$$= \frac{1}{\operatorname{vol}(G)} \int_{\mathfrak{g}^*} \prod_{a=1}^{\dim G} \frac{d\phi^a}{2\pi} e^{-\frac{1}{4iT}(\phi^a)^2} \int_{\mathcal{M}} \frac{\omega^n}{n!} e^{-\phi^a \otimes H^a}$$

$$= \frac{1}{\operatorname{vol}(G)} \lim_{s\to\infty} \int_{\mathfrak{g}^*} \prod_{a=1}^{\dim G} \frac{d\phi^a}{2\pi} e^{-\frac{1}{4iT}(\phi^a)^2} \int_{\mathcal{M}} \frac{\omega^n}{n!} e^{-\phi^a \otimes H^a - sD_{\mathbf{g}}\lambda}$$

(3.125)

where $\lambda \in \Lambda_G^1 \mathcal{M}$ and we have applied the localization principle to the Duistermaat-Heckman type integral over $\mathcal{M}$ on the right-hand side of the first equality in (3.125). The localization 1-form λ is chosen just as before using a G-invariant metric on $\mathcal{M}$ and the Hamiltonian vector field associated with the square of the moment map. In Chapter 8 we shall outline how the formal infinite-dimensional generalization of this last localization formula can be used to evaluate the partition function of 2-dimensional Yang-Mills theory [171].

Finally, we point out the work of Jeffrey and Kirwan [78] who rigorously derived, in certain special cases, the contribution to Z_G from the reduced phase space $\mathcal{M}_0 = H^{-1}(0)/G$ in (3.122). Let $H_C \subset G$ be the Cartan subgroup of G, and assume that the fixed points p of the induced H_C-action on $\mathcal{M}$ are isolated and non-degenerate. Then for any equivariantly-closed differential form α of degree $\dim \mathcal{M}_0$ in $\Lambda_G \mathcal{M}$, we have the so-called residue formula [78]

$$\int_{\mathcal{M}_0} \alpha|_{\mathcal{M}_0} = \sum_{p \in \mathcal{M}_{H_C}} \operatorname{Res}\left[e^{-\phi^a \otimes H^a(p)} \frac{\left(\prod_\beta \beta\right) \alpha^{(0)}(p)}{e(p)} \right] \qquad (3.126)$$

where β are the roots associated to $H_C \subset G$ (the eigenvalues of the generators of H_C in the adjoint representation of G), and Res is Jeffrey-Kirwan-Kalkman residue, defined as the coefficient of $\frac{1}{\phi}$ where ϕ is the element of the symmetric algebra $S(\mathfrak{g}^*)$ representing the induced H_C-action on $\mathcal{M}$ (see [78] and [82] for its precise definition). This residue, whose explicit form was computed by Kalkman [82] (for some more recent results see [135, 163]), depends on the fixed-point set $\mathcal{M}_{H_C}$ of the H_C-action on $\mathcal{M}$ and it can be expressed in terms of the weight determinants $e(p)$ in (3.64) of the H_C-action and the values $H(p) = \phi^a \otimes H^a(p)$. It is in forms similar to (3.126) that the first non-abelian generalizations of the Duistermaat-Heckman theorem due to Guilleman and Prato appeared [63]. The residue formula can explicitly be used to obtain information about the cohomology ring of the reduced phase space $\mathcal{M}_0$ above [78, 82]. This is particularly useful in applications to topological gauge theories (see Chapter 8).

3.9 The Wu Localization Formula

The final generalization of the Duistermaat-Heckman theorem that we shall present here is an interesting application, due to Wu [173], of Witten's localization formula in the form (3.125) when applied to a global $U(1)$-action on $\mathcal{M}$. This yields a localization formula for Hamiltonians which are not themselves the associated symplectic moment map, but are functionals of such an observable H. This is accomplished via the localization formula

$$Z_{U(1)}(T) = \int_{\mathcal{M}} \frac{\omega^n}{n!} \, e^{iTH^2}$$

$$= \left(-\frac{1}{4\pi i T}\right)^{1/2} \lim_{s \to \infty} \int_0^{2\pi} d\phi \; e^{-\frac{1}{4iT}\phi^2} \int_{\mathcal{M}} \frac{\omega^n}{n!} \, e^{-\phi \otimes H - s D_{\mathbf{u}(1)}\lambda}$$
(3.127)

The final integral on the right-hand side of (3.127) is just that which appears in the canonical localization integral (2.128) used in the derivation of the Duistermaat-Heckman formula. Working this out just as before and performing the resulting Gaussian ϕ-integral yields Wu's localization formula for circle actions [173]

$$Z_{U(1)}(T) = \frac{(2\pi)^n}{(n-1)!} \sum_{p \in \mathcal{M}_V} \frac{1}{e(p)} \int_0^\infty ds \; s^{n-1} \, e^{iT(s+|H(p)|)^2} + \int_{\mathcal{M}_0} e^{\omega + iF/4T} \bigg|_{\mathcal{M}_0}$$
(3.128)

where $F = dA$ is the curvature of an abelian gauge connection on the (nontrivial) principal $U(1)$-bundle $H^{-1}(0) \to \mathcal{M}_0$. The formula (3.128) can be used to determine the symplectic volume of the Marsden-Weinstein reduced phase space $\mathcal{M}_0$ [173]. This gives an alternative localization for Hamiltonians which themselves do not generate an isometry of some metric g on $\mathcal{M}$, but are quadratic in such isometry generators. As we shall see, the path integral generalizations of Wu's formula are rather important for certain physical problems.

4. Quantum Localization Theory for Phase Space Path Integrals

In quantum mechanics there are not too many path integrals that can be evaluated explicitly and exactly, while the analog of the stationary phase approximation, i.e. the semi-classical approximation, can usually be obtained quite readily. In this Chapter we shall investigate the possibility of obtaining some path integral analogs of the Duistermaat-Heckman formula and its generalizations. A large class of examples where one has an underlying equivariant cohomology which could serve as a structure responsible for localization is provided by phase space path integrals, i.e. the direct loop space analogs of (3.52). Of course, as path integrals in general are mathematically awkward objects, the localization formulas that we will obtain in this way are not really definite predictions but rather suggestions for what kind of results to expect. Because of the lack of rigor that goes into deriving these localization formulas it is perhaps surprising then that some of these results are not only conceptually interesting but also physically reasonable.

Besides these there are many other field-theoretic analogies with the functional integral generalization of the Duistermaat-Heckman theorem, the common theme being always some underlying geometrical or topological structure which is ultimately responsible for localization. We have already mentioned one of these in the last Chapter, namely the Witten localization formula which is in principle the right framework to apply equivariant localization to a cohomological formulation of 2-dimensional quantum Yang-Mills theory (see Chapter 8). Another large class of quantum models for which the Duistermaat-Heckman theorem seems to make sense is $N = \frac{1}{2}$ supersymmetric quantum mechanics [8]. This formal application, due to Atiyah and Witten, was indeed the first encouraging evidence that such a path integral generalization of the rigorous localization formulas of the last Chapter exists. Strictly speaking though, this example really falls into the category of the Berline-Vergne localization of Section 2.6 as the free loop space of a configuration manifold is not quite a symplectic manifold in general [21]. More generally, the Duistermaat-Heckman localization can be directly generalized to the infinite-dimensional case within the Lagrangian formalism, if the loop space defined over the *configuration* space has on it a natural symplectic structure. This is the case, for example, for geodesic motion on a Lie group manifold, where the space of based loops is a Kähler manifold [142] and the

78   4. Quantum Localization Theory for Phase Space Path Integrals

stationary phase approximation is well-known to be exact [38, 146]. This formal localization has been carried out by Picken [139, 140].

We will discuss these specific applications in more detail, but we are really interested in obtaining some version of the equivariant localization formulas available which can be applied to non-supersymmetric models and when the partition functions cannot be calculated directly by some other means. The Duistermaat-Heckman theorem in this context would now express something like the exactness of the one-loop approximation to the path integral. These functional integral formulas, and their connections to the finite-dimensional formulas of Chapter 3, will be discussed at length in this Chapter. The formal techniques we shall employ throughout use ideas from supersymmetric and topological field theories, and indeed we shall see how to interpret an arbitrary phase space path integral quite naturally both as a supersymmetric and as a topological field theory partition function. In the Hamiltonian approach to localization, therefore, topological field theories fit quite naturally into the loop space equivariant localization framework. As we shall see, this has deep connections with the integrability properties of these models. In all of this, the common mechanism will be a fundamental cohomological nature of the model which can be understood in terms of a supersymmetry allowing one to deform the integrand without changing the integral.

4.1 Phase Space Path Integrals

We begin this Chapter by deriving the quantum mechanical path integral for a bosonic quantum system with no internal degrees of freedom. For simplicity, we shall present the calculation for $n = 1$ degree of freedom in Darboux coordinates on $\mathcal{M}$, i.e. we essentially carry out the calculation on the plane $\mathbb{R}^2$. The extension to $n > 1$ will then be immediate, and then we simply add the appropriate symplectic quantities to obtain a canonically-invariant object on a general symplectic manifold $\mathcal{M}$ to ensure invariance under transformations which preserve the density of states.

To transform the classical theory of the last Chapter into a quantum mechanical one (i.e. to 'quantize' it), we replace the phase space coordinates (p, q) with operators $(\hat{p}, \hat{q})$ which obey an operator algebra that is obtained by replacing the Poisson algebra of the Darboux coordinates (3.18) by allowing the commutator bracket of the basis operators $(\hat{p}, \hat{q})$ to be simply equal to the Poisson brackets of the same objects as elements of the Poisson algebra of C^∞-functions on the phase space, times an additional factor of $i\hbar$ where $\hbar$ is Planck's constant,

$$[\hat{p}, \hat{q}] = i\hbar \qquad (4.1)$$

The operators $(\hat{p}, \hat{q})$ with the canonical commutation relation (4.1) make the space of C^∞-functions on $\mathcal{M}$ into an infinite-dimensional associative operator algebra called the Heisenberg algebra[1]. This algebra can be represented on the space $L^2(q)$ of square integrable functions of the configuration space coordinate q by letting the operator $\hat{q}$ act as multiplication by q and $\hat{p}$ as the derivative

$$\hat{p} = i\hbar \frac{\partial}{\partial q} \tag{4.2}$$

This representation of the Heisenberg algebra is called the Schrödinger picture and the elements of the Hilbert space $L^2(q)$ are called the wavefunctions or physical quantum states of the dynamical system[2].

The eigenstates of the (Hermitian) position and momentum operators are denoted by the usual Dirac bra-ket notation

$$\hat{q}|q\rangle = q|q\rangle \quad , \quad \hat{p}|p\rangle = p|p\rangle \tag{4.3}$$

These states are orthonormal,

$$\langle q|q'\rangle = \delta(q - q') \quad , \quad \langle p|p'\rangle = \delta(p - p') \tag{4.4}$$

and they obey the momentum and position space completeness relations

$$\int_{-\infty}^{\infty} dp\, |p\rangle\langle p| = \int_{-\infty}^{\infty} dq\, |q\rangle\langle q| = \mathbf{1} \tag{4.5}$$

with $\mathbf{1}$ the identity operator on the respective space. In the representation (4.2) on L^2-functions the momentum and configuration space representations are related by the usual Fourier transformation

$$|q\rangle = \int_{-\infty}^{\infty} \frac{dp}{\sqrt{2\pi\hbar}}\, e^{-ipq/\hbar}|p\rangle \tag{4.6}$$

which identifies the matrix element

$$\langle p|q\rangle = \langle q|p\rangle^* = \frac{1}{\sqrt{2\pi\hbar}}\, e^{-ipq/\hbar} \tag{4.7}$$

The basis operators have the matrix elements

$$\langle p|\hat{q}|q\rangle = q\langle p|q\rangle \quad , \quad \langle p|\hat{p}|q\rangle = p\langle p|q\rangle = i\hbar\frac{\partial}{\partial q}\langle p|q\rangle \tag{4.8}$$

All observables (i.e. real-valued C^∞-functions of (p,q)) now become Hermitian operators acting on the Hilbert space. In particular, the Hamiltonian of

---
[1] More precisely, the operators $(\hat{p}, \hat{q})$ generate the universal enveloping algebra of an extended affine Lie algebra which is usually identified as the Heisenberg algebra.
[2] Strictly speaking, these function spaces should be properly defined as distribution spaces in light of the discussion which follows.

the dynamical system now becomes a Hermitian operator $\hat{H} \equiv H(\hat{p},\hat{q})$ with the matrix elements

$$\langle p|\hat{H}|q\rangle = H(p,q)\langle p|q\rangle = H(p,q)\frac{e^{-ipq/\hbar}}{\sqrt{2\pi\hbar}} \tag{4.9}$$

and the eigenvalues of this operator determine the energy levels of the physical system.

The time evolution of any quantum operator is determined by the quantum mapping above of the Hamilton equations of motion (3.32). In particular, the time evolution of the position operator is determined by

$$\dot{\hat{q}}(t) = \frac{1}{i\hbar}\left[\hat{q},\hat{H}\right] \tag{4.10}$$

which may be solved formally by

$$\hat{q}(t) = e^{i\hat{H}t/\hbar}\hat{q}(0)\,e^{-i\hat{H}t/\hbar} \tag{4.11}$$

so that the time evolution is determined by a unitary transformation of the position operator $\hat{q}(0)$. In the Schrödinger representation, we treat the operators as time-independent quantities using the unitary transformation (4.11) and consider the time-evolution of the quantum states. The configuration of the system at a time t is defined using the unitary time-evolution operator in (4.11) acting on an initial configuration $|q\rangle$ at time $t=0$,

$$|q,t\rangle = e^{i\hat{H}t/\hbar}|q\rangle \tag{4.12}$$

which is an eigenstate of (4.11) for all t.

An important physical quantity is the quantum propagator

$$\mathcal{K}(q',q;T) = \langle q',T|q,0\rangle = \langle q'|\,e^{-i\hat{H}T/\hbar}|q\rangle \tag{4.13}$$

which, according to the fundamental principles of quantum mechanics [101], represents the probability of the system evolving from a state with configuration q to one with configuration q' in a time interval T. The propagator (4.13) satisfies the Schrödinger wave equation

$$i\hbar\frac{\partial}{\partial T}\mathcal{K}(q',q;T) = \hat{H}\mathcal{K}(q',q;T) \tag{4.14}$$

where the momentum operators involved in the Hamiltonian $\hat{H}$ on the right-hand side of (4.14) are represented in the Schrödinger polarization (4.2). The Schrödinger equation is to be solved with the Dirac delta-function initial condition

$$\mathcal{K}(q',q;T=0) = \delta(q'-q) \tag{4.15}$$

The function $\mathcal{K}(q',q;T)$ acts as an integration kernel which determines the time-evolution of the wavefunctions as

$$\Psi(q';T) = \int_{-\infty}^{\infty} dq\, \mathcal{K}(q',q;T)\Psi(q;0) \qquad (4.16)$$

where $\Psi(q;t) \equiv \langle q,t|\Psi\rangle$ are the time-dependent configuration space representations of the physical states $|\Psi\rangle$ of the system. Thus the propagator represents the fundamental quantum dynamics of the system and the stationary state solutions to the Schrödinger equation (4.14) determine the energy eigenvalues of the dynamical system.

The phase space path integral provides a functional representation of the quantum propagator in terms of a 'sum' over continuous trajectories on the phase space. It is constructed as follows [147]. Between the initial and final configurations q and q' we introduce $N-1$ intermediate configurations $q_0, \ldots, q_N$ with $q_0 \equiv q$ and $q_N \equiv q'$, and each separated by the time interval

$$\Delta t = T/N \qquad (4.17)$$

Introducing intermediate momenta $p_1, \ldots, p_N$ and inserting the completeness relations

$$\int_{-\infty}^{\infty} dq_{j-1}\, dq_j\, dp_j\, |q_j\rangle\langle q_j|p_j\rangle\langle p_j|q_{j-1}\rangle\langle q_{j-1}| = 1 \quad , \quad j=1,\ldots,N \qquad (4.18)$$

into the matrix element (4.13) we obtain

$$\mathcal{K}(q',q;T) = \langle q'|(\,\mathrm{e}^{-i\hat{H}\Delta t/\hbar})^N|q\rangle$$

$$= \int_{-\infty}^{\infty} \prod_{j=1}^{N} dq_{j-1}\, dq_j\, dp_j\, \langle q'|q_j\rangle\langle q_j|\,\mathrm{e}^{-i\hat{H}\Delta t/\hbar}|p_j\rangle\langle p_j|q_{j-1}\rangle\langle q_{j-1}|q\rangle \qquad (4.19)$$

$$= \int_{-\infty}^{\infty} \prod_{j=1}^{N-1} dq_j \prod_{j=1}^{N} \frac{dp_j}{2\pi\hbar}\, \exp\left\{\frac{i}{\hbar}\sum_{i=1}^{N}\left(p_i\frac{q_i-q_{i-1}}{\Delta t} - H(p_i,q_i)\right)\Delta t\right\}$$

$$\times \delta(q_0 - q)\delta(q_N - q')$$

where we have used the various identities quoted above. In the limit $N\to\infty$, or equivalently $\Delta t \to 0$, the discrete points (p_j, q_j) describe paths $(p(t), q(t))$ in the phase space between the configurations q and q', and the sum in (4.19) becomes the continuous limit of a Riemann sum representing a discretized time integration. Then (4.19) becomes

$$\mathcal{K}(q',q;T) = \lim_{N\to\infty} \int_{-\infty}^{\infty} \prod_{j=1}^{N-1} dq_j \prod_{j=1}^{N} \frac{dp_j}{2\pi\hbar}\, \exp\left\{\frac{i}{\hbar}\int_0^T dt\, (p(t)\dot{q}(t) - H(p,q))\right\}$$

$$\times \delta(q(0)-q)\delta(q(T)-q')$$

$$(4.20)$$

82   4. Quantum Localization Theory for Phase Space Path Integrals

Note that the argument of the exponential in (4.20) is just the classical action of the dynamical system, because its integrand is the usual Legendre transformation between the Lagrangian and Hamiltonian descriptions of the classical dynamics [55]. Notice also that, in light of the Heisenberg uncertainty principle $\Delta q \Delta p \sim 2\pi\hbar$, the normalization factors $2\pi\hbar$ there can be physically interpreted as the volume of an elementary quantum state in the phase space.

The integration measure in (4.20) formally gives an integral over all phase space paths defined in the time interval $[0, T]$. This measure is denoted by

$$[dp\, dq] \equiv \lim_{N\to\infty} \prod_{j=1}^{N} \frac{dp_j}{2\pi\hbar} \prod_{j=1}^{N-1} dq_j \equiv \prod_{t\in[0,T]} \frac{dp(t)}{2\pi\hbar} dq(t) \qquad (4.21)$$

and it is called the Feynman measure. The last equality means that it is to be understood as a 'measure' on the infinite-dimensional functional space of phase space trajectories $(p(t), q(t))$, where for each fixed time slice $t \in [0, T]$, $dp(t)\, dq(t)$ is ordinary Riemann-Lebesgue measure. Being an infinite-dimensional quantity, it is not rigorously defined, and some special care must be taken to determine the precise meaning of the limit $N \to \infty$ above. This has been a topic of much dispute over the years and we shall make no attempt in this Book to discuss the ill-defined ambiguities associated with the Feynman measure. Many rigorous attempts at formulating the path integral have been proposed in constructive quantum field theory. For instance, it is possible to give the limit (4.21) a somewhat precise meaning using the so-called Lipschitz functions of order $\frac{1}{2}$ which assumes that the paths which contribute in (4.20) grow no faster than $\mathcal{O}(\sqrt{t})$ (these functional integrals are called Wiener integrals)[3]. We shall at least assume that the integration measure (4.21) is supported on C^∞ phase space paths and that the quantum mechanical propagator given by (4.20) is a tempered distribution, i.e. it can diverge with at most a polynomial growth. This latter restriction on the path integral is part of the celebrated Wightman axioms for quantum field theory which allows one to at least carry out certain formal rigorous manipulations from the theory of distributions.

However, a physicist will typically proceed without worry and succeed in extracting a surprising amount of information from formulas such as (4.20) without the need to investigate in more detail the implications of the limit $N \to \infty$ above. To actually carry out functional integrations such as (4.20) one uses formal functional analogs of the usual rules of Riemann-Lebesgue integration in the straightforward sense, where all time integrals are treated as continuous sums on the functional space (i.e. the time parameter t is regarded as a continuous index).

---

[3] Note that the transition from the multiple integral representation in (4.19) to the representation (4.20),(4.21) in terms of phase space paths requires that these trajectories can at least be approximated by piecewise-linear functions.

If we set $q = q'$ and integrate over all q, then the left-hand side of (4.20) yields

$$\int_{-\infty}^{\infty} dq \ \langle q| \ e^{-i\hat{H}T/\hbar} |q\rangle \equiv \text{tr} \, \| \, e^{-i\hat{H}T/\hbar} \| = \sum dE \ e^{-iET/\hbar} \quad (4.22)$$

where E are the energy eigenvalues of $\hat{H}$ and the symbol $\|\cdot\|$ will be used to emphasize that the matrix of interest is considered as an infinite dimensional one over either the Hilbert space of physical states or the functional trajectory space. On the other hand, the right-hand side of (4.20) becomes

$$Z(T) = \int [dp \, dq] \ \exp\left\{\frac{i}{\hbar} \int_0^T dt \ (p(t)\dot{q}(t) - H(p,q))\right\} \delta(q(0) - q(T)) \quad (4.23)$$

which is called the quantum partition function. From (4.22) we see that the quantum partition function describes the spectrum of the quantum Hamiltonian of the dynamical system and that the poles of its Fourier transform

$$G(E) = \int_0^{\infty} dT \ e^{iET/\hbar} Z(T) \quad (4.24)$$

give the bound state spectrum of the system [101]. The quantity (4.24) is none other than the energy Green's function $G(E) = \text{tr}\|(E - \hat{H})^{-1}\|$ which is associated with the Schrödinger equation (4.14). Thus the quantum partition function is in some sense the fundamental quantity which describes the quantum dynamics (i.e. the energy spectrum) of a Hamiltonian system.

Finally, the generalization to an arbitrary symplectic manifold $(\mathcal{M}, \omega)$ of dimension $2n$ is immediate. The factor $p\dot{q}$ becomes simply $p_\mu \dot{q}^\mu$ in higher dimensions, and, in view of (3.19), the canonical form of this is $\theta_\mu(x)\dot{x}^\mu$ in an arbitrary coordinate system on $\mathcal{M}$. Likewise, the phase space measure $dp \wedge dq$ according to (3.23) should be replaced by the canonically-invariant Liouville measure (3.22). Thus the quantum partition function for a generic dynamical system $(\mathcal{M}, \omega, H)$ is defined as

$$Z(T) = \int_{L\mathcal{M}} [d\mu_L(x)] \ e^{iS[x]} = \int_{L\mathcal{M}} [d^{2n}x] \prod_{t \in [0,T]} \sqrt{\det \|\omega(x(t))\|} \ e^{iS[x]} \quad (4.25)$$

where

$$S[x] = \int_0^T dt \ (\theta_\mu(x)\dot{x}^\mu - H(x)) \quad (4.26)$$

is the classical action of the Hamiltonian system. Here and in the following we shall set $\hbar \equiv 1$ for simplicity, and the functional integration in (4.25) is taken over the loop space $L\mathcal{M}$ of $\mathcal{M}$, i.e. the infinite-dimensional space of paths $x(t) : [0, T] \to \mathcal{M}$ obeying periodic boundary conditions $x^\mu(0) = x^\mu(T)$. Although much of the formalism which follows can be applied to path integrals over the larger trajectory space of all paths, we shall find it convenient to deal mostly with the loop space over the phase space. The partition function (4.25) can be regarded as the formal infinite-dimensional analog of the classical integral (3.52), or, as mentioned before, the prototype of a topological field theory functional integral regarded as a $(0 + 1)$-dimensional quantum field theory. In the latter application the discrete index sums over μ contain as well integrals over the manifold on which the fields are defined.

Notice that the symplectic potential θ appearing in (4.25),(4.26) is only locally defined, and so some care must be taken in defining (4.25) when ω is not globally exact. We shall discuss this procedure later on. Note also that the Liouville measure in (4.25), which is defined by the last equality in (4.21), differs from that of (4.23) in that in the latter case there is one extra momentum integration in the phase space Feynman measure (4.21), so that the endpoints are fixed and we integrate over all intermediate momenta. Thus one must carefully define appropriate boundary conditions for the integrations in (4.25) for the Schrödinger path integral measure in order to maintain a formal analogy between the finite and infinite dimensional cases. We shall elaborate on this point in the next Chapter. Further discussion of this and the proper discretizations and ordering prescriptions that are needed to define the functional integrations that appear above can be found in [147] and [93].

4.2 Example: Path Integral Derivation of the Atiyah-Singer Index Theorem

As we did at the start of Chapter 2 above, we shall motivate the formal manipulations that will be carried out on the phase space path integral (4.25) with an explicit example which captures the essential ideas we shall need. At the same time, this particular example sets the stage for the analogies with topological field theory functional integrals which will follow and will serve as a starting point for some of the applications which will be discussed in Chapter 8. We will consider the derivation, via the evaluation of a path integral for supersymmetric quantum mechanics [5, 48], of the Atiyah-Singer index theorem which expresses the fact that the analytical index of a Dirac operator is a topological invariant of the background fields in the quantum field theory in which it is defined. This theorem and its extensions have many uses in quantum field theory, particularly for the study of anomalies and the fractional fermion number of solitons [41, 125, 158].

Consider a Dirac operator $i\slashed{\nabla}$ on an even-dimensional compact orientable Riemannian manifold $\mathcal{M}$ with metric g of Minkowski signature,

4.2 Example: Path Integral Derivation of the Atiyah-Singer Index Theorem

$$i\slashed{\nabla} \equiv i\gamma^\mu \nabla_\mu \equiv i\gamma^\mu \left(\partial_\mu + \frac{1}{8}\omega_{\mu jk}[\gamma^j,\gamma^k] + iA_\mu\right) \quad (4.27)$$

Here $\gamma^\mu(x)$ are the Dirac matrices which generate the Clifford algebra of $\mathcal{M}$,

$$\gamma^\mu\gamma^\nu + \gamma^\nu\gamma^\mu = 2g^{\mu\nu}(x) \quad (4.28)$$

and A_μ is a connection on a principal fiber bundle $E \to \mathcal{M}$ (i.e. a gauge field). We shall assume for simplicity that the structure group of the principal bundle is $G = U(1)$, so that A is a connection on a line bundle $L \to \mathcal{M}$. The spin-connection $\omega^j_{\mu i}$ is defined as follows. At each point $x \in \mathcal{M}$ we introduce a local basis of orthonormal tangent vectors $e^i_\mu(x)$, called a vielbein, where μ labels the basis components in $T\mathcal{M}$ and $i = 1,\ldots,\dim\mathcal{M}$ parametrizes the fibers of $T\mathcal{M}$ (i.e. the local rotation index in the tangent space). Orthonormality means that $g^{\mu\nu}(x)e^i_\mu(x)e^j_\nu(x) = \eta^{ij}$ is the flat Minkowski metric in $T\mathcal{M}$, or equivalently

$$\eta_{ij}e^i_\mu(x)e^j_\nu(x) = g_{\mu\nu}(x) \quad (4.29)$$

In this vielbein formalism, $\gamma^i \equiv e^i_\mu(x)\gamma^\mu(x)$ in (4.27) and the spin-connection (i.e. connection on the spin bundle $S\mathcal{M}$ of $\mathcal{M}$, defined by the $\dim\mathcal{M}$-dimensional spinor representations of the local Lorentz group of the tangent bundle) is

$$\omega^i_{\mu j} = e^i_\nu(\partial_\mu E^\nu_j + \Gamma^\nu_{\mu\lambda}E^\lambda_j) \quad (4.30)$$

and $E^\mu_i(x)$ are the inverse vielbein fields, i.e. $E^\mu_i e^j_\mu = \delta^j_i$. The spin-connection (4.30) is a gauge field of the local Lorentz group of the tangent bundle, i.e. under a local Lorentz transformation $e^i_\mu(x) \to \Lambda^i_j(x)e^j_\mu(x)$, $\Lambda(x) \in SO(2n-1,1)$, on the frame bundle of $\mathcal{M}$, the gauge field ω_μ transforms in the usual way as $\omega_\mu \to \Lambda\omega_\mu\Lambda^{-1} - \partial_\mu\Lambda\cdot\Lambda^{-1}$. It is defined so that the covariant derivative in (4.27) coincides with the Levi-Civita-Christoffel connection, i.e. $\nabla_\mu e^i_\nu \equiv \partial_\mu e^i_\nu - \Gamma^\lambda_{\mu\nu}e^i_\lambda + \omega^i_{\mu j}e^j_\nu = 0$. The covariant derivative in (4.27) is in general regarded as a connection on the bundle $T\mathcal{M}\otimes S\mathcal{M}\otimes L$ which together define the twisted spin complex of $\mathcal{M}$ (the 'twisting' being associated with the presence of the gauge field A).

The chiral representation of the Dirac matrices is that in which the chirality matrix $\gamma^c \equiv i\gamma^1\gamma^2\cdots\gamma^{2n}$, with the properties $(\gamma^c)^2 = \mathbf{1}$ and $(\gamma^c)^\dagger = \gamma^c$, is diagonal. Since the Dirac operator (4.27) commutes with γ^c, in this representation of the Clifford algebra (4.28) these 2 operators can be written in the block forms

$$\gamma^c = \begin{pmatrix} 1 & 0 \\ 0 & -1 \end{pmatrix}, \quad i\slashed{\nabla} = \begin{pmatrix} 0 & \mathcal{D} \\ \mathcal{D}^\dagger & 0 \end{pmatrix} \quad (4.31)$$

The analytical index of $i\slashed{\nabla}$ is then defined as the difference between the dimensions of the kernel and co-kernel of the elliptic operator $\mathcal{D}$,

$$\mathrm{index}(i\slashed{\nabla}) \equiv \dim\ker\mathcal{D} - \dim\mathrm{coker}\,\mathcal{D} = \dim\ker\mathcal{D} - \dim\ker\mathcal{D}^\dagger \quad (4.32)$$

86   4. Quantum Localization Theory for Phase Space Path Integrals

In the chiral representation (4.31), the Dirac spinors, i.e. the solutions Ψ of the Dirac equation

$$i\slashed{\nabla}\Psi = \begin{pmatrix} 0 & \mathcal{D} \\ \mathcal{D}^\dagger & 0 \end{pmatrix}\begin{pmatrix} \Psi_+ \\ \Psi_- \end{pmatrix} = E\Psi \qquad (4.33)$$

are determined by their positive and negative chirality spin components $\Psi_\pm$. Since the zero mode solutions, $E = 0$, satisfy $\mathcal{D}^\dagger\Psi_+ = \mathcal{D}\Psi_- = 0$, the index (4.32) is just the difference between the number of positive and negative chirality zero-mode solutions of the Dirac equation (4.33), i.e.

$$\mathrm{index}(i\slashed{\nabla}) = \mathrm{Tr}_{E=0}\gamma^c \qquad (4.34)$$

Moreover, since $[i\slashed{\nabla}, \gamma^c] = 0$, the chirality operator provides a one-to-one mapping between positive and negative non-zero energy states. Thus the index (4.34) can be written as a trace over the full Hilbert space $\mathcal{H}$ spanned by the Dirac spinors as

$$\mathrm{index}(i\slashed{\nabla}) = \mathrm{tr}_{\mathcal{H}}\|\gamma^c\,\mathrm{e}^{-T(\mathcal{D}\mathcal{D}^\dagger + \mathcal{D}^\dagger\mathcal{D})}\| \qquad (4.35)$$

where we have used the fact that the spinors satisfying the eigenvalue equation (4.33) also obey the Schrödinger equations

$$\mathcal{D}\mathcal{D}^\dagger\Psi_+ = E^2\Psi_+ \quad , \quad \mathcal{D}^\dagger\mathcal{D}\Psi_- = E^2\Psi_- \qquad (4.36)$$

The parameter $T > 0$ in (4.35) regulates the operator trace.

The representation (4.34) of the index of a Dirac operator is known as the Witten index [22, 166, 167]. We can identify the positive and negative chirality spinors as bosons and fermions, respectively, and then the chirality operator can be written as $\gamma^c = (-1)^F$ where F is the fermion (or ghost) number operator. The operator $\mathcal{D}^\dagger$ can then be identified with a supersymmetry generator Q, which provides a mapping between fermions and bosons, associated with a supersymmetric theory with Hamiltonian given by the graded BRST commutator (see (4.35) above)

$$H = \{Q, Q^\dagger\} \geq 0 \qquad (4.37)$$

as is standard in a supersymmetric model. This is equivalent to the statement above that bosonic and fermionic states of non-zero energy in the supersymmetric theory are always paired. Since $QQ^\dagger$ and $Q^\dagger Q$ are positive-definite Hermitian operators, the zero modes $|0\rangle$ of H are supersymmetric, $Q|0\rangle = Q^\dagger|0\rangle = 0$, and thus they provide a (trivial) 1-dimensional representation of the supersymmetry. Small perturbations of the background gravitational and gauge fields g and A may excite the $E = 0$ states, but bosonic and fermionic states must always be lifted in pairs. Consequently, the Witten index

$$\mathrm{index}(i\slashed{\nabla}) = \mathrm{tr}_{\mathcal{H}}\|(-1)^F\,\mathrm{e}^{-TH}\| \equiv \mathrm{str}\|\mathrm{e}^{-TH}\| \qquad (4.38)$$

4.2 Example: Path Integral Derivation of the Atiyah-Singer Index Theorem

is a topological invariant that is independent of the choice of spin and gauge connections. Here str denotes the supertrace and, since $\mathcal{M}$ is compact by assumption, there are only finitely many modes which contribute in (4.38). Thus all these quantities are independent of the parameter T. In the low temperature limit $(T \to \infty)$ only zero modes contribute to (4.38) according to their chirality.

The physical relevence of the Witten index is immediate. As the zero energy states in general need not be paired, the non-vanishing of the Witten index (4.38) implies that there is at least one zero energy state which is then an appropriate supersymmetric ground state of the underlying supersymmetric theory. Thus the non-vanishing of (4.38) is a sufficient condition for the presence of supersymmetric ground states. Conversely, a necessary criterion for dynamical supersymmetry breaking is that $\text{Tr}_{E=0}(-1)^F$ should vanish.

Using the standard path integral techniques of the last Section, it is straightforward to write down a path integral representation of (4.38) [5]. The collection of all fields Φ will clearly involve both bosonic and fermionic degrees of freedom which will be connected by a supersymmetry, i.e. the appropriate path integral representation will be that of a supersymmetric field theory. Furthermore, the integral over the function space of fields on $\mathcal{M}$ will be restricted to fields which satisfy periodic boundary conditions for both the space and time coordinates, $\Phi(t+T) = \Phi(t)$. This restriction is necessary for the pairing of states discussed above, and the reason for this condition in the time direction for the fermionic fields is because of the presence of the Klein operator $(-1)^F$ in the supertrace. The path integral representation of the index (4.38) is then [5, 48]

$$\text{index}(i\nabla\!\!\!\!/\,) = \int_{L\mathcal{M} \otimes L\Lambda^1 \mathcal{M}} [d^{2n}x] \, [d^{2n}\psi] \, e^{iT I_{1/2}[x,\psi]} \qquad (4.39)$$

where $\psi^\mu(t)$ are anticommuting periodic paths on $\mathcal{M}$ (which, according to Section 2.6, can be taken to lie in $L\Lambda^1\mathcal{M}$) with path integration defined using functional analogs of the Berezin integration rules discussed in Section 2.6. The action in (4.39) is that of $N = \frac{1}{2}$ (Dirac) supersymmetric quantum mechanics, i.e. the invariant action for a spinning particle in background gravitational and gauge fields[4]

$$I_{1/2}[x,\psi] = \oint_{S^1} d\tau \left(\frac{1}{2} g_{\mu\nu} \dot{x}^\mu \dot{x}^\nu + \dot{x}^\mu A_\mu + \frac{1}{2} g_{\mu\nu} \psi^\mu \nabla_\tau \psi^\nu - \frac{1}{2} \psi^\mu F_{\mu\nu} \psi^\nu \right)$$

(4.40)

---

[4] In general an N-component supersymmetric model contains N fermion chiral conjugate pairs $(\bar{\psi}_i, \psi_i)$ with $2N$ associated superpartner bosonic fields F_i, and N corresponding supersymmetry charges $(Q_i^\dagger, Q_i)$ which mix the fields with their superpartners.

where ψ^μ are the Grassmann (superpartner) coordinates for the particle configurations $x^\mu \in \mathcal{M}$, and we have rescaled the time by T so that time integrations lie on the unit circle S^1. Here

$$\nabla_\tau W^\mu(x(\tau)) = \partial_\tau W^\mu(x(\tau)) + \Gamma^\mu_{\nu\lambda}(x(\tau))\dot{x}^\nu(\tau)W^\lambda(x(\tau)) \tag{4.41}$$

is the covariant derivative along the loop $x(\tau)$ induced by the Riemannian connection ∇ on $\mathcal{M}$, and $F_{\mu\nu} = \partial_\mu A_\nu - \partial_\nu A_\mu$ is the gauge field strength tensor. In (4.40), the particle current $\dot{x}^\mu$ is minimally coupled to the gauge field A_μ, and its spinor degrees of freedom couple to the electromagnetic field of A_μ by the usual Pauli magnetic moment interaction. The action (4.40) has the (infinitesimal) supersymmetry

$$\mathcal{S}x^\mu(\tau) = \psi^\mu(\tau) \quad , \quad \mathcal{S}\psi^\mu(\tau) = \dot{x}^\mu(\tau) \tag{4.42}$$

The action (4.40) arises from the standard supersymmetric non-linear (Wess-Zumino) sigma-model, and we shall see in Chapter 8 how to write this supersymmetric model in a more conventional fashion using superspace coordinates and superfields.

Let us briefly describe how one arrives at the path integral representation (4.39) (see [71] for details). The local Dirac algebra of the spin bundle is represented by the anticommutator $[\psi^i, \psi^j]_+ = \eta^{ij}$, so that the Clifford algebra of $\mathcal{M}$ is represented is represented as

$$[\psi^\mu, \psi^\nu]_+ = g^{\mu\nu}(x) \tag{4.43}$$

where as always $\psi^\mu = E_i^\mu \psi^i$. The zero-mode equation for the Dirac operator $i\slashed{\nabla}$ can therefore be realized as the graded constraint equation

$$\mathcal{S} \equiv \psi^\mu \left(\partial_\mu + \frac{1}{4}\omega_{\mu ij}\psi^i\psi^j + iA_\mu \right) = 0 \tag{4.44}$$

where the supersymmetry generator $\mathcal{S}$ associated with (4.27) generates the (graded) $N = \frac{1}{2}$ supersymmetry algebra[5]

$$\{\mathcal{S}, \mathcal{S}\} = H = g^{\mu\nu}\left(\partial_\mu + \frac{1}{4}\omega_{\mu ij}\psi^i\psi^j + iA_\mu\right)\left(\partial_\nu + \frac{1}{4}\omega_{\nu k\ell}\psi^k\psi^\ell + iA_\nu\right)$$

$$+ \frac{1}{2}\psi^\mu F_{\mu\nu}\psi^\nu$$

$$\{\mathcal{S}, H\} = \{H, H\} = 0 \tag{4.45}$$

In arriving at (4.45) we have used the various symmetry properties of the Riemann curvature tensor. Notice that the Hamiltonian H vanishes on physical (supersymmetric) ground states, so that there are no local propagating

---
[5] See Appendix A for the convention for the graded commutator $\{\cdot, \cdot\}$.

4.2 Example: Path Integral Derivation of the Atiyah-Singer Index Theorem

degrees of freedom and the model can only describe the global topological characteristics of the manifold $\mathcal{M}$, i.e. this supersymmetric model defines a topological field theory.

The constraint algebra (4.45) contains first class constraints, i.e. it defines a closed algebra between H and $\mathcal{S}$ (see Appendix A) such that H is supersymmetric under the infinitesimal supersymmetry transformations generated by $\mathcal{S}$. The constraints $H = \mathcal{S} = 0$ ensure the reparametrization invariance of the trajectories $x^\mu(\tau)$. It is straightforward to now construct the BRST gauge fixed path integral associated with this constraint algebra (in the proper time gauge). Since the various ghost degrees of freedom only couple to world line quantities and not to the metric structure of $\mathcal{M}$, (4.39) coincides with the canonical BRST gauge-fixed path integral describing the propagation of a Dirac particle on the configuration space $\mathcal{M}$ (with the identification $p_\mu \sim \partial_\mu$ as the canonical momentum conjugate to x^μ). The gauge-fixed quantum action (4.40) is written only modulo the ghost field and other contributions that decouple from the background metric of $\mathcal{M}$, as these fields only contribute to the overall normalization in (4.39). The necessity to use periodic boundary conditions in the path integral follows from the identification $\psi^i \sim \gamma^c\gamma^i$ of the Dirac matrices.

There are several ways to evaluate explicitly the supersymmetric path integral (4.39). The traditional method is to exploit the T-independence and use, in the high-temperature limit $(T \to 0)$, either a heat kernel expansion of the trace in (4.35) [41] or a normal coordinate expansion to evaluate the partition function (4.39) [5, 48]. Here, however, we wish to emphasize the observation of Atiyah and Witten [8] (and the later generalizations to twisted Dirac operators by Bismut [23, 24] and Jones and Petrack [80]) that the path integral for $N = \frac{1}{2}$ supersymmetric quantum mechanics admits a formal equivariant cohomological structure on the superloop space $L\mathcal{M} \otimes L\Lambda^1\mathcal{M}$. To see how this structure arises, we introduce a geometric framework for manipulating the path integral (4.39). These geometric manipulations will be the starting point for the general analyses of generic phase space path integrals which will follow. Given any functional $F[x]$ of closed paths in the loop space $L\mathcal{M}$, we define functional differentiation, for which functional integration is the anti-derivative thereof, by the rule

$$\frac{\delta}{\delta x^\mu(\tau)} F[x(\tau')] = \delta(\tau - \tau') F'[x(\tau')] \qquad (4.46)$$

and the rules for functional differentiation of the periodic Grassmann-valued paths by the anti-commutator

$$\left[\frac{\delta}{\delta \psi^\mu(\tau)}, \psi^\nu(\tau')\right]_+ = \delta^\nu_\mu \delta(\tau - \tau') \qquad (4.47)$$

The crucial point is that the fermionic part of the supersymmetric action (4.40) is bilinear in the fermion fields so that the functional Berezin integration induces a determinant factor $\det^{1/2}\|\hat{\Omega}\|$ which makes the remaining

90   4. Quantum Localization Theory for Phase Space Path Integrals

integration over $L\mathcal{M}$ in (4.39) resemble the phase space path integral (4.25). More precisely, the loop space fermionic bilinear form appearing in (4.40) is

$$\hat{\Omega}[x,\psi] = \oint_{S^1} d\tau \, \frac{1}{2}\psi^\mu(\tau)\left(g_{\mu\nu}\nabla_\tau - F_{\mu\nu}(x(\tau))\right)\psi^\nu(\tau) \quad (4.48)$$

which, after Berezin integration, induces a loop space Liouville measure $[d^{2n}x]\sqrt{\det\|\hat{\Omega}\|}$. Introducing the nilpotent graded derivative operator

$$D = \oint_{S^1} d\tau \, \psi^\mu(\tau)\frac{\delta}{\delta x^\mu(\tau)} \quad (4.49)$$

we see that (4.48) can be expressed as a D-exact quantity

$$\hat{\Omega}[x,\psi] = D\hat{\Sigma}[x,\psi] \quad (4.50)$$

where

$$\hat{\Sigma}[x,\psi] = \oint_{S^1} d\tau \, \{g_{\mu\nu}(x(\tau))\dot{x}^\nu(\tau) + A_\mu(x(\tau))\}\,\psi^\mu(\tau) \equiv \oint_{S^1} d\tau \, \hat{\Sigma}_\mu(x(\tau))\psi^\mu(\tau) \quad (4.51)$$

The functional (4.48) can be interpreted as a loop space symplectic structure. Strictly speaking though, it is properly termed a 'pre-symplectic' structure because although it is D-closed, $D\hat{\Omega}[x,\psi] = 0$, it is not necessarily non-degenerate on the loop space. It is this interpretation of supersymmetric theories in general that makes infinite-dimensional generalizations of the equivariant localization formalisms of Chapter 3 very powerful tools.

In particular, the $N = \frac{1}{2}$ supersymmetry (4.42) can be represented by a loop space equivariant derivative operator. To see this, introduce the nilpotent graded contraction operator

$$I_{\dot{x}} = \oint_{S^1} d\tau \, \dot{x}^\mu(\tau)\frac{\delta}{\delta\psi^\mu(\tau)} \quad (4.52)$$

and define the corresponding graded equivariant exterior derivative operator

$$D_{\dot{x}} = D + I_{\dot{x}} = \oint_{S^1} d\tau \, \left(\psi^\mu(\tau)\frac{\delta}{\delta x^\mu(\tau)} + \dot{x}^\mu(\tau)\frac{\delta}{\delta\psi^\mu(\tau)}\right) \quad (4.53)$$

Then the supersymmetry (4.42) is immediately recognized as the action of the derivative $D_{\dot{x}} \sim S$ on $L\mathcal{M} \otimes L\Lambda^1\mathcal{M}$. The square of $D_{\dot{x}}$ is the generator of time translations on the superloop space $L\mathcal{M} \otimes L\Lambda^1\mathcal{M}$,

$$D_{\dot{x}}^2 = \oint_{S^1} d\tau \, \left(\dot{x}^\mu(\tau)\frac{\delta}{\delta x^\mu(\tau)} + \dot{\psi}^\mu(\tau)\frac{\delta}{\delta\psi^\mu(\tau)}\right) = \oint_{S^1} d\tau \, \frac{d}{d\tau} \quad (4.54)$$

4.2 Example: Path Integral Derivation of the Atiyah-Singer Index Theorem

so that its action on a loop space functional $W[x, \psi]$ is

$$D_{\dot{x}}^2 W[x,\psi] = \oint_{S^1} d\tau \, \frac{d}{d\tau} W[x,\psi] = W[x(1),\psi(1)] - W[x(0),\psi(0)] \quad (4.55)$$

Consequently, (4.53) is a nilpotent operator provided we restrict to single-valued loop space functionals $W[x, \psi]$. Hence the action of $N = \frac{1}{2}$ supersymmetric quantum mechanics defines an equivariant structure on $L\mathcal{M} \otimes L\Lambda^1\mathcal{M}$ and on the basis of the general arguments of Chapter 2 we expect its path integral to localize to an integral over $\mathcal{M}$, the zero locus of the vector field $\dot{x}^\mu(\tau)$ (i.e. the constant paths $x(t) = x(0) \in \mathcal{M} \; \forall t$). This is well-known to be the case [5, 48].

In fact, the full action (4.40) is $D_{\dot{x}}$-exact,

$$I_{1/2}[x,\psi] = \oint_{S^1} d\tau \, \hat{\Sigma}_\mu(x(\tau))\dot{x}^\mu(\tau) + \hat{\Omega}[x,\psi] = I_{\dot{x}}\hat{\Sigma}[x,\psi] + \hat{\Omega}[x,\psi] \equiv D_{\dot{x}}\hat{\Sigma}[x,\psi] \quad (4.56)$$

and its bosonic part resembles the general phase space action functional (4.26) with $H \equiv 0$ there. As mentioned before, the vanishing of the Hamiltonian is the topological feature of such supersymmetric field theories. Now the equivariant localization principle applied to the case at hand would imply on its own that the path integral

$$\text{index}(i\nabla\!\!\!\!/\,) = \int_{L\mathcal{M} \otimes L\Lambda^1\mathcal{M}} [d^{2n}x] \, [d^{2n}\psi] \; e^{iTD_{\dot{x}}\hat{\Sigma}[x,\psi]} \quad (4.57)$$

is formally independent of the parameter T, and thus it manifestly localizes onto $\dot{x}^\mu(\tau) = 0$ (for $T \to \infty$). Of course, we cannot simply set $T = 0$ in (4.57) because the bosonic integration would yield ∞ while the fermionic one would give 0, leading to an ill-defined quantity. In any case, if we think of the coefficient T in front of the action as Planck's constant $\hbar$, then this is just another way of seeing that the semi-classical approximation is exact. The T-independence can be understood from the point of view that if we differentiate the right-hand side of (4.57) with respect to T, then we obtain the vacuum expectation value $\langle 0|D_{\dot{x}}\hat{\Sigma}|0\rangle$ in the supersymmetric quantum field theory above. If the vacuum itself is invariant under the $N = \frac{1}{2}$ supersymmetry of the model, then $D_{\dot{x}}|0\rangle = 0$ and the vacuum expectation value of this operator vanishes. It is these same sorts of arguments which establish the topological invariance of BRST-exact path integrals (also known as cohomological or Witten-type topological field theories) in general [22]. The above connection between the formalisms of the previous Chapters and the Atiyah-Singer index theorem is the usual intimate connection between standard supersymmetric models (for instance those which arise in the Duistermaat-Heckman interpretation of the quantum mechanics of spin [156, 4]) and equivariant cohomology [26].

We now use the fact that (4.57) can be evaluated for $T \to \infty$ and use a trick similar to that in Section 3.7. We introduce based loops on $L\mathcal{M} \otimes L\Lambda^1 \mathcal{M}$,

$$x^\mu(\tau) = x_0^\mu + \hat{x}^\mu(\tau) \quad , \quad \psi^\mu(\tau) = \psi_0^\mu + \hat{\psi}^\mu(\tau) \tag{4.58}$$

with $(x_0, \psi_0) \in \mathcal{M} \otimes \Lambda^1 \mathcal{M}$ the constant modes of the fields and $(\hat{x}, \hat{\psi})$ the non-constant fluctuations about these zero-modes, and define the path integral measure by[6]

$$[d^{2n}x]\,[d^{2n}\psi] \equiv d^{2n}x_0\, d^{2n}\psi_0 \prod_{\tau \in S^1} d^{2n}\hat{x}(\tau)\, d^{2n}\hat{\psi}(\tau) \tag{4.59}$$

We then rescale the non-constant modes as

$$\hat{x}^\mu(\tau) \to \hat{x}^\mu(\tau)/\sqrt{T} \quad , \quad \hat{\psi}^\mu(\tau) \to \hat{\psi}^\mu(\tau)/\sqrt{T} \tag{4.60}$$

With this rescaling, we find after some algebra that in the limit $T \to \infty$ the action (4.40) becomes

$$T \cdot I_{1/2}[x, \psi] \stackrel{T \to \infty}{\longrightarrow} \oint_{S^1} d\tau \left(\frac{1}{2} g_{\mu\nu}(x_0) \dot{\hat{x}}^\mu(\tau) \dot{\hat{x}}^\nu(\tau) + \frac{1}{2} \hat{\psi}^i(\tau) \eta_{ij} \partial_\tau \hat{\psi}^j(\tau) \right.$$

$$\left. - \frac{1}{2} \psi_0^\mu F_{\mu\nu}(x_0) \psi_0^\nu + \frac{1}{2} R_{ij\mu\nu}(x_0) \psi_0^i \psi_0^j \dot{\hat{x}}^\mu(\tau) \dot{\hat{x}}^\nu(\tau) \right)$$

$$+ \mathcal{O}(1/\sqrt{T}) \tag{4.61}$$

where we have Taylor expanded the quantities in (4.40) about (x_0, ψ_0) using (4.58) and (4.60). The Jacobians for the scaling by $\sqrt{T}$ cancel out from the bosonic and fermionic integration measures in (4.59), and the remaining functional integrations over non-constant modes are Gaussian. This illustrates the strong role that supersymmetry plays here in reducing the complicated integrations in (4.39) to Gaussian ones.

Evaluating these Gaussian integrations in (4.39) leads to

$$\text{index}(i\slashed{\nabla}) = \int_{\mathcal{M} \otimes \Lambda^1 \mathcal{M}} d^{2n}x_0\, d^{2n}\psi_0\; e^{\frac{i}{4\pi} F_{\mu\nu}(x_0) \psi_0^\mu \psi_0^\nu}$$

$$\times \left(\text{det}' \| \delta_\nu^\mu \partial_\tau - R_\nu^\mu(x_0, \psi_0) \| \right)^{-1/2} \tag{4.62}$$

where we have ignored (infinite) constant factors arising from the Gaussian functional integrations and normalized the $U(1)$ connection. Here the prime on the determinant means that it is taken over the fluctuation modes with periodic boundary conditions (i.e. the determinant with zero modes excluded).

---

[6] In Section 4.6 we shall be a bit more precise about this decomposition over a general superloop space $L\mathcal{M} \otimes L\Lambda^1 \mathcal{M}$. For now, we are just concerned with using this to evaluate the index.

4.2 Example: Path Integral Derivation of the Atiyah-Singer Index Theorem

The exponential factor in (4.62) is immediately seen to be the (ordinary) Chern character ch(F) of the given complex line bundle $L \to \mathcal{M}$, while the functional determinant coincides (modulo overall signs to be discussed below) with the Euler form of the normal bundle to $\mathcal{M}$ in $L\mathcal{M}$ (this is the bundle spanned by the non-constant modes of $x(\tau)$). Thus (4.62) coincides with a formal application of the degenerate Duistermaat-Heckman integration formula (3.118) (more precisely the degenerate version of the Berline-Vergne theorem) to the infinite-dimensional integral (4.39).

Finally, we discuss how to calculate the Euler form in (4.62). A regularization scheme in general must always be chosen to evaluate infinite-dimensional determinants [102]. Notice first that here the infinite-dimensional Pfaffian arising from the fermionic integration cancels from the result of the infinite-dimensional Gaussian integral over the bosonic fluctuation modes. Thus, just as in the finite-dimensional case, the sign dependence of the Pfaffian gets transfered to the inverse square root of the determinant. The spectral asymmetry associated with the sign of the infinite-dimensional Pfaffian (see (3.62)) has to be regulated and is given by the Atiyah-Patodi-Singer eta-invariant [41, 158] of the Dirac operator $\partial_\tau - R$,

$$\eta(\partial_\tau - R) = \lim_{s \to 0} \oint d\lambda\, \mathrm{sgn}(\lambda)|\lambda|^{-s} + \dim\,\ker(\partial_\tau - R)$$
$$= \lim_{s \to 0} \frac{1}{\Gamma\left(\frac{s+1}{2}\right)} \int_0^\infty dt\, t^{(s-1)/2}\,\mathrm{tr}\left\|(\partial_\tau - R)\,\mathrm{e}^{-t(\partial_\tau - R)^2}\right\| \quad (4.63)$$

where the integration (and/or sum) is over all non-zero eigenvalues λ of $\partial_\tau - R$ and

$$\Gamma(x) = \int_0^\infty dt\, t^{x-1}\,\mathrm{e}^{-t}\ ,\qquad x > 0 \quad (4.64)$$

is the Euler gamma-function.

Next, we evaluate the determinant using standard supersymmetry regularizations [5, 48] for first-order differential operators defined on a circle. The most convenient such choice is Riemann zeta-function regularization. The non-constant single-valued eigenfunctions of the operator ∂_τ on S^1 are $\mathrm{e}^{2\pi i k\tau}$, where k are non-zero integers. Since the matrix R is antisymmetric, it can be skew-diagonalized into n 2×2 skew-diagonal blocks $R^{(j)}$ with skew eigenvalues λ_j, where $j = 1,\ldots,n$. For each such block $R^{(j)}$, we get the formal contribution to the determinant in (4.129),

$$\mathrm{det}'\|\partial_\tau - R^{(j)}\| = \prod_{k \neq 0} (2\pi i k + \lambda_j)(2\pi i k - \lambda_j)$$
$$= g(\lambda_j/2\pi i) g(-\lambda_j/2\pi i) \prod_{k \neq 0} (2\pi i)^2 \quad (4.65)$$

94   4. Quantum Localization Theory for Phase Space Path Integrals

where we have defined the function $g(z)$ as the formal product

$$g(z) = \prod_{k \neq 0}(k+z) \tag{4.66}$$

We can determine the regulated form of the function $g(z)$ by examining its logarithmic derivative $g'(z)/g(z)$ [48]. This is, as a function of $z \in \mathbb{C}$, a function with simple poles of residue 1 at $z = k$ a non-zero integer. Thus we take $g'(z)/g(z) = \pi \cot \pi z - 1/z + b$ and integrating this we get

$$g(z) = \sin \pi z \; e^{bz}/\pi z \tag{4.67}$$

where we have normalized $g(z)$ so that $g(0) = 1$. The arbitrary phase in (4.67) appears because the zeroes of the function (4.66) occur at $z = k \in \mathbb{Z}$ which determine it up to a function without zeroes, i.e. an exponential function. When substituted into (4.65), it is related to the sign of the determinant, and hence to the eta-invariant (4.63). In certain instances (see Section 5.4) it is necessary to make a specific choice for the regularization of this phase [102] (i.e. a choice for b). In our case here, however, the phase b will cancel out explicitly in (4.65) and so we can neglect its effect.

The infinite prefactor in (4.65) is regularized using the Riemann zeta-function

$$\zeta(s) = \sum_{k=1}^{\infty} \frac{1}{k^s} \tag{4.68}$$

which is finite for $s \geq 0$ with $\zeta(0) = -1/2$ [60]. We find that

$$\prod_{k \neq 0}(2\pi i)^2 = \prod_{k>0}(2\pi i)^4 = (2\pi i)^{4(\sum_{k=1}^{\infty} \frac{1}{k^s})|_{s=0}} = (2\pi i)^{4\zeta(0)} = (2\pi i)^{-2} \tag{4.69}$$

and thus the block contribution (4.65) to the functional determinant in (4.62) is

$$\det' \|\partial_\tau - R^{(j)}\| = \frac{1}{\pi^2}\left(\frac{\sin\frac{i\lambda_j}{2}}{\lambda_j}\right)^2 = \left(\frac{1}{2\pi i}\right)^2 \det\left[\frac{\sinh \frac{1}{2} R^{(j)}}{\frac{1}{2} R^{(j)}}\right] \tag{4.70}$$

Multiplying the blocks together we see that the fluctuation determinant appearing in (4.62) is just given by the ordinary $V = 0$ $\hat{A}$-genus (2.96) with respect to the curvature R, and thus the index is

$$\text{index}(i\nabla\!\!\!\!/\,) = \int_M \text{ch}(F) \wedge \hat{A}(R) \tag{4.71}$$

The result (4.71) is the celebrated Atiyah-Singer index theorem for a twisted spin complex [41]. We see then that a formal application of the Berline-Vergne theorem yields the well-known Atiyah-Singer geometrical representation of the index of $i\displaystyle{\not}\nabla$. This result can be generalized to include the coupling of fermions to a non-abelian gauge field A on a vector bundle $E \to \mathcal{M}$. Now the functional $\hat{\Omega}[x,\psi]$ above is no longer closed ($F = dA + [A \stackrel{\wedge}{,} A]/2$ obeys the Bianchi identity), but the construction above can still be carried through using the coadjoint orbit representation of the structure group of the principal fiber bundle [3, 71] (see Section 5.1). The above representation of the Witten index in terms of a supersymmetric path integral can also be generalized to other differential operators, not just the Dirac operator (4.27). For instance, the Witten index for the DeRham exterior derivative operator d describes the DeRham complex of the manifold $\mathcal{M}$ [166]. The index is now the Euler characteristic of $\mathcal{M}$, the supersymmetric path integral is that of $N = 1$ (DeRham) supersymmetric quantum mechanics, and the localization formula reproduces the Gauss-Bonnet-Chern theorem. An equivariant generalization then yields the Poincaré-Hopf theorem of classical Morse theory [166]. We shall elaborate on some of these ideas, as well as how they extend to infinite-dimensional cases relevant to topological field theories, in Chapter 8.

Finally, we point out that the equivariant cohomological interpretation above is particularly well-suited to reproduce the Callias-Bott index theorems [70], i.e. the analog of the Atiyah-Singer index theorem for a Dirac operator on an odd-dimensional non-compact manifold [41]. The supersymmetric model is now that of $N = 1$ supersymmetric quantum mechanics with background monopole and soliton configurations. The trace over zero modes representing the Witten index can in this case be infinite-dimensional, and it is simply not true that the partition function is independent of the parameter T (the index being obtained for $T \to \infty$) so that one cannot simplify matters by taking the $T \to 0$ limit. The canonical realization of the $N = 1$ supersymmetry by considering an equivariant structure over an extended superloop space, defined by a larger mixing of bosonic and fermionic coordinates, preserves the contributions of the zero modes which would otherwise be lost [107, 108] and the localization tricks used above become directly applicable. We shall discuss this a bit more in Chapter 8. Furthermore, the index in these cases can be computed from a higher-dimensional Atiyah-Singer index theorem by introducing a simple first class constraint (i.e. one that is a symmetry of the Hamiltonian, or equivalently a constant of the motion) that eliminates the extra dimensions. We refer to [70] for more details about this approach to index theorems in general.

4.3 Loop Space Symplectic Geometry and Equivariant Cohomology

The example of the last Section has shown that a formal generalization of symplectic geometry and equivariant cohomology to the loop space of a physical problem can result in a (correct) localization formula in the same spirit

96   4. Quantum Localization Theory for Phase Space Path Integrals

as those of Chapters 2 and 3. The localization principle in this context was just a manifestation of the supersymmetry of that model. It has also provided us with some important functional space tools that will be used throughout this Book, as well as hints on how to proceed to loop space generalizations of the results of the earlier Chapters. Following these lessons we have learned, we shall now focus on developing some geometric methods of determining quantum partition functions of generic (not necessarily supersymmetric) dynamical systems. Given the formulation of the path integral in Section 4.1 above on a general symplectic manifold, we wish to treat the problem of its exact evaluation within the geometric context of Chapter 3. As exemplified by the example of the previous Section, for this we need a formulation of exterior and symplectic differential geometry on the loop space $L\mathcal{M}$ over the phase space $\mathcal{M}$. This will ultimately lead to a formal, infinite-dimensional generalization of the equivariant localization priniciple for path integrals, and thus formal conditions and methods for evaluating exactly these functional integrations which in general are far more difficult to deal with than their classical counterparts. As with the precise definition of the functional integrals above, we shall be rather cavalier here about the technicalities of infinite-dimensional manifolds. The loop space $L\mathcal{M} \to \mathcal{M}$ is an infinite-dimensional vector bundle – the fiber over a point $x \in \mathcal{M}$ is the space of all loops $x(t)$ based at x, $x(T) = x(0) = x$, which is an infinite-dimensional non-abelian group with group multiplication of loops $(x_1 x_2)(t)$ defined by first traversing the loop $x_1(t)$, and then the loop $x_2(-t)$ in the opposite direction. These quantities should therefore be properly defined using Sobolev completions of the infinite-dimensional groups and spaces involved. This can always be done in an essentially straightforward and routine manner [22].

We define the exterior algebra $L\Lambda\mathcal{M}$ of the loop space by lifting the Grassmann generators η^μ of $\Lambda\mathcal{M}$ to anti-commuting periodic paths $\eta^\mu(t)$ which generate $L\Lambda\mathcal{M}$ and which are to be identified as the basis $dx^\mu(t)$ of loop space 1-forms. With this, we can define loop space differential k-forms

$$\alpha = \int_0^T dt_1 \cdots dt_k \, \frac{1}{k!} \alpha_{\mu_1 \cdots \mu_k}[x; t_1, \ldots, t_k] \eta^{\mu_1}(t_1) \cdots \eta^{\mu_k}(t_k) \qquad (4.72)$$

and the loop space exterior derivative is defined by lifting the exterior derivative of the phase space $\mathcal{M}$,

$$d_L = \int_0^T dt \, \eta^\mu(t) \frac{\delta}{\delta x^\mu(t)} \qquad (4.73)$$

The loop space symplectic geometry is determined by a loop space symplectic 2-form

$$\Omega = \int_0^T dt \, dt' \, \frac{1}{2} \Omega_{\mu\nu}[x; t, t'] \eta^\mu(t) \eta^\nu(t') \qquad (4.74)$$

4.3 Loop Space Symplectic Geometry and Equivariant Cohomology

which is closed
$$d_L \Omega = 0 \tag{4.75}$$
or in local coordinates $x^\mu(t)$ on $L\mathcal{M}$,

$$\frac{\delta}{\delta x^\mu(t)} \Omega_{\nu\lambda}[x;t',t''] + \frac{\delta}{\delta x^\nu(t)} \Omega_{\lambda\mu}[x;t',t''] + \frac{\delta}{\delta x^\lambda(t)} \Omega_{\mu\nu}[x;t',t''] = 0 \tag{4.76}$$

Thus we can apply the infinite-dimensional version of Poincaré's lemma to represent Ω locally in terms of the exterior derivative of a loop space 1-form

$$\vartheta = \int_0^T dt\ \vartheta_\mu[x;t]\eta^\mu(t) \tag{4.77}$$

as
$$\Omega = d_L \vartheta \tag{4.78}$$

We further assume that (4.74) is non-degenerate, i.e. the matrix $\Omega_{\mu\nu}[x;t,t']$ is invertible on the loop space.

The canonical choice of symplectic structure on $L\mathcal{M}$ which coincides with the loop space Liouville measure introduced in (4.25) is that which is induced from the symplectic structure of the phase space,

$$\Omega_{\mu\nu}[x;t,t'] = \omega_{\mu\nu}(x(t))\delta(t-t') \tag{4.79}$$

which is diagonal in its loop space indices t, t'. We shall use similar liftings of other quantities from the phase space to the loop space. In this way, elements $\alpha(x)$ of $L\Lambda_x \mathcal{M}$ (or $LT_x\mathcal{M}$) at a loop $x \in L\mathcal{M}$ are regarded as deformations of the loop, i.e. as elements of $\Lambda\mathcal{M}$ (or $T\mathcal{M}$) restricted to the loop $x^\mu(t)$ such that $\alpha[x;t] \in \Lambda_{x(t)}\mathcal{M}$ (or $T_{x(t)}\mathcal{M}$). This means that these vector bundles over $L\mathcal{M}$ are infinite-dimensional spaces of sections of the pull-back of the phase space bundles to $[0,T]$ by the map $x(t) : [0,T] \to \mathcal{M}$. In particular, we define loop space canonical transformations as loop space changes of variable $F[x(t)]$ that leave Ω invariant. These are the transformations of the form

$$\vartheta \xrightarrow{F} \vartheta_F = \vartheta + d_L F \tag{4.80}$$

Thus in the context of the loop space symplectic geometry determined by (4.79), the quantum partition function is an integral over the infinite-dimensional symplectic manifold $(L\mathcal{M}, \Omega)$ with the loop space Liouville measure there determined by the canonically-invariant closed form on $L\mathcal{M}$ given by exterior products of Ω with itself,

$$[d\mu_L(x)] = [d^{2n}x]\ \sqrt{\det \|\Omega\|} \tag{4.81}$$

The loop space Hamiltonian vector field associated with the action (4.26) has components

98   4. Quantum Localization Theory for Phase Space Path Integrals

$$V_S^\mu[x;t] = \int_0^T dt'\, \Omega^{\mu\nu}[x;t,t']\frac{\delta S[x]}{\delta x^\nu(t')} = \dot{x}^\mu(t) - V^\mu(x(t)) \quad (4.82)$$

with $V^\mu = \omega^{\mu\nu}\partial_\nu H$ as usual the Hamiltonian vector field on $\mathcal{M}$. The zeroes of V_S

$$L\mathcal{M}_S = \{x(t) \in L\mathcal{M} : V_S[x(t)] = 0\} \quad (4.83)$$

are the extrema of the action (4.26) and coincide with the classical trajectories of the dynamical system, i.e. the solutions of the classical Hamilton equations of motion. The loop space contraction operator with respect to a loop space vector field $W^\mu[x;t]$ is given by

$$i_W = \int_0^T dt\, W^\mu[x;t]\frac{\delta}{\delta\eta^\mu(t)} \quad (4.84)$$

Thus we can define a loop space equivariant exterior derivative

$$Q_W = d_L + i_W \quad (4.85)$$

whose square is the Lie derivative along the loop space vector field W,

$$Q_W^2 = d_L i_W + i_W d_L = \mathcal{L}_W = \int_0^T dt\left(W^\mu\frac{\delta}{\delta x^\mu} + \partial_\nu W^\mu \eta^\nu \frac{\delta}{\delta\eta^\mu}\right) \quad (4.86)$$

When $W = V_S$ is the loop space Hamiltonian vector field, we shall for ease denote the corresponding operators above as $i_{V_S} \equiv i_S$, etc.

The partition function can be written as in the finite-dimensional case using the functional Berezin integration rules to absorb the determinant factor into the exponential in terms of the anti-commuting periodic fields $\eta^\mu(t)$,

$$\begin{aligned}Z(T) &= \int_{L\mathcal{M}\otimes L\Lambda^1\mathcal{M}} [d^{2n}x]\,[d^{2n}\eta]\, \exp\left\{iS[x] + \frac{i}{2}\int_0^T dt\, \omega_{\mu\nu}(x(t))\eta^\mu(t)\eta^\nu(t)\right\} \\ &= \int_{L\mathcal{M}\otimes L\Lambda^1\mathcal{M}} [d^{2n}x]\,[d^{2n}\eta]\, e^{i(S[x]+\Omega[x,\eta])}\end{aligned}$$
(4.87)

so that in this way $Z(T)$ is written in terms of an augmented action $S+\Omega$ on the super-loop space $L\mathcal{M} \otimes L\Lambda^1\mathcal{M}$. From this we can now formally describe the S^1-equivariant cohomology of the loop space.

The operator Q_S is nilpotent on the subspace

$$L\Lambda_S\mathcal{M} = \{\alpha \in L\Lambda\mathcal{M} : \mathcal{L}_S\alpha = 0\} \quad (4.88)$$

of equivariant loop space functionals. The loop space observable $S[x]$ defines the loop space Hamiltonian vector field through

$$d_L S = -i_S \Omega \tag{4.89}$$

from which it follows that the integrand of the quantum partition function (4.87) is equivariantly closed,

$$Q_S(S + \Omega) = (d_L + i_S)(S + \Omega) = 0 \tag{4.90}$$

and so the augmented action $S + \Omega$ can be locally represented as the equivariant exterior derivative of a 1-form $\hat{\vartheta}$,

$$S + \Omega = Q_S \hat{\vartheta} = \int_0^T dt \left(V_S^\mu \hat{\vartheta}_\mu + \frac{1}{2} \Omega_{\mu\nu} \eta^\mu \eta^\nu \right) \tag{4.91}$$

From (4.90) we find that

$$Q_S^2 \hat{\vartheta} = \mathcal{L}_S \hat{\vartheta} = 0 \tag{4.92}$$

and so $\hat{\vartheta}$ lies in the subspace (4.88). If Φ_S is some globally defined loop space 0-form with

$$\mathcal{L}_S(d_L \Phi_S) = 0 \tag{4.93}$$

then we see that $\hat{\vartheta}$ is not unique but the augmented action (4.91) is invariant under the loop space canonical transformation

$$\hat{\vartheta} \to \hat{\vartheta} + d_L \Phi_S \tag{4.94}$$

Thus the partition function (4.87) has a very definite interpretation in terms of the loop space equivariant cohomology $H_S(L\mathcal{M})$ determined by the operator Q_S on $L\Lambda_S \mathcal{M}$.

4.4 Hidden Supersymmetry and the Loop Space Localization Principle

The fact that the integrand of the partition function above can be interpreted in terms of a loop space equivariant cohomology suggests that we can localize it by choosing an appropriate representative of the loop space equivariant cohomology class determined by the augmented action $S + \Omega$. However, the arguments which showed in the finite-dimensional cases that the partition function integral is invariant under such topological deformations cannot be straightforwardly applied here since there is no direct analog of Stokes' theorem for infinite-dimensional manifolds. Nonetheless, the localization priniciple can be established by interpreting the equivariant cohomological structure on $L\mathcal{M}$ as a "hidden" supersymmetry of the quantum theory. In this way one has a sort of Stokes' theorem in the form of a Ward identity associated with this supersymmetry (as was the case in Section 4.2 above), where we interpret the fundamental localization property (2.126) as an infinitesimal change of variables in the integral. The partition function (4.87) can be interpreted as a BRST gauge-fixed path integral [22] with the $\eta^\mu(t)$ viewed as fermionic ghost fields and $x^\mu(t)$ as the fundamental bosonic fields of the model. The supersymmetry is suggested by the ungraded structure of Q_S on $L\Lambda_S\mathcal{M}$ which maps even-degree, commuting loop space forms (bosons) into odd-degree, anti-commuting forms (fermions)[7]. Since the fermion fields $\eta^\mu(t)$ appear by themselves without a conjugate partner, this determines an $N = \frac{1}{2}$ supersymmetry. The $N = \frac{1}{2}$ supersymmetry algebra $Q_S^2 = \mathcal{L}_S$ implies that Q_S is a supersymmetry charge on the subspace $L\Lambda_S\mathcal{M}$, and the augmented action is supersymmetric, $Q_S(S + \Omega) = 0$. Thus here $L\Lambda_S\mathcal{M}$ coincides with the BRST complex of physical (supersymmetric) states, and the BRST transformations of the fundamental bosonic fields $x^\mu(t)$ and their superpartners $\eta^\mu(t)$ are given by the action of the infinitesimal supersymmetry generator Q_S [8]

$$Q_S x^\mu(t) = \eta^\mu(t) \quad , \quad Q_S \eta^\mu(t) = V_S^\mu[x;t] \tag{4.95}$$

This formal identification of the equivariant cohomological structure as a hidden supersymmetry allows one to interpret the (non-supersymmetric) quantum theory as a supersymmetric or topological field theory. It was Blau, Keski-Vakkuri and Niemi [30] who pointed out that a quite general localization principle could be formulated for path integrals using rather formal functional techniques introduced in the BRST quantization of first class constrained systems [118]. In these theories a BRST transformation produces

---

[7] In this interpretation the form degree can be thought of as a ghost number, so that the physical observables of the system (i.e. those with ghost number 0) are the smooth functions on $\mathcal{M}$. Furthermore, at this point it is useful to recall the analogy between $Q_S = d_L + i_S$ and the gauge-covariant derivative in a gauge theory for the following analogies with BRST quantization of gauge theories. See Appendix A for a brief review of some of the ideas of BRST quantization.

[8] In supersymmetric quantum field theories the BRST transformations of operators and fields are represented by a graded BRST commutator $\{Q_S, \cdot\}$. This commutator in the case at hand can be represented by the Poisson structure of the phase space as follows. We introduce periodic trajectories $\lambda_\mu(t)$ in $L\mathcal{M}$ conjugate to $x^\mu(t)$ and anticommuting periodic paths $\bar\eta_\mu(t)$ conjugate to $\eta^\mu(t)$, i.e.

$$\{\lambda_\mu(t), x^\nu(t')\}_\omega = \{\bar\eta_\mu(t), \eta^\nu(t')\}_\omega = \delta_\mu^\nu \delta(t-t')$$

which are to be identified as the Poisson algebra realization of the operators $\lambda_\mu(t) \sim \frac{\delta}{\delta x^\mu(t)}$ and $\bar\eta_\mu(t) \sim \frac{\delta}{\delta \eta^\mu(t)}$ acting in the usual way. This gives a Poisson bracket realization of the actions of the operators d_L and i_S, and then the action of Q_S is represented by the BRST commutator $\{Q_S, \cdot\}_\omega$. In the following, one can keep in mind this representation which maintains a complete formal analogy with supersymmetric theories.

4.4 Hidden Supersymmetry and the Loop Space Localization Principle

a super-Jacobian on the super-loop space $L\mathcal{M} \otimes L\Lambda^1\mathcal{M}$ whose corrections are related to anomalies and BRST supersymmetry breaking. The arguments below are therefore valid *provided* that the Q_S-supersymmetry above is not broken in the quantum theory.

The argument for infinite-dimensional localization proceeds as follows. Consider the 1-parameter family of phase space path integrals

$$\mathcal{Z}(\lambda) = \int_{L\mathcal{M} \otimes L\Lambda^1\mathcal{M}} [d^{2n}x]\, [d^{2n}\eta]\, e^{i(S[x]+\Omega[x,\eta]+\lambda Q_S\psi[x,\eta])} \quad (4.96)$$

where $\lambda \in \mathbb{R}$ and $\psi \in L\Lambda_S^1\mathcal{M}$ is a gauged fermion field which is homotopic to 0 under the supersymmetry transformation generated by Q_S (i.e. $\psi \equiv \psi_{s=1}$, where ψ_s, $s \in [0,1]$, is a 1-parameter family of gauge fermions with $Q_S^2\psi_s = 0$ and $\psi_{s=0} = 0$). As in the finite-dimensional case, we wish to establish the λ-independence of this path integral, i.e. that (4.96) depends only on the BRST cohomology class determined by the augmented action, so that a choice of $\lambda \neq 0$ amounts to a choice of representative of $S + \Omega$ in its loop space equivariant cohomology class and different choices of non-trivial representatives then lead to the desired localization schemes. Consider an infinitesimal variation $\lambda \to \lambda + \delta\lambda$ of the argument of (4.96), i.e. let $\psi \to \psi + \delta\psi$ with

$$\delta\psi = \delta\lambda \cdot \psi \quad (4.97)$$

and consider the infinitesimal supersymmetry transformation on the super-loop space parametrized by the gauge fermion $\delta\psi \in L\Lambda_S^1\mathcal{M}$,

$$\begin{aligned} x^\mu \to \bar{x}^\mu &= x^\mu + \delta x^\mu = x^\mu + \delta\psi \cdot Q_S x^\mu = x^\mu + \delta\psi \cdot \eta^\mu \\ \eta^\mu \to \bar{\eta}^\mu &= \eta^\mu + \delta\eta^\mu = \eta^\mu + \delta\psi \cdot Q_S\eta^\mu = \eta^\mu + \delta\psi \cdot V_S^\mu \end{aligned} \quad (4.98)$$

Since $Q_S(S + \Omega) = \mathcal{L}_S\psi = 0$, the argument of the path integral (4.96) is BRST-invariant.

However, the corresponding super-Jacobian arising in the Feynman measure in (4.96) on $L\mathcal{M} \otimes L\Lambda^1\mathcal{M}$ is non-trivial and it has precisely the same functional form as that in a standard BRST transformation [118]. The pertinent super-Jacobian here is given by the super-determinant

$$[d^{2n}\bar{x}]\, [d^{2n}\bar{\eta}] = \text{sdet} \left\| \begin{array}{cc} \frac{\delta\bar{x}}{\delta x} & \frac{\delta\bar{x}}{\delta\eta} \\ \frac{\delta\bar{\eta}}{\delta x} & \frac{\delta\bar{\eta}}{\delta\eta} \end{array} \right\| [d^{2n}x]\, [d^{2n}\eta] \quad (4.99)$$

and the path integral (4.96) is invariant under arbitrary smooth changes of variables. For infinitesimal $\delta\lambda$, the identity

$$\text{tr}\log\|A\| = \log\det\|A\| \quad (4.100)$$

implies that the super-determinant in (4.99) can be computed in terms of the super-trace, the super-loop space sum of the diagonal entries in (4.99), as $\text{sdet}\|A\| = 1 + \text{str}\|A\|$. This gives

$$[d^{2n}\bar{x}]\,[d^{2n}\bar{\eta}] = \left\{1 + \int_0^T dt\, \left(\frac{\delta}{\delta x^\mu}(\delta\psi)\eta^\mu - \frac{\delta}{\delta \eta^\mu}(\delta\psi)V_S^\mu\right)\right\}[d^{2n}x]\,[d^{2n}\eta]$$

$$= \left\{1 - \int_0^T dt\, \left(\eta^\mu \frac{\delta}{\delta x^\mu} + V_S^\mu \frac{\delta}{\delta \eta^\mu}\right)\delta\psi\right\}[d^{2n}x]\,[d^{2n}\eta]$$

$$= (1 - Q_S\delta\psi)[d^{2n}x]\,[d^{2n}\eta] \sim e^{-\delta\lambda \cdot Q_S\psi}[d^{2n}x]\,[d^{2n}\eta]$$
(4.101)

Thus substituting the change of variables (4.98) with super-Jacobian (4.101) into the path integral (4.96) we immediately see that

$$\mathcal{Z}(\lambda) = \mathcal{Z}(\lambda - \delta\lambda) \tag{4.102}$$

which establishes the independence of the path integral (4.96) under homotopically trivial deformations which live in the subspace (4.88). This proof of the λ-independence (or the ψ-independence more generally) of (4.96) is a specialization of the Fradkin-Vilkovisky theorem [13, 14, 118] to the supersymmetric theory above, which states that local supersymmetric variations of gauge fermions in a supersymmetric BRST gauge-fixed path integral leave it invariant. Indeed, the addition of the BRST-exact term $Q_S\psi$ can be regarded as a gauge-fixing term (the reason why ψ is termed here a 'gauge fermion') which renormalizes the theory but leaves it invariant under these perturbative deformations. The addition of this term to the action of the quantum theory above is therefore regarded as a topological deformation, in that it does not change the value of the original partition function which is the $\lambda \to 0$ limit of (4.96) above. This is consistent with the general ideas of topological field theory, in which a supersymmetric BRST-exact action is known to have no local propagating degrees of freedom and so can only describe topological invariants of the underlying space. We shall discuss these more topological aspects of BRST-exact path integrals, also known as Witten-type topological field theories [22], in due course. In any case, we can now write down the loop space localization principle

$$Z(T) = \lim_{\lambda \to \infty} \int_{L\mathcal{M} \otimes L\Lambda^1\mathcal{M}} [d^{2n}x]\,[d^{2n}\eta]\, e^{i(S[x] + \Omega[x,\eta] + \lambda Q_S\psi[x,\eta])} \tag{4.103}$$

so that the quantum partition function localizes onto the zeroes of the gauge fermion field ψ.

Given the localization property (4.103) of the quantum theory, we would now like to pick a suitable representative ψ making the localization manifest.

4.4 Hidden Supersymmetry and the Loop Space Localization Principle

As in the finite dimensional cases, the localizations of interest both physically and mathematically are usually the fixed point locuses of loop space vector fields W on $L\mathcal{M}$. To translate this into a loop space differential form, we introduce a metric tensor G on the loop space and take ψ to be the associated metric-dual form

$$\psi = \int_0^T dt\, dt'\, G_{\mu\nu}[x;t,t']W^\mu[x;t]\eta^\nu(t') \tag{4.104}$$

of the loop space vector field W. The supersymmetry condition $\mathcal{L}_S\psi = 0$ is then equivalent to the Killing equation $\mathcal{L}_S G = 0$ and the additional requirement $\mathcal{L}_S W = 0$ on W [9], where

$$\mathcal{L}_S W = \int_0^T dt \left(\frac{d}{dt} - \mathcal{L}_{V(x(t))}\right) W[x;t] \tag{4.105}$$

In principle there are many useful choices for W obeying such a restriction, but we shall be concerned mostly with those which can be summarized in

$$W^\mu[x;t] = r\dot{x}^\mu(t) - sV^\mu(x(t)) \tag{4.106}$$

where the parameters r, s are chosen appropriate to the desired localization scheme.

As for the metric in (4.104), there are also in principle many possibilities. However, there only seems to be 1 general class of loop space metric tensors to which general arguments and analyses can be applied. To motivate these, we note first that the equivariant exterior derivative Q_S can be written as

$$Q_S = Q_{\dot{x}} - i_V = d_L + i_{\dot{x}} - i_V \tag{4.107}$$

and the square of the operator $Q_{\dot{x}}$ is just the generator of time translations

$$Q_{\dot{x}}^2 = \mathcal{L}_{\dot{x}} = \int_0^T dt \left(\dot{x}^\mu(t)\frac{\delta}{\delta x^\mu(t)} + \dot{\eta}^\mu(t)\frac{\delta}{\delta \eta^\mu(t)}\right) = \int_0^T dt\, \frac{d}{dt} \tag{4.108}$$

This operator arises when we assume that the loop space Hamiltonian vector field generates an S^1-flow on the loop space, parametrized by a parameter $\tau \in [0,1]$ so that the flow is $x^\mu(t) \to x^\mu(t;\tau)$ with $x^\mu(t;0) = x^\mu(t;1)$, such

---

[9] We also require that the combination (4.104) be such that it determines a homotopically trivial element as above, so that it introduces no extra topological effects into the path integral (4.103) when evaluated on contractable loops. For the most part, we shall be rather cavalier about this requirement and discuss it only towards the end of this Book.

that in the selected loop space coordinates $x^\mu(t)$ the flow parameter τ also shifts the loop (time) parameter $t \to t + \tau$. In this case we have

$$V_S^\mu[x;t] = \left.\frac{\partial x^\mu(t;\tau)}{\partial \tau}\right|_{\tau=0} = \dot{x}^\mu(t) \tag{4.109}$$

and the supersymmetry transformation (4.95) becomes

$$Q_{\dot{x}} x^\mu(t) = \eta^\mu(t) \quad , \quad Q_{\dot{x}} \eta^\mu(t) = \dot{x}^\mu(t) \tag{4.110}$$

which we recall is the infinitesimal supersymmetry discussed in Section 4.2 above. In particular, the effective action is now (locally) of the functional form

$$S + \Omega = \int_0^T dt \left(\theta_\mu(x) \dot{x}^\mu + \frac{1}{2} \omega_{\mu\nu}(x(t)) \eta^\mu \eta^\nu \right) = (d_L + i_{\dot{x}}) \vartheta = Q_{\dot{x}} \hat{\vartheta} \tag{4.111}$$

and the topological invariance of the quantum theory, i.e. the invariance of (4.111) under BRST-deformations by elements ψ of the subspace $L\Lambda_S^1 \mathcal{M}$, is according to (4.108) determined by *arbitrary* globally defined single-valued functionals on $L\mathcal{M}$, i.e. $\psi(0) = \psi(T)$. This form of the $U(1)$-equivariant cohomology on the loop space is called the model-independent circle action.

We shall therefore demand that the localization functionals in (4.104) be invariant under the model-independent S^1-action on $L\mathcal{M}$ (i.e. rigid rotations $x(t) \to x(t + \tau)$ of the loops). This requires that the loop space metric tensor above obey $\mathcal{L}_{\dot{x}} G = 0$, or equivalently that $G_{\mu\nu}[x;t,t'] = G_{\mu\nu}[x;t-t']$ is diagonal in its loop space indices. Since the quantum theory is to describe the dynamics of a given Hamiltonian system for which we know the underlying manifold $\mathcal{M}$, the best way to pick the Riemannian structure on $L\mathcal{M}$ is to lift a metric tensor g from $\mathcal{M}$ so that G takes the ultra-local form

$$G_{\mu\nu}[x;t,t'] = g_{\mu\nu}(x(t))\delta(t-t') \tag{4.112}$$

and its action on loop space vector fields is given by

$$G(V_1, V_2) = \int_0^T dt \, g_{\mu\nu}(x(t)) V_1^\mu[x;t] V_2^\nu[x;t] \tag{4.113}$$

Because of the reparametrization invariance of the integral (4.113), the metric tensor G is invariant under the canonical flow on $L\mathcal{M}$ generated by $\dot{x}$. The Lie derivative condition on G is then equivalent to the Lie derivative condition (2.111) with respect to the Hamiltonian vector field V on $\mathcal{M}$. Thus infinite-dimensional localization requires as well that the phase space $\mathcal{M}$ admit a globally-defined $U(1)$-invariant Riemannian structure on $\mathcal{M}$ with respect to the *classical* dynamics of the given Hamiltonian system. As discussed before, the condition that the Hamiltonian H generates an isometry of a metric g on $\mathcal{M}$ (through the induced Poisson structure on $(\mathcal{M}, \omega)$) is a very restrictive condition on the Hamiltonian dynamics. Essentially it means that H must be related to the global action (2.36) of a group G on $\mathcal{M}$, so that the classical mechanics generates a very large degree of symmetry. As mentioned before, the infinite-dimensional results above, in particular the evaluation of the super-Jacobian in (4.101), are as reliable as the corresponding calculations in standard BRST quantization, provided that the boundary conditions in (4.96) are also supersymmetric. Provided that the assumptions on the classical properties of the Hamiltonian are satisfied (as for the finite-dimensional cases), the above derivation will stand correct unless the supersymmetry $Q_S^2 = \mathcal{L}_S$ is broken in the quantum theory, for instance by a scale anomaly in the rescaling of the metric $G_{\mu\nu} \to \lambda \cdot G_{\mu\nu}$ above. See Appendices A and B for the precise correspondence between BRST quantization, equivariant cohomology and localization.

4.5 The WKB Localization Formula

We shall now begin examining the various types of localization formulas that can be derived from the general principles of the last Section. The first infinite-dimensional localization formula that we shall present is the formal generalization of the Duistermaat-Heckman integration formula, whose derivation follows the loop space versions of the steps used in Sections 2.6 and 3.3. We assume that the action S has (finitely-many) isolated and non-degenerate critical trajectories, so that the zero locus (4.83) consists of isolated classical loops in $L\mathcal{M}$, i.e. we assume that the determinant of the associated Jacobi fields arising from a second-order variation of S is non-vanishing on these classical trajectories. Under these assumptions, we set $r = s = 1$ in (4.106), so that

$$\psi = \int_0^T dt \, g_{\mu\nu} V_S^\mu \eta^\nu$$

$$Q_S \psi = \int_0^T dt \, \left[g_{\mu\nu} V_S^\mu V_S^\nu + \eta^\mu \left(g_{\mu\nu} \partial_t - g_{\nu\lambda} \partial_\mu V^\lambda + V_S^\lambda \partial_\mu g_{\nu\lambda} \right) \eta^\nu \right]$$

(4.114)

Proceeding just as in the finite-dimensional case, the evaluation of the localization integral (4.103) gives

$$Z(T) \sim \int_{L\mathcal{M}} [d^{2n}x]\, \sqrt{\det \|\Omega\|} \sqrt{\det \|\delta V_S\|}\, \delta(V_S)\, e^{iS[x]}$$

$$\sim \int_{L\mathcal{M}} [d^{2n}x]\, \sqrt{\det \|\Omega\|} \sqrt{\det \|\delta^\mu_\nu \partial_t - \partial_\nu(\omega^{\mu\lambda}\partial_\lambda H)\|}$$ (4.115)

$$\times \delta(\dot{x}^\mu - \omega^{\mu\nu}\partial_\nu H)\, e^{iS[x]}$$

$$\sim \sum_{x(t) \in L\mathcal{M}_S} \frac{\sqrt{\det \|\omega(x(t))\|}\, e^{iS[x]}}{\sqrt{\det \|\delta^\mu_\nu \partial_t - \partial_\nu(\omega^{\mu\lambda}\partial_\lambda H)\|}}$$

where here and in the following the symbol $\sim$ will be used to signify the absorption of infinite prefactors into the determinants which arise from the functional Gaussian integrations. The functional determinant in the denominator of (4.115) can be evaluated in the same manner as in Section 4.2 above with $R \to dV$ there and the eigenfunctions of ∂_t are now the periodic functions on $[0, T]$ instead of S^1 so that the eigenvalues are replaced as $2\pi i k \to 2\pi i k/T$ in (4.65). The result can be written in terms of the Dirac $\hat{A}$-genus of the tangent bundle of $\mathcal{M}$

$$\det \|\delta^\mu_\nu \partial_t - \partial_\nu V^\mu\| = \left(\frac{T}{2\pi i}\right)^{2n} \det\left[\frac{\sinh \frac{T}{2} dV}{\frac{T}{2} dV}\right] = \hat{A}(T \cdot dV)^{-2} \quad (4.116)$$

using the ordinary moment map for the $U(1)$-action here.

This result is the famous WKB approximation to the partition function [147], except that it is summed over all classical paths and not just those which minimize the action S. If we reinstate the factors of $\hbar$, then it is formally the leading term of the stationary phase expansion of the partition function in powers of $\hbar$ as $\hbar \to 0$. The limit $\hbar \to 0$ is called the classical limit of the quantum mechanics problem above, since then according to (4.1) the operators $\hat{p}$ and $\hat{q}$ behave as ordinary commuting c-numbers as in the classical theory. For $\hbar \to 0$ we can naturally evaluate the path integral by the stationary-phase method discussed in Section 3.3, i.e. we expand the trajectories $x(t) = x_0(t) + \delta x(t)$ in the action, with $x_0(t) \in L\mathcal{M}_S$ and $\delta x(t)$ the fluctuations about the classical paths $x_0(t)$ with $\delta x(0) = \delta x(T) = 0$, and then carry out the leading Gaussian functional integration over these fluctuations. Indeed, this was the way Feynman originally introduced the path integral to describe quantum mechanics as a sum over trajectories which fluctuate around the classical paths of the system. This presentation of quantum mechanics thus leads to the dynamical Hamilton action principle of classical mechanics [55], i.e. the classical paths of motion of a dynamical system are those which minimize the action, as a limiting case. If the classical trajectories were unique, then we would only obtain the factors $e^{iS[x]/\hbar}$ above as $\hbar \to 0$. Quantum mechanics can then be interpreted as implying fluctuations (the one-loop determinant factors in (4.115)) around these classical trajectories.

We should point out here that the standard WKB formulas are usually given for configuration space path integrals where the fluctuation determinant $(\det \|L_S(x(t))\|)^{-1/2}$ appearing in (4.115) is the so-called Van Vleck determinant which is essentially the Hessian of S in configuration space coordinates q. Here the determinant is the functional determinant of the Jacobi operator which arises from the usual Legendre transformation to phase space coordinates (p,q). This operator is important in the Hamilton-Jacobi theory of classical mechanics [6, 55], and this determinant can be interpreted as the density of classical trajectories. The result (4.115) and the assumptions that went into deriving it, such as the non-vanishing of the determinant of the Jacobi fields and the existence of an invariant phase space metric, are certainly true for the classic examples in quantum mechanics and field theory where the semi-classical approximation is known to be exact, such as for the propagator of a particle moving on a group manifold [38, 139, 146]. The above localization principle yields sufficient, geometric conditions for when a given path integral is given exactly by its WKB approximation, and it therefore has the possibility of expanding the set of quantum systems for which the Feynman path integral is WKB exact and localizes onto the classical trajectories of the system.

4.6 Degenerate Path Integrals and the Niemi-Tirkkonen Localization Formula

There are many instances in which the WKB approximation is unsuitable for a quantum mechanical path integral, such as a dynamical system whose classical phase space trajectories coalesce at some point. It is therefore desirable to seek alternative, more general localization formulas which can be applied to larger classes of quantum systems. Niemi and Palo [121] have investigated the types of degeneracies that can occur for phase space path integrals and have argued that for Hamiltonians which generate circle actions the classical trajectories can be characterized as follows. In general, the critical point set of the action S with non-constant periodic solutions $x^\mu(T) = x^\mu(0) = x_0^\mu$ lie on a compact submanifold $L\mathcal{M}_S$ of the phase space $\mathcal{M}$. In this context, $L\mathcal{M}_S$ is refered to as the moduli space of T-periodic classical solutions and it is in general a non-isolated set for only some discrete values of the propagation time T. For generic values of T the periodic solutions with $x^\mu(T) = x^\mu(0) = x_0^\mu$ exist only if x_0^μ lies on the critical submanifold $\mathcal{M}_V$ of the Hamiltonian H. Then the classical equations of motion reduce to $\dot{x}^\mu = V^\mu = \omega^{\mu\nu}\partial_\nu H = 0$ and so the moduli space $L\mathcal{M}_S$ coincides with the critical point set $\mathcal{M}_V \subset \mathcal{M}$. Notice that in this case the functional determinants involving the symplectic 2-form in (4.115) cancel out and one is left with only the regularized determinant in (4.116). We shall see some specific examples of this later on.

With this in mind we can derive a loop space analog of the degenerate Duistermaat-Heckman formula of Section 3.7. We decompose $L\mathcal{M}$ and $L\Lambda^1\mathcal{M}$

into classical modes and fluctuations about the classical solutions and scale the latter by $1/\sqrt{\lambda}$,

$$x^\mu(t) = \bar{x}^\mu(t) + x_f^\mu(t)/\sqrt{\lambda} \quad , \quad \eta^\mu(t) = \bar{\eta}^\mu(t) + \eta_f^\mu(t)/\sqrt{\lambda} \qquad (4.117)$$

where $\bar{x}(t) \in L\mathcal{M}_S$ are the solutions of the classical equations of motion, i.e. $V_S^\mu(\bar{x}(t)) = \dot{\bar{x}}^\mu - \omega^{\mu\nu}(\bar{x})\partial_\nu H(\bar{x}) = 0$, and $\bar{\eta}^\mu(t) \sim d\bar{x}^\mu(t) \in \Lambda^1 L\mathcal{M}_S$ span the kernel of the loop space Riemann moment map,

$$(\Omega_S)_{\mu\nu}(\bar{x})\bar{\eta}^\nu = 0 \qquad (4.118)$$

where

$$\Omega_S = d_L\psi = \int_0^T dt \, \frac{\delta}{\delta x^\mu}\left(g_{\nu\lambda}V_S^\lambda\right)\eta^\mu\eta^\nu \qquad (4.119)$$

with ψ given in (4.114). In particular, this implies that $\bar{\eta}^\mu(t)$ are Jacobi fields, i.e. they obey the fluctuation equation

$$\left(\delta_\nu^\mu \partial_t - \partial_\nu V^\mu(\bar{x})\right)\bar{\eta}^\nu = 0 \qquad (4.120)$$

The fluctuation modes in (4.117) obey the boundary conditions $x_f^\mu(0) = x_f^\mu(T) = 0$ and $\eta_f^\mu(0) = \eta_f^\mu(T) = 0$.

The super-loop space measure with this decomposition is then

$$[d^{2n}x]\,[d^{2n}\eta] = d^{2n}\bar{x}(t)\,d^{2n}\bar{\eta}(t)\prod_{t\in[0,T]} d^{2n}x_f(t)\,d^{2n}\eta_f(t) \qquad (4.121)$$

where as usual the change of variables (4.117) has unit Jacobian because the determinants from the bosonic and fermionic fluctuations cancel (this is the powerful manifestation of the "hidden" supersymmetry in these theories). The calculation now proceeds analogously to that in Section 3.7, so that evaluating the Gaussian integrals over the fluctuation modes localizes the path integral to a finite-dimensional integral over the moduli space $L\mathcal{M}_S$ of classical solutions,

$$Z(T) \sim \int_{L\mathcal{M}_S} d^{2n}\bar{x}(t)\,\frac{\sqrt{\det\omega(\bar{x})}\,e^{iS[\bar{x}]}}{\text{Pfaff}\|\delta_\nu^\mu\partial_t - (\mu_S)_\nu^\mu(\bar{x}) - R_\nu^\mu(\bar{x})\|\big|_{\mathcal{N}L\mathcal{M}_S}} \qquad (4.122)$$

where $\mu_S = g^{-1}\cdot\Omega_S$ and R is as usual the Riemann curvature 2-form of the metric g evaluated on $L\mathcal{M}_S$. In (4.122) the Pfaffian is taken over the fluctuation modes $x_f^\mu(t)$ about the classical trajectories $\bar{x}^\mu(t) \in L\mathcal{M}_S$ (i.e. along the normal bundle $\mathcal{N}L\mathcal{M}_S$ in $L\mathcal{M}$), and the measure there is an invariant measure over the moduli space of classical solutions which is itself a symplectic manifold. The localization formula (4.122) is the loop space version of the degenerate localization formula (3.118) in which the various factors can be

4.6 The Niemi-Tirkkonen Localization Formula

interpreted as loop space extensions of the equivariant characteristic classes. In particular, in the limit where the solutions to the classical equations of motion $V_S^\mu(x(t)) = 0$ become isolated and non-degenerate paths the integration formula (4.122) reduces to the standard WKB localization formula (4.115).

However, the degenerate localization formula (4.122) is hard to use in practise because in general the moduli space of classical solutions has a complicated, T-dependent structure, i.e. it is usually a highly non-trivial problem to solve the classical equations of motion for the T-periodic classical trajectories of a dynamical system[10]. We would therefore like to obtain alternative degenerate localization formulas which are applicable independently of the structure of the moduli space $L\mathcal{M}_S$ above. Given the form of (4.122), we could then hope to obtain a localization onto some sort of equivariant characteristic classes of the manifold $\mathcal{M}$. The first step in this direction was carried out by Niemi and Tirkkonen in [127]. Their localization formula can be derived by setting $s = 0$, $r = 1$ in (4.106) so that

$$\psi = \int_0^T dt \ g_{\mu\nu} \dot{x}^\mu \eta^\nu$$

$$Q_S \psi = \int_0^T dt \ \left[g_{\mu\nu} \dot{x}^\mu (\dot{x}^\nu - V^\nu) + \eta^\mu \left(g_{\mu\nu} \partial_t + \dot{x}^\lambda g_{\mu\rho} \Gamma^\rho_{\lambda\nu} \right) \eta^\nu \right]$$

(4.123)

Here the zero locus of the vector field (4.106) consists of the constant loops $\dot{x}^\mu = 0$, i.e. points on $\mathcal{M}$, so that the canonical localization integral will reduce to an integral over the finite-dimensional manifold $\mathcal{M}$ (as in Section 4.2), rather than a sum or integral over the moduli space of classical solutions as above.

To evaluate the right-hand side of (4.103) with (4.123), we use the standard trick. We decompose $L\mathcal{M}$ and $L\Lambda^1 \mathcal{M}$ into constant modes and fluctuation modes and scale the latter by $1/\sqrt{\lambda}$,

$$x^\mu(t) = x_0^\mu + \hat{x}^\mu(t)/\sqrt{\lambda} \quad , \quad \eta^\mu(t) = \eta_0^\mu + \hat{\eta}^\mu(t)/\sqrt{\lambda} \qquad (4.124)$$

where

$$x_0^\mu = \frac{1}{T} \int_0^T dt \ x^\mu(t) \quad , \quad \eta_0^\mu = \frac{1}{T} \int_0^T dt \ \eta^\mu(t)$$

$$\partial_t x_0^\mu = \partial_t \eta_0^\mu = 0 \quad , \quad \int_0^T dt \ \hat{x}^\mu(t) = \int_0^T dt \ \hat{\eta}^\mu(t) = 0$$

(4.125)

---

[10] Some features of the space of T-periodic classical trajectories for both energy conserving and non-conserving Hamiltonian systems have been discussed recently by Niemi and Palo in [119, 122].

The decomposition (4.124) is essentially a Fourier decomposition in terms of some complete sets of states $\{x_k^\mu(t)\}_{k\in\mathbb{Z}}$ and $\{\eta_k^\mu(t)\}_{k\in\mathbb{Z}}$, so that

$$\hat{x}^\mu(t) = \sum_{k\neq 0} s_k^\mu x_k^\mu(t) \quad , \quad \hat{\eta}^\mu(t) = \sum_{k\neq 0} \sigma_k^\mu \eta_k^\mu(t) \tag{4.126}$$

and the Feynman measure in the path integral is then defined just as before as

$$[d^{2n}x]\,[d^{2n}\eta] = d^{2n}x_0\,d^{2n}\eta_0 \prod_{t\in[0,T]} d^{2n}\hat{x}(t)\,d^{2n}\hat{\eta}(t)$$
$$= d^{2n}x_0\,d^{2n}\eta_0 \prod_{k\neq 0} d^{2n}s_k\,d^{2n}\sigma_k \tag{4.127}$$

With the rescaling in (4.124) of the fluctuation modes, the gauge fixing term $Q_S\psi$ is

$$Q_S\psi = \int_0^T dt\,\left[x^\mu\left((\Omega_V)_{\mu\nu}\partial_t - g_{\mu\nu}\partial_t^2\right)x^\nu + \frac{1}{2}R_{\mu\nu}x^\mu\dot{x}^\nu + \hat{\eta}^\mu g_{\mu\nu}\partial_t\hat{\eta}^\nu\right]$$
$$+\mathcal{O}(1/\sqrt{\lambda})$$
(4.128)

where we have integrated by parts over t and used the periodic boundary conditions.

In (4.128) we see the appearence of the equivariant curvature of the Riemannian manifold $(\mathcal{M},g)$. Since Ω_V and R there act on the fluctuation modes, as usual they can be interpreted as forming the equivariant curvature of the normal bundle of $\mathcal{M}$ in $L\mathcal{M}$. With the above rescaling the fluctuation and zero modes decouple in the localization limit $\lambda \to \infty$, just as before. The integrations over the fluctuations are as usual Gaussian, and the result of these integrations is

$$Z(T) \sim \int_\mathcal{M} \mathrm{ch}_V(-iT\omega) \wedge \left(\mathrm{det}'\|\delta_\nu^\mu\partial_t - (R_V)_\nu^\mu\|\right)^{-1/2} \tag{4.129}$$

This form of the partition function is completely analogous to the degenerate localization formula of Section 3.7, and it is also similar to the formula (4.122), except that now the domain of integration has changed from the moduli space $L\mathcal{M}_S$ of classical solutions to the entire phase space $\mathcal{M}$. This makes the formula (4.129) much more appealing, in that there is no further reference to the T-dependent submanifold $L\mathcal{M}_S$ of $\mathcal{M}$. Note that (4.129) differs from the classical partition function for the dynamical system $(\mathcal{M},\omega,H)$ by a one-loop determinant factor which can be thought of as encoding the information due to quantum fluctuations. The classical Boltzmann weight e^{-iTH} comes from evaluating the action $S[x]$ on the constant loops $x \in \mathcal{M} \subset L\mathcal{M}$, so that

(4.129) is another sort of semi-classical localization of the Feynman path integral.

The fluctuation determinant in (4.129) can be again evaluated just as in Section 4.2 and it now yields the equivariant $\hat{A}$-genus (2.96) with respect to the equivariant curvature R_V. The localization formula (4.129) is therefore

$$Z(T) \sim \int_{\mathcal{M}} \text{ch}_V(-iT\omega) \wedge \hat{A}_V(TR) \tag{4.130}$$

This is the Niemi-Tirkkonen localization formula [127] and it expresses the quantum partition function as a (finite-dimensional) integral over the phase space $\mathcal{M}$ of equivariant characteristic classes in the $U(1)$-equivariant cohomology generated by the Hamiltonian vector field V on $\mathcal{M}$. The huge advantage of this formula over the localization formula of the last Section is that no assumptions appear to have gone into its derivation (other than the standard localization constraints). It thus applies not only to the cases covered by the WKB localization theorem, but also to those where the WKB approximation breaks down (e.g. when classical paths coalesce in $L\mathcal{M}$). Indeed, being a localization onto time-independent loops it does not detect degenerate types of phase space trajectories that a dynamical system may possess.

In fact, the localization formula (4.130) can be viewed as an integral over the equivariant generalization of the Atiyah-Singer index density of a Dirac operator with background gravitational and gauge fields, and it therefore represents a sort of equivariant generalization of the Atiyah-Singer index theorem for a twisted spin complex. Indeed, when $H, V \to 0$ the effective action in the canonical localization integral is

$$(S + \Omega + \lambda Q_S \psi)|_{H=V=0}$$
$$= \int_0^T dt \left[\lambda g_{\mu\nu} \dot{x}^\mu \dot{x}^\nu + \theta_\mu \dot{x}^\mu + \lambda g_{\mu\nu} \eta^\mu \nabla_t \eta^\nu + \frac{1}{2} \eta^\mu \omega_{\mu\nu} \eta^\nu \right] \tag{4.131}$$

On the other hand, the left-hand side of the localization formula (4.103) becomes

$$Z(T)|_{H=0} = \text{tr} \| e^{-i\hat{H}T} \| \Big|_{T=0} = \dim \mathcal{H}_\mathcal{M} \tag{4.132}$$

which is an integer representing the dimension of the free Hilbert space associated with $S(H = 0)$ and which can therefore only describe the topological characteristics of the manifold $\mathcal{M}$. Recalling the discussion of Section 4.2, we see that the action (4.131) is the supersymmetric action for a bosonic field $x^\mu(t)$ and its Dirac fermion superpartner field $\eta^\mu(t)$ in the background of a gauge field θ_μ and a gravitational field $g_{\mu\nu}$, i.e. the action of $N = \frac{1}{2}$ Dirac supersymmetric quantum mechanics. Moreover, the integer (4.132) coincides with the $V = 0$ limit of (4.130) which is the ordinary Atiyah-Singer index for a twisted spin complex (the 'twisting' here associated with the usual symplectic line bundle $L \to \mathcal{M}$). Thus the localization formalism here is just

112   4. Quantum Localization Theory for Phase Space Path Integrals

a more general case of the localization example of Section 4.2 above which reproduced quite beautifully the celebrated Atiyah-Singer index theorem[11]. We shall describe some more of these cohomological field theoretical aspects of equivariant localization in Section 4.10, and the connections between the localization formalism and other supersymmetric quantum field theories in Chapter 8. The equivariant cohomological structure of these theories is consistent with the topological nature of supersymmetric models (the basic topological field theories – see Section 4.10 below) and they always yield certain topological invariants of the underlying manifolds such as the Atiyah-Singer index.

4.7 Connections with the Duistermaat-Heckman Integration Formula

In this Section we shall point out some relations between the path integral localization formulas derived thus far and their relations to the finite-dimensional Duistermaat-Heckman formula. Since the localization formulas are all derived from the same fundamental geometric constraints, one would expect that, in some limits at least, they are all related to each other. In particular, when 2 localization formulas hold for a certain quantum mechanical path integral, they must both coincide somehow. We can relate the various localization formulas by noting that the integrand of (4.130) is an equivariantly closed differential form on $\mathcal{M}$ (being an equivariant characteristic class) with respect to the finite-dimensional equivariant cohomology defined by the ordinary Cartan derivative $D_V = d + i_V$. Thus we can apply the Berline-Vergne theorem (in degenerate form – compare with Section 3.7) of Section 2.6 to localize the equivariant Atiyah-Singer index onto the critical points of the Hamiltonian H to obtain

$$Z(T) \sim \int_{\mathcal{M}_V} \frac{\mathrm{ch}_V(-iT\omega)}{E_V(R)|_{\mathcal{N}_V}} \wedge \hat{A}_V(TR) \bigg|_{\mathcal{M}_V} \qquad (4.133)$$

so that a (degenerate) Hamiltonian gives a localization of the path integral onto $\mathcal{M}_V$ in terms of the equivariant Chern class restricted to $\mathcal{M}_V$, and the equivariant Euler class and $\hat{A}$-genus of the normal bundle $\mathcal{N}_V$. Note that this differs from the finite-dimensional localization formula (3.118) only in the appearance of the equivariant $\hat{A}$-genus which arises from the evaluation of the temporal determinants which occur. This factor therefore encodes the quantum fluctuations about the classical values, and its appearance is

---

[11] This analogy, as well as the localization of the quantum partition function in general, requires that boundary conditions for the path integral be selected which respect the pertinent supersymmetry. We shall say more about this requirement later on.

4.7 Connections with the Duistermaat-Heckman Integration Formula

quite natural according to the general supersymmetry arguments above (as the Dirac $\hat{A}$-genus quite frequently arises from supersymmetric field theory path integrals). Furthermore, the localization formula (4.133) follows from the moduli space formula (4.122) for certain values of the propagation time T (see the discussion at the beginning of the last Section).

The connection between the WKB and Niemi-Tirkkonen localization formulas is now immediate if we assume that the critical point set $\mathcal{M}_V$ of the Hamiltonian consists of only isolated and non-degenerate points (i.e. the Hamiltonian H is a Morse function). Then in the canonical localization vector field (4.106) we can set $r = 0$ and $s = -1$ so that

$$\psi = i_V g = \int_0^T dt \, g_{\mu\nu} V^\mu \eta^\nu$$

$$Q_S \psi = \int_0^T dt \, \left[\frac{1}{2} (\Omega_V)_{\mu\nu} \eta^\mu \eta^\nu + V^\mu g_{\mu\nu} (\dot{x}^\nu - V^\nu) \right]$$

(4.134)

We use the rescaled decomposition (4.124) again which decouples the zero modes from the fluctuation modes because of the "hidden" supersymmetry. The Gaussian integrations over the fluctuation modes then yields

$$Z(T) \sim \int_{\mathcal{M}} d^{2n}x_0 \, e^{-iTH} \sqrt{\frac{\det \Omega_V}{\det'\|\partial_t - \Omega_V\|}} \, \delta(V) \sim \sum_{p \in \mathcal{M}_V} \frac{e^{-iTH(p)}}{\sqrt{\det \Omega_V}} \hat{A}(T\Omega_V)$$

(4.135)

where the (ordinary) Dirac $\hat{A}$-genus arises from evaluating the temporal determinant in (4.135) as described before and we recall that $\Omega_V(p) = 2dV(p) = 2\omega^{-1}(p)\mathcal{H}(p)$ at a critical point $p \in \mathcal{M}_V$. Thus under these circumstances we can localize the partition function path integral onto the time-independent classical trajectories of the dynamical system, yielding a localization formula that differs from the standard Duistermaat-Heckman formula (3.63) only by the usual quantum fluctuation term.

The localization formula (4.135) of course also follows directly from the degenerate formula (4.133) in the usual way, and it can be shown [85] to also follow from the WKB formula (4.115) using the Weinstein action invariant [26, 165] which probes the first cohomology group of the symplectomorphism group of the symplectic manifold (i.e. the diffeomorphism subgroup of canonical transformations). This latter argument requires that $\mathcal{M}$ is compact, the classical trajectories are non-intersecting and each classical trajectory can be contracted to a critical point of H through a family of classical trajectories (for instance when $H^1(\mathcal{M}; \mathbb{R}) = 0$), and that the period T is such that the boundary condition $x^\mu(0) = x^\mu(T)$ admits only constant loops as solutions to the classical equations of motion. The localization onto the critical points of the Hamiltonian is not entirely surprising, since as discussed at the beginning of the last Section for Hamiltonian circle actions on $\mathcal{M}$ the zero locuses $L\mathcal{M}_S$ and $\mathcal{M}_V$ in general coincide. Drawing from the analogy of (4.135) with the Duistermaat-Heckman theorem (i.e. that the equivariant Atiyah-Singer index (4.130) is given exactly by its stationary phase approximation), one can, in particular, in this case conclude from Kirwan's theorem that the Hamiltonian H is a perfect Morse function that admits only even Morse indices [85]. We see therefore that localization formulas and various Morse theoretic arguments (such as Kirwan's theorem) follow (formally) for path integrals in exactly the same way that they followed for ordinary finite-dimensional phase space integrals.

4.8 Equivariant Localization and Quantum Integrability

For the remainder of this Chapter we shall discuss some more formal features of the localization formalism for path integrals, as well as some extensions of it. We have shown in Chapter 3 that there is an intimate connection between classical integrability and the localization formalism for dynamical systems. With this in mind, we can use the localization formalism to construct an alternative, geometric formulation of the problem of *quantum* integrability [47, 120] (in the sense that the quantum partition function can be evaluated exactly) which differs from the usual approaches to this problem [35]. As in Section 3.6 we consider a generic integrable Hamiltonian which is a functional $H = H(I)$ of action variables I^a which are in involution as in (3.87). From the point of view of the localization constraints above, the condition that H generates a circle action which is an isometry of some Riemannian geometry on $\mathcal{M}$ means that the action variables I^a generate the Cartan subalgebra of the associated isometry group of $(\mathcal{M}, g)$ in its Poisson bracket realization on $(\mathcal{M}, \omega)$.

For such a dynamical system, we use a set of generating functionals $J^a(t)$ to write the quantum partition function as

$$Z(T) = \exp\left(-i \int_0^T dt\, H\left[\frac{\delta}{i\delta J(t)}\right]\right) \int_{L\mathcal{M}} [d^{2n}x]\, \sqrt{\det \|\Omega\|}$$

$$\times \exp\left\{i \int_0^T dt\, (\theta_\mu \dot{x}^\mu - J^a I^a)\right\}\Bigg|_{J=0}$$

(4.136)

To evaluate the path integral in (4.136), we consider an infinitesimal variation of its action

$$\delta(\theta_\mu \dot{x}^\mu - J^a I^a) = \delta x^\mu (\omega_{\mu\nu} \dot{x}^\nu - J^a \partial_\mu I^a) \qquad (4.137)$$

with the infinitesimal Poisson bracket variation

$$\delta x^\mu = \epsilon^a \{I^a, x^\mu\}_\omega = -\epsilon^a \omega^{\mu\nu} \partial_\nu I^a \qquad (4.138)$$

4.8 Equivariant Localization and Quantum Integrability

where ϵ^a are infinitesimal coordinate-independent parameters. The transformation (4.137),(4.138) corresponds to the leading order infinitesimal limit of the canonical transformation

$$x^\mu \to x^\mu + \left\{ e^{-\epsilon^a I^a}, x^\mu\, e^{\epsilon^a I^a} \right\}_\omega = x^\mu + \epsilon^a \{x^\mu, I^a\}_\omega \qquad (4.139)$$
$$+ \frac{1}{2}\epsilon^a \epsilon^b \left\{ \{x^\mu, I^a\}_\omega, I^b \right\}_\omega + \ldots$$

and it gives

$$\delta(\theta_\mu \dot{x}^\mu - J^a I^a) = -\dot{\epsilon}^a I^a \qquad (4.140)$$

after an integration by parts over time. Since the Liouville measure in (4.136) is invariant under canonical transformations, it follows that the only effect of the variation (4.140) on the loop space coordinates in (4.136) is to shift the external sources as $J^a \to J^a + \dot{\epsilon}^a$. Note that if we identify $J^a(t)$ as the temporal component A_0^a of a gauge field then this shift has the same functional form as a time-dependent abelian gauge transformation [120]. Thus if for some reason the quantum theory breaks the invariance of the Liouville measure under these coordinate transformations, we would expect to be able to relate the non-trivial Jacobian that arises to conventional gauge anomalies [151].

Thus if we Fourier decompose the fields $J^a(t)$ into their zero modes J_0^a and fluctuation modes $\hat{J}^a(t)$ as in (4.124), we can use this canonical transformation to 'gauge' away the time-dependent parts of J^a in (4.136) so that the path integral there depends only on the constant modes J_0^a of the generating functionals and the partition function is given by

$$Z(T) = \exp\left(-iTH\left[\frac{1}{i}\frac{\partial}{\partial J_0}\right]\right) \int_{L\mathcal{M}} [d^{2n}x]\, \sqrt{\det \|\Omega\|}$$
$$\times \exp\left\{ i\int_0^T dt\, (\theta_\mu \dot{x}^\mu - J_0^a I^a) \right\}\bigg|_{J_0=0} \qquad (4.141)$$

Since the Hamiltonian $J_0^a I^a$ in the action in (4.141) generates an abelian group action on $\mathcal{M}$, we can localize it using the Niemi-Tirkkonen formula (4.130) to arrive at

$$Z(T) \sim \exp\left(-iTH\left[\frac{1}{i}\frac{\partial}{\partial J_0}\right]\right) \int_\mathcal{M} \mathrm{ch}_{J_0^a I^a}(-iT\omega) \wedge \hat{A}_{J_0^a I^a}(TR)\bigg|_{J_0=0} \qquad (4.142)$$

and so the path integral now localizes to a derivative expansion of equivariant characteristic classes. The localization formula (4.142) is valid for *any* integrable Hamiltonian system whose conserved charges $J_0^a I^a$ generate a global isometry on $\mathcal{M}$, and consequently the localization formalism can be used to establish the exact quantum solvability of generic integrable models.

116   4. Quantum Localization Theory for Phase Space Path Integrals

Indeed, there are several non-trivial examples of integrable models where the WKB localization formula (4.115) is known to be valid, and this has led to the conjecture that for a large class of integrable field theories a "proper" version of the semi-classical approximation should yield a reliable reproduction of the features of the exact quantum theory [176]. The formula (4.142) is one such candidate, and thus it yields an explicit realization of this conjecture. However, one may also hope that the localization principle of Section 4.4 could be used to derive weaker versions of the localization formulas above for some dynamical systems which are not necessarily completely integrable [85] (in the sense that the localization formalism above does not carry through). For this, we consider a Hamiltonian with $r < n$ conserved charges I^a which are in involution as in (3.87),(3.88), and which have the classical equations of motion $\dot{I}^a = 0$. We then set

$$\psi = \int_0^T dt\ I^a \partial_\mu I^a \eta^\mu \quad , \quad Q_S\psi = \int_0^T dt\ \left(\dot{I}^a\right)^2 \qquad (4.143)$$

in the canonical localization integral (4.103). The cohomological relation $Q_S^2\psi = \mathcal{L}_S\psi = 0$ follows from the involutary property of the charges I^a. Then the right-hand side of (4.103) yields a localization of the path integral onto the constant values of the conserved charges I^a,

$$Z(T) \sim \int_{L\mathcal{M}} [d^{2n}x]\ \sqrt{\det \|\Omega\|}\ \prod_{a=1}^r \delta(\dot{I}^a)\ e^{iS[x]} \qquad (4.144)$$

The formula (4.144) is a weaker version of the above localization formulas which is valid for any non-integrable system that admits conserved charges. It can be viewed as a quantum generalization of the classical reduction theorem [7] which states that conserved charges in involution reduce the dynamics onto the symplectic subspace of the original phase space determined by the constant (classical) values of the integrals of motion I^a. When H is completely integrable this subspace coincides with the invariant Liouville tori discussed in Section 3.6. Thus even when there are corrections to the various localization formulas above (e.g. the WKB approximation), the supersymmetry arguments of Section 4.4 can be used to derive weaker versions of the localization formulas. Notice that, as anticipated, the localization formula (4.144) does not presume any isometric structure on the phase space (see the discussion of Section 3.6). Equivariant cohomology might therefore provide a natural geometric framework for understanding quantum integrability, and the localization formulas associated with general integrable models represent equivariant characteristic classes of the phase space. For more details about this and other connections between equivariant localization and integrability, see [47, 84, 85].

4.9 Localization for Functionals of Isometry Generators

In the last Section we considered a particular class of Hamiltonians which were functionals of action variables and we were able to derive a quite general localization formula for these dynamical systems. It is natural to explore now whether or not localization formulas could be derived for Hamiltonians which are more general types of functionals. We begin with the case where the Hamiltonian of a dynamical system is an *a priori* arbitrary functional $\mathcal{F}(H)$ of an observable H which generates an abelian isometry through the Hamiltonian equations for H in the usual sense. Thus we want to evaluate the path integral [128]

$$Z(T|\mathcal{F}(H)) = \int_{L\mathcal{M}} [d^{2n}x] \sqrt{\det\|\Omega\|} \exp\left\{i\int_0^T dt \ (\theta_\mu \dot{x}^\mu - \mathcal{F}(H))\right\} \quad (4.145)$$

We shall see that such path integrals are important for certain physical applications. Note, however, that although such functionals may seem arbitrary, we must at least require that $\mathcal{F}(H)$ be a semi-bounded functional of the observable H [159]. Otherwise, a Wick rotation off of the real time axis to imaginary time may produce a propagator $\text{tr}\|e^{-iT\mathcal{F}(H)}\|$ which is not a tempered distribution and thus eliminating any rigorous attempts to make the path integral a well-defined mathematical entity.

The formalism used to treat path integrals such as (4.145) is the auxilliary field formalism for supersymmetric theories [71, 107, 108] which enables one to relate the loop space equivariant cohomology determined by the derivative Q_S to the more general model-independent S^1 loop space formalism, i.e. that determined by the equivariant exterior derivative $Q_{\dot{x}}$. We recall from Section 4.4 that in this formulation the path integral action is BRST-exact, as required for supersymmetric field theories. Here the auxilliary fields that are introduced turn out to coincide with those used to formulate generic Poincaré supersymmetric theories in terms of the model-independent S^1 loop space equivariant cohomology which renders their actions BRST-exact. These supersymmetric models will be discussed in Chapter 8.

To start, we *assume* that there is a function $\phi(\xi)$ such that $\mathcal{F}(H)$ is a Gaussian functional integral transformation of it,

$$\exp\left(-i\int_0^T dt \ \mathcal{F}(H)\right) = \int_{L\mathbb{R}} [d\xi] \exp\left\{i\int_0^T dt \ \left(\frac{1}{2}\xi^2 - \phi(\xi)H\right)\right\} \quad (4.146)$$

Because of the local integrability of $\mathcal{F}(H)$, locally such a function $\phi(\xi)$ can always be constructed, but there may be obstructions to constructing $\phi(\xi)$ globally on the loop space $L\mathcal{M}$, for the reasons discussed before. The transformation $\xi \to \phi$ which maps the Gaussian in ξ to a non-linear functional of ϕ is just the Nicolai transformation in supersymmetry theory [22, 116], i.e. the

118   4. Quantum Localization Theory for Phase Space Path Integrals

change of variables that maps the bosonic part of the supersymmetric action into a Gaussian such that the Jacobian for this change of variables coincides with the determinant obtained by integrating over the bilinear fermionic part of the supersymmetric action. This observation enables one to explicitly construct a localization for the path integral (4.145).

Notice that when $\mathcal{F}(H)$ is either linear or quadratic in the observable H, the Nicolai transform ξ is directly related to the functional Fourier transformation of $\mathcal{F}(H)$,

$$\exp\left(-i\int_0^T dt\ \mathcal{F}(H)\right) = \int_{L\mathbb{R}} [d\phi]\ \exp\left(-i\int_0^T dt\ \hat{F}(\phi)\right) \exp\left(-i\int_0^T dt\ \phi H\right) \tag{4.147}$$

However, for more complicated functionals $\mathcal{F}(H)$ this connection is less straightforward. In particular, if we change variables $\xi \to \phi$ in the Gaussian transformation (4.146), we find

$$\exp\left(-i\int_0^T dt\ \mathcal{F}(H)\right) = \int_{L\mathbb{R}} [d\phi] \prod_{t\in[0,T]} \xi'(\phi)\ \exp\left\{i\int_0^T dt\ \left(\frac{1}{2}\xi^2(\phi) - \phi H\right)\right\} \tag{4.148}$$

so that the effect of this transformation is to isolate the isometry generator H and make it contribute linearly to the effective action in (4.145) (as we did in the last Section). This allows one to localize (4.145) using the general prescriptions of Section 4.4 above.

Substituting (4.148) into (4.145), we then carry out the same steps which led to the Niemi-Tirkkonen localization formula (4.130). However, now there is an auxilliary, time-dependent field ϕ which appears in the path integral action which must be incorporated into the localization procedure. These fields appear in the terms ϕH above and are therefore interpreted as the *dynamical generators* of $S(\mathbf{u(1)}^*)$. We introduce a superpartner η for the auxilliary field ϕ whose Berezin integration absorbs the Jacobian factor in (4.148). The path integral (4.145) thus becomes a functional integral over an extended superloop space. As discussed in Appendix B, one can now introduce an extended BRST-operator incorporating the super-multiplet (ϕ, η) such that the partition function is evaluated with a BRST-exact action whose argument lies in the BRST-complex of physical states and, as was the case in Section 4.2, the Niemi-Tirkkonen localization onto constant modes becomes manifest. This extended BRST-operator is the so-called Weil differential whose cohomology defines the BRST model for the $U(1)$-equivariant cohomology [99, 129]. This more sophisticated technique is required whenever the basis elements ϕ^a of the symmetric algebra $S(\mathbf{g}^*)$ are made dynamical and are integrated out, as is the case here.

We shall not enter into the cumbersome details of this extended superspace evaluation of (4.145), but merely refer to [128] for the details (see also

4.9 Localization for Functionals of Isometry Generators

Appendix B for a sketch of the idea). The final result is the integration formula

$$Z(T|\mathcal{F}(H)) \sim \int_{-\infty}^{\infty} d\phi_0 \; \xi_0'(\phi_0) \; e^{iT\xi_0^2(\phi_0)/2} \int_{\mathcal{M}} \mathrm{ch}_{\phi_0 V}(-iT\omega) \wedge \hat{A}_{\phi_0 V}(TR) \tag{4.149}$$

where ϕ_0 are the zero modes of the auxilliary field ϕ. (4.149) is valid (formally) for any semi-bounded functional $\mathcal{F}(H)$ of an isometry generator H on $\mathcal{M}$. Thus even for functionals of Hamiltonian isometry generators the localization formula is a relatively simple expression in terms of equivariant characteristic classes. The only computational complication in these formulas is the identification of the function $\xi(\phi)$ (or the functional Fourier transform $\hat{F}(\phi)$). We note that when $\mathcal{F}(H) = H$, we have $\phi(\xi) = 1$ and (4.149) reduces consistently to the Niemi-Tirkkonen localization formula (4.130). In the important special case $\mathcal{F}(H) = H^2$, we find $\phi(\xi) = \xi$ (i.e. $\hat{F}(\phi) = \phi^2$) and the localization formula (4.149) becomes

$$Z(T|H^2) \sim \int_{-\infty}^{\infty} d\phi_0 \; e^{iT\phi_0^2/2} \int_{\mathcal{M}} \mathrm{ch}_{\phi_0 V}(-iT\omega) \wedge \hat{A}_{\phi_0 V}(TR) \tag{4.150}$$

which is the formal path integral generalization of the Wu localization formula (3.128).

In fact, the above dynamical treatment of the multipliers ϕ suggests a possible non-abelian generalization of the localization formulas and hence a path integral generalization of the Witten localization formula of Section 3.8 [161]. At the same time we generalize the localization formalism of Section 4.8 above to the case where the Hamiltonian is a functional of the generators of the full isometry group of $(\mathcal{M}, g)$, and not just simply the Cartan subgroup thereof. We consider a general non-abelian Hamiltonian moment map (3.30) where the component functions H^a are assumed to generate a Poisson algebra realization of the isometry group G of some Riemannian metric g on $\mathcal{M}$. As mentioned in Section 3.8, when the ϕ^a are fixed we are essentially in the abelian situation above and this case will be discussed in more detail in what follows. Here we assume that the multipliers ϕ^a are time-dependent and we integrate over them in the path integral following the same prescription for equivariant integration introduced in Section 3.8. This corresponds to modelling the G-equivariant cohomology of $\mathcal{M}$ in the Weil algebra using the BRST formalism [129, 161] (see Appendix B). When the ϕ^a are fixed parameters, the action functional (4.26) generates the action of S^1 on $L\mathcal{M}$ in the model independent circle action described in Section 4.4 above. However, when the ϕ^a are dynamical quantities, S generates the action of the semi-direct product $LG \rtimes S^1$, where the action of S^1 corresponds to translations of the loop parameter t and $LG = C^\infty(S^1, G)$ is the loop group of the isometry group G. These actions are generated, respectively, by the loop space vector

120   4. Quantum Localization Theory for Phase Space Path Integrals

fields

$$V_{S^1} = \int_0^T dt \, \dot{x}^\mu(t) \frac{\delta}{\delta x^\mu(t)}$$

$$V_{LG} = \int_0^T dt \, \phi^a(t) \omega^{\mu\nu}(x(t)) \left(\frac{\delta}{\delta x^\nu(t)} H^a \right) \frac{\delta}{\delta x^\mu(t)} \equiv \int_0^T dt \, \phi^a(t) V^a(t) \quad (4.151)$$

The commutator algebra of the vector fields (4.151) is that of $LG \otimes S^1$ on $L\mathcal{M}$,

$$[V_{S^1}, V_{LG}] = \int_0^T dt \, \dot{\phi}^a H^a \quad , \quad [V^a(t), V^b(t')] = f^{abc} V^c(t) \delta(t-t') \quad (4.152)$$

The equivariant extension of the symplectic 2-form Ω on $L\mathcal{M}$ is therefore $S + \Omega$.

If the multipliers ϕ^a (now regarded as local coordinates on $L\mathbf{g}^*$) are integrated over directly, then the isometry functions H^a become constraints because the ϕ^a appear linearly in the action and so act as Lagrange multipliers. In this case we are left with a topological quantum theory (i.e. there are no classical degrees of freedom) with vanishing classical action, in parallel to the finite-dimensional case of Section 3.8. Alternatively, we can add a functional $F = F(\phi^a)$ to the argument of the exponential term in the partition function such that the quantity $S + \Omega + F$ is equivariantly closed. We then introduce a non-abelian generalization of the procedure outlined above [161] (see Appendix B for details). Introducing an extended equivariant BRST operator Q_T for the semi-direct product action of $LG \otimes S^1$ on $L\mathcal{M}$ (the non-abelian version of that above), it turns out that $S + \Omega + F$ is equivariantly closed with respect to Q_T only for either $F = 0$ or $F = \frac{1}{2}(\phi^a)^2$, where the latter is the invariant polynomial corresponding to the quadratic Casimir element of G. Note that this is precisely the choice that was made in our definition of equivariant integration in Section 3.8. As shown in Appendix B, within this framework we can reproduce loop space generalizations of the cohomological formulation of Section 3.2 for the Hamiltonian dynamics. The rest of the localization procedure now carries through parallel to that above and in the Niemi-Tirkkonen localization, and it yields the localization formula [161]

$$Z(T) \sim \int_{\mathbf{g}^*} \prod_{a=1}^{\dim G} d\phi_0^a \, e^{iT(\phi_0^a)^2/2} \int_{\mathcal{M}} \mathrm{ch}_{\phi_0^a V^a}(-iT\omega) \wedge \hat{A}_{\phi_0^a V^a}(TR) \quad (4.153)$$

which is a non-abelian version of the quadratic localization formula (4.150) and is the path integral generalization of the Witten localization formula presented in Section 3.8 [12]. Notice that the primary difference between this non-abelian localization and its abelian counterpart is that in the latter the functional $F(\phi)$ is *a priori* arbitrary.

4.10 Topological Quantum Field Theories

In this last Section of this Chapter we return to the case where the dual basis elements of $S(\mathbf{g}^*)$ are fixed numbers. We wish to study the properties of the quantum theory when the effective action is BRST-exact as in (4.91) locally on the loop space [85, 123]. In this case the quantum theory is said to be topological, in that there are no local physical degrees of freedom and the remaining partition function can only describe topological invariants of the space on which it is defined [22]. We shall see this explicitly below, and indeed we have already seen hints of this in the expressions for the path integral in terms of equivariant characteristic classes above. To get a flavour for this, we first consider a quantum theory that admits a model independent circle action globally on the loop space, i.e. whose loop space Hamiltonian vector field generates a global constant velocity $U(1)$ action on $L\mathcal{M}$, so that its action functional is given locally by (4.111). In this case, the determinant that appears in the denominator of the WKB localization formula (4.115) is

$$\det \|\delta^2 S\| \Big|_{\dot{x}=0} = \det \|\delta(\Omega \cdot \dot{x})\| \Big|_{x=x_0} = \det \|\Omega \partial_t\| \Big|_{x=x_0} \quad (4.154)$$

where the localization is now onto the constant loops $x_0 \in \mathcal{M}$. Since the determinants on the right-hand side of (4.115) now cancel modulo the factor $\det \|\partial_t\|$, only the zero modes of ∂_t can contribute. Thus the (degenerate) WKB localization formula in this case becomes

$$Z(T) \sim \int_{\mathcal{M}} d^{2n}x_0 \sqrt{\det \|\Omega|_{\partial_t=0}\|} \Big|_{x=x_0} \quad (4.155)$$

and only the zero modes of the symplectic 2-form contribute. Since this path integral yields the topological Witten index of the corresponding supersymmetric theory [167], the localization formula identifies the loop space characteristic class which corresponds to the Witten index of which the ensuing Atiyah-Singer index counts the zero modes of the associated Dirac operator. This is one of the new insights gained into supersymmetric theories from the equivariant localization formalism. (4.155) is a purely cohomological representative of the manifold $\mathcal{M}$ which contains no physical information.

Next, consider the more general case of an equivariantly-exact action (4.91). Note that this is precisely the solution to the problem of solving

---

[12] The procedure outlined above could also be employed in the discussion of the Witten localization formalism in Section 3.8. This has been implicitly done in carrying out the equivariant integrations there (see Appendix B).

122   4. Quantum Localization Theory for Phase Space Path Integrals

the loop space equivariant Poincaré lemma for $S+\Omega$. If we assume that the symplectic potential is invariant under the global $U(1)$-action on $\mathcal{M}$, as in (3.26), then the Hamiltonian is given by $H = i_V\theta$ and the loop space 1-form $\hat{\vartheta}$ in (4.91) is given by

$$\hat{\vartheta} = \int_0^T dt\, \theta_\mu(x(t))\eta^\mu(t) \qquad (4.156)$$

The loop space localization principle naively implies that the resulting path integral should be trivial. Indeed, since the 1-form (4.156) lies in the subspace (4.88), the partition function can be written as

$$Z(T) = \int_{L\mathcal{M}\otimes L\Lambda^1\mathcal{M}} [d^{2n}x]\,[d^{2n}\eta]\, e^{i\lambda Q_S \hat{\vartheta}} \qquad (4.157)$$

and it is independent of the parameter $\lambda \in \mathbb{R}$. In particular, it should be independent of the action S.

However, the above argument for the triviality of the path integral assumes that θ is homotopic to 0 in the subspace (4.88) under the supersymmetry generated by Q_S, i.e. that (4.91) holds globally for all loops. For the remainder of this Chapter we will assume that the manifold $\mathcal{M}$ is simply connected, so that $H^1(\mathcal{M};\mathbb{R}) = 0$. Then the above argument presumes that the second DeRham cohomology group $H^2(\mathcal{M};\mathbb{R}) = 0$ is trivial. If this is not the case, then one must be careful about arguing the λ-independence of the path integral (4.157). Consider the family of symplectic 2-forms

$$\omega^{(\lambda)} = \lambda d\theta = \lambda\omega \qquad (4.158)$$

associated with the action in (4.157). We consider a closed loop $\gamma(x)$ in the phase space $\mathcal{M}$ parametrized by the periodic trajectory $x(t) : [0,T] \to \mathcal{M}$. Since by assumption $\gamma(x)$ is the boundary of a 2-surface Σ_1 in $\mathcal{M}$, Stokes' theorem implies that the kinetic term $\theta^{(\lambda)} = \lambda\theta$ in (4.157) can be written as

$$\int_0^T dt\, \theta_\mu^{(\lambda)}(x(t))\dot{x}^\mu(t) = \oint_{\gamma(x)} \theta^{(\lambda)} = \int_{\Sigma_1} \omega^{(\lambda)} \qquad (4.159)$$

For consistency of the path integral (4.157), which is expressed as a sum over closed loops in $\mathcal{M}$, the phase (4.159) must be independent of the representative surface Σ_1 spanning $\gamma(x)$, owing to the topological invariance of the partition function $Z(T)$ over $L\mathcal{M}$. Thus if we introduce another surface Σ_2 (of opposite orientation to Σ_1) with boundary $\gamma(x)$ and let Σ be the closed surface (sphere) which is divided into 2 halves Σ_1 and Σ_2 by $\gamma(x)$, then we have

4.10 Topological Quantum Field Theories

$$e^{i\oint_\Sigma \omega^{(\lambda)}} = e^{i\int_{\Sigma_1}\omega^{(\lambda)}} e^{-i\int_{\Sigma_2}\omega^{(\lambda)}} \qquad (4.160)$$

and consequently the integral of $\omega^{(\lambda)}$ over any closed orientable surface Σ in $\mathcal{M}$ must satisfy a version of the Dirac or Wess-Zumino-Witten (flux) quantization condition [168]

$$\frac{1}{2\pi}\oint_\Sigma \omega^{(\lambda)} = \frac{\lambda}{2\pi}\oint_\Sigma \omega \in \mathbb{Z} \qquad (4.161)$$

This means that $\omega^{(\lambda)}$ is an integral element of $H^2(\mathcal{M};\mathbb{R})$, i.e. it defines an integer cohomology class in $H^2(\mathcal{M};\mathbb{Z})$, which is possible only for certain discrete values of $\lambda \in \mathbb{R}$. It follows that a continuous variation $\delta\lambda$ of λ cannot leave the path integral (4.157) invariant and it depends non-trivially on the localization 1-form $\psi \equiv \vartheta$ and thus also on the action S.

Thus the path integral (4.157) defines a consistent quantum theory only when the symplectic 2-form (4.158) defines an integral curvature on $\mathcal{M}$. However, if we introduce a variation $\theta \to \theta + \delta\theta$ of the symplectic potential in (4.157) corresponding to a variation $\omega \to \omega + \delta\omega$ with $\delta\omega = d\alpha$ a trivial element of $H^2(\mathcal{M};\mathbb{R})$ in the subspace (4.88), then the localization principle implies that the path integral remains unchanged (using Stokes' theorem for $\delta\omega$ in (4.161)). Thus the path integral depends only on the cohomology class of ω in $H^2(\mathcal{M};\mathbb{R})$, not on the particular representative $\omega = d\theta$, which means that the partition function (4.157) determines a cohomological topological quantum field theory on the phase space $\mathcal{M}$.

Furthermore, we note that within the framework of the Niemi-Tirkkonen localization formula, the BRST-exact term $Q_S(\lambda\psi + \hat{\vartheta})$, with $\hat{\vartheta}$ given by (4.156) and ψ given in (4.123), gives the effective action in the canonical localization integral (4.96). We saw earlier that the $Q_{\dot{x}}$-exact piece of this action corresponds to the Atiyah-Singer index of a Dirac operator $i\not{\nabla}$ in the background of a $U(1)$ gauge field θ_μ and a gravitational field $g_{\mu\nu}$. The remaining terms there, given by the i_V-exact pieces, then coincide with the terms that one expects in a supersymmetric path integral representation of the infinitesimal Lefschetz number (also known as a character index or equivariant G-index)

$$\text{index}_H(i\not{\nabla};T) = \lim_{\lambda\to\infty} \text{tr}\| e^{iTH}(e^{-\lambda\mathcal{D}^\dagger\mathcal{D}} - e^{-\lambda\mathcal{D}\mathcal{D}^\dagger})\| \qquad (4.162)$$

generated by the Hamiltonian H [21, 23, 24, 120, 127]. This follows from arguments similar to those in Section 4.2 which arrived at the supersymmetric path integral representation of the (ordinary) Atiyah-Singer index. In general, when the Dirac operator is invariant under the action of the isometry group G on $\mathcal{M}$, $[V^a, i\not{\nabla}] = 0$, then the eigenstates of $i\not{\nabla}$ which correspond to a fixed eigenvalue E define a representation of the Lie algebra of G. It is possible to show just as before that the right-hand side of the Lefschetz number (4.162) is independent of $\lambda \in \mathbb{R}^+$, and, therefore, when either $\mathcal{D}^\dagger\mathcal{D}$ or $\mathcal{D}\mathcal{D}^\dagger$ has no

124   4. Quantum Localization Theory for Phase Space Path Integrals

zero modes, we can take the limit $\lambda \to 0$ there and only the zero modes of $i\slashed{V}$ contribute to (4.162). Consequently, the equivariant index coincides with the character

$$\text{index}_H(i\slashed{V}; T) = \text{str}_R \, e^{iTH} \tag{4.163}$$

of the (reducible or irreducible) representation R of the Cartan element H of G determined by the zero modes of $i\slashed{V}$.

Consequently, in the case of Hamiltonian systems for which $\mathcal{L}_V g = \mathcal{L}_V \theta = 0$, the Niemi-Tirkkonen localization formula (4.130) reproduces the Lefschetz fixed point formulas of Bismut [23, 24] and Atiyah, Bott and Singer [41], provided that boundary conditions for the path integral have been properly selected. Thus a purely bosonic theory can be related to the properties of a (functional) Dirac operator defined in the canonical phase space of the bosonic theory, and this analogy leads one to the hope that the above localization prescriptions can be made quite rigorous in a number of interesting infinite-dimensional cases. Note also that the path integral (4.157) has the precise form of a Witten-type or cohomological quantum field theory, which is characterized by a classical action which is BRST-exact with the BRST charge Q_S representing gauge and other symmetries of the classical theory. These types of topological field theories are known to have partition functions which are given exactly by their semi-classical approximation – more precisely, they admit Nicolai maps which trivialize the action and restrict to the moduli space of classical solutions [22]. Thus the topological and localization properties of supersymmetric and topological field theories find their natural explanation within the framework of loop space equivariant localization.

Of course, the above results rely heavily on the G-invariance condition (3.26) for the symplectic potential θ. In the general case, we recall from Section 3.2 that we have the relation (3.47) which holds locally in a neighbourhood $\mathcal{N}$ in $\mathcal{M}$ away from the critical points of H and in which $\omega = d\theta$. In this case, (3.47) gives a solution to the equivariant Poincaré lemma and although the action is locally BRST-exact, globally the quantum theory is non-trivial and may not be given exactly by a semi-classical approximation. Then the path integral (4.96) has the form of a gauge-fixed topological field theory, otherwise known as a Schwarz-type or quantum topological field theory [22], with $\hat{Q}_S$ the BRST charge representing the gauge degrees of freedom. With $\hat{\vartheta}$ as in (4.156), the loop space equivariant symplectic 2-form can be written in the neighbourhood $L\mathcal{N}$ as

$$S + \Omega = Q_S(\hat{\vartheta} + d_L F) - \int_0^T dF(x(t)) \tag{4.164}$$

and the path integral can be represented locally as

$$Z(T) = \int_{L\mathcal{M} \otimes L\Lambda^1 \mathcal{M}} [d^{2n}x]\,[d^{2n}\eta] \; e^{iQ_S(\hat{\vartheta}+d_L F) - i \oint_{\gamma(x)} dF} \tag{4.165}$$

4.10 Topological Quantum Field Theories

If we assume that $\mathcal{M}$ is simply connected, so that $H^1(\mathcal{M};\mathbb{R}) = 0$, then, by Stokes' theorem, the dF term in (4.165) can be ignored for closed trajectories on the phase space[13]. Since from (4.164) we have

$$\mathcal{L}_S(\hat{\vartheta} + d_L F) = Q_S(S + \Omega) = 0 \tag{4.166}$$

it follows that $\hat{\vartheta} + d_L F \in L\Lambda_S^1 \mathcal{M}$ and the effective classical action $S + \Omega$ is equivariantly-exact in the neighbourhood $L\mathcal{N}$.

The non-triviality of the path integral now depends on the non-triviality that occurs when the local neighbourhoods $\mathcal{N}$ above are patched together. In particular, we can invoke the above argument to conclude that the partition function (4.165) depends only on the cohomology class of ω in $H^2(\mathcal{M};\mathbb{R})$, in addition to the critical point set of the action S. Thus the partition function in the general case locally determines a cohomological topological quantum field theory. From the discussion of Section 3.6 we see that this is consistent with the fact that the theory is locally integrable outside of the critical point set of H. We recall also from that discussion that in a neighbourhood $\mathcal{N}$ where action-angle variables can be introduced and where H does not have any critical points, we can construct an explicit realization of the function F above and hence an explicit realization of the topological quantum theory (4.165). For integrable models where action-angle variables can be defined almost everywhere on the phase space $\mathcal{M}$, the ensuing theory is topological, i.e. it can be represented by a topological action of the form (4.164) almost everywhere on the loop space $L\mathcal{M}$. Notice that all of the above arguments stem from the assumption that $H^1(\mathcal{M};\mathbb{R}) = 0$. In Chapter 6 we shall encounter a cohomological topological quantum field theory defined on a multiply-connected phase space which obeys all of the equivariant localization criteria. We also remark that in the general case, when ω is not globally exact, the Wess-Zumino-Witten prescription above for considering the action (4.26) in terms of surface integrals as in (4.159) makes rigorous the definition of the partition function on a general symplectic manifold, a point which up until now we have ignored for simplicity. In this case the required consistency condition (4.161) means that ω itself defines an integral curvature, which is consistent with the usual ideas of geometric quantization [172]. We shall see how this prescription works on a multiply-connected phase space in Chapter 6.

---

[13] This term is analogous to the instanton term $F \wedge F$ in 4-dimensional Yang-Mills theory which can be represented in terms of a locally exact form and is therefore non-trivial only for space-times which have non-contractable loops [151].

5. Equivariant Localization on Simply Connected Phase Spaces: Applications to Quantum Mechanics, Group Theory and Spin Systems

When the phase space $\mathcal{M}$ of a dynamical system is compact, the condition that the Hamiltonian vector field V generate a global isometry of some Riemannian geometry on $\mathcal{M}$ automatically implies that its orbits must be closed circles (see ahead Section 5.2). This feature is usually essential for the finite-dimensional localization theorems, but within the loop space localization framework, where the arguments for localization are based on formal supersymmetry arguments on the infinite-dimensional manifold $L\mathcal{M}$, the flows generated by V need not be closed and indeed many of the formal arguments of the last Chapter will still apply to non-compact group actions. For instance, if we wanted to apply the localization formalism to an n-dimensional potential problem, i.e. on the non-compact phase space $\mathcal{M} = \mathbb{R}^{2n}$, then we would expect to be allowed to use a Hamiltonian vector field which generates non-compact global isometries. As we have already emphasized, the underlying feature of quantum equivariant localization is the interpretation of an equivariant cohomological structure of the model as a supersymmetry among the physical, auxilliary or ghost variables. But as shown in Section 4.3, this structure is exhibited quite naturally by *arbitrary* phase space path integrals, so that, under the seemingly weak conditions outlined there, this formally results in the equivariant localization of these path integrals. This would in turn naively imply the exact computability of any phase space path integral.

Of course, we do not really expect this to be the case, and there is therefore the need to explore the loop space equivariant localization formalism in more detail to see precisely what sort of dynamical systems will localize. In this Chapter we shall explore the range of applicability of the equivariant localization formulas [40, 159] by presenting a more detailed analysis of the meaning and implications of the required localization symmetries, and we shall work out numerous explicit mathematical and physical applications of the formalisms of the previous Chapters. As we shall see, the global isometry condition on the Hamiltonian dynamics is a very restrictive one, essentially meaning that H is related to a global group action (2.36). The natural examples of such situations are the harmonic oscillator and free particle Hamiltonians on $\mathbb{R}^{2n}$ (the trivial Gaussian, free field theories), and the quantization of spin [117] (i.e. the height function on the sphere), or more generally the quantization of the coadjoint orbits of Lie groups [4, 26, 85, 128, 156, 159] and

128   5. Equivariant Localization on Simply Connected Phase Spaces

the equivalent Kirillov-Kostant geometric quantization of homogeneous phase space manifolds [2, 3]. Indeed, the exactness of the semi-classical approximation (or the Duistermaat-Heckman formula) for these classes of phase space path integrals was one of the most important inspirations for the development of quantum localization theory and these systems will be extensively studied in this Chapter, along with some generalizations of them. We shall see that the Hamiltonian systems whose phase space path integrals can be equivariantly localized essentially all fall into this general framework, and that the localization formulas in these cases always represent important group-theoretical invariants called characters, i.e. the traces $\mathrm{tr}_R g = \mathrm{tr}_R \, \mathrm{e}^{c^a X^a}$ evaluated in an irreducible representation R of a group G which are invariant under similarity transformations representing equivalent group representations, and they reproduce, in certain instances, some classical formulas for these characters [87]. In our case the group G will be the group of isometries of a Riemannian structure on $\mathcal{M}$.

As it is essentially the isometry group G that determines the integrable structure of the Hamiltonian system in the equivariant localization framework, we shall study the localization formalism from the point of view of what the possible isometries can be for a given phase space manifold. A detailed analysis of this sort will lead to a geometrical characterization of the integrable dynamical systems from the viewpoint of localization and will lead to topological field theoretical interpretations of integrability, as outlined in Section 4.10. It also promises deeper insights into what one may consider to be the geometrical structure of the quantum theory. This latter result is a particularly interesting characterization of the quantum theory because the partition functions considered are all *ab initio* independent of any Riemannian geometry on the underlying phase space (as are usually the classical and quantum mechanics). Nonetheless, we shall see that for a given Riemannian geometry, the localizable dynamical systems depend on this geometry in such a way so that they determine Hamiltonian isometry actions.

Strictly speaking, most of this general geometric analysis in this Chapter and the next will only carry through for a 2-dimensional phase space. The reason for this is that the topological and geometrical classifications of Riemann surfaces is a completely solved problem from a mathematical point of view. We may therefore invoke this classification scheme to in turn classify the Hamiltonian systems which fit the localization framework. Such a neat mathematical characterization of higher dimensional manifolds is for the most part an unsolved problem (although much progress has been made over the last 7 years or so in the classification of 3- and 4-manifolds), so that a classification scheme such as the one that follows does not generalize to higher-dimensional models. We shall, however, illustrate how these situations generalize to higher dimensions via some explicit examples which will show that the 2-dimensional classifications do indeed tell us about the properties of general localizable dynamical systems. In particular, we shall

5. Equivariant Localization on Simply Connected Phase Spaces

see that from certain points of view all the localizable Hamiltonians represent "generalized" harmonic oscillators, a sort of feature that is anticipated from the previous integrability arguments and the local forms of Hamiltonians which generate circle actions. These seemingly trivial behaviours are a reflection of the large degree of symmetry that is the basis for the large reduction of the complicated functional integrals to Gaussian ones. We will also analyse in full detail the localization formulas of the last Chapter, which will therefore give explicit examples of the cohomological and integrable models that appear quite naturally in (loop space) equivariant localization theory. This analysis will also provide *new* integrable quantum systems, as we shall see, which fall into the class of the generalized localization formulas (e.g. the Niemi-Tirkkonen formula (4.130)), but not the more traditional WKB approximation. Such examples represent a major, non-trivial advance of localization theory and illustrate the potential usefulness of the localization formulas as reliable calculational tools.

At the same time we can address some of the issues that arise when dealing with phase space path integrals, which are generally regarded as rather disreputable because of the unusual discretization of momentum and configuration paths that occurs (in contrast to the more conventional configuration space (Lagrangian) path integral [147]). For instance, we recall from Section 4.1 that the general identification between the Schrödinger picture path integral and loop space Liouville measures was done rather artificially, basically by drawing an analogy between them. For a generic phase space path integral to represent the actual energy spectrum of the quantum Hamiltonian, one would have to carry out the usual quantization of generic Poisson brackets $\{x^\mu, x^\nu\}_\omega = \omega^{\mu\nu}(x)$. However, unlike the Heisenberg canonical commutation relations (4.1), the Lie algebra generated by this procedure is not necessarily finite-dimensional (for $\mathcal{M}$ compact) and so the representation problem has no straightforward solution when the phase space is not a cotangent bundle $\mathcal{M} \otimes \Lambda^1 \mathcal{M}$ [96], as is the case for a Euclidean phase space. This approach is therefore hopelessly complicated and in general hardly consistent. One way around this, as we shall see, is to use instead *coherent state* path integrals. This enables one to obtain the desired identification above while maintaining the original phase space path integral, and therefore at the same time keeping a formal analogy between the finite-dimensional and loop space localization formulas. Furthermore, because of their classical properties, coherent states are particularly well-suited for semi-classical studies of quantum dynamics. We shall see that all the localizable dynamical systems in 2-dimensions have phase space path integrals that can be represented in terms of coherent states, thus giving an explicit evaluation of the quantum propagator and the connection with some of the conventional coadjoint orbit models.

In this Chapter we shall in addition confine our attention to the case of a simply-connected phase space, leaving the case where $\mathcal{M}$ can have non-contractible loops for the next Chapter. In both cases, however, we shall focus on the construction of localizable Hamiltonian systems starting from a *generic* phase space metric, which will illustrate explicitly the geometrical dependence of these dynamical systems and will therefore give a further probe into the geometrical nature of (quantum) integrability. In this way, we will get a good general idea of what sort of phase space path integrals will localize and a detailed description of the symmetries responsible for localization, as well as what sort of topological field theories the localization formulas will represent.

5.1 Coadjoint Orbit Quantization and Character Formulas

There is a very interesting class of cohomological quantum theories which arise quite naturally within the framework of equivariant localization. These will set the stage for the discussion of this Chapter wherein we shall focus on the generic equivariant Hamiltonian systems with simply connected phase spaces and thus present numerous explicit examples of the localization formalism. For a (compact or non-compact) semi-simple Lie group G (i.e. one whose Lie algebra $\mathbf{g}$ has no abelian invariant subalgebras), we are interested in the coadjoint action of G on the coset space $\mathcal{M}_G = G/H_C = \{gh_C : g \in G\}$, where $H_C \sim (S^1)^r$ is the Cartan subgroup of G. The coset obtained by quotienting a Lie group by a maximal torus is often called a 'flag manifold'. The coadjoint orbit

$$O_{\Lambda'} = \{\mathrm{Ad}^*(g)\Lambda' : g \in G\} \simeq \mathcal{M}_G \quad , \quad \Lambda' \in \mathbf{h}^* \tag{5.1}$$

is the orbit of maximal dimensionality of G. Here $\mathrm{Ad}^*(g)\Lambda'$ denotes the coadjoint action of G on Λ', i.e.

$$(\mathrm{Ad}^*(g)\Lambda')(\gamma) = \Lambda'(g^{-1}\gamma g) \quad , \quad \forall \gamma \in \mathbf{g} \tag{5.2}$$

and $\mathbf{h}$ is the Cartan subalgebra of $\mathbf{g}$. The natural isomorphism in (5.1) between the flag manifold $\mathcal{M}_G = G/H_C$ and the coadjoint orbit $O_{\Lambda'}$ is $gH_C \to \mathrm{Ad}^*(g)\Lambda'$ with the maximal torus H_C identified as the stabilizer group of the point $\Lambda' \in \mathbf{h}^*$. We assume henceforth that $H^1(G) = H^2(G) = 0$. There is a natural G-invariant symplectic structure on the coadjoint orbit (5.1) which is defined by the Kirillov-Kostant 2-form [2, 3]. This 2-form at the point $\Lambda \in \mathbf{g}^*$ is given by

$$\omega_\Lambda = \frac{1}{2}\Lambda([\mathcal{T} \stackrel{\wedge}{,} \mathcal{T}]) \tag{5.3}$$

where $\mathcal{T}$ is a 1-form with values in the Lie algebra $\mathbf{g}$ which satisfies the equation

$$d\Lambda(\gamma) = \mathrm{ad}^*(\mathcal{T})\Lambda(\gamma) \equiv \Lambda([\gamma, \mathcal{T}]) \quad , \quad \forall \gamma \in \mathbf{g} \tag{5.4}$$

5.1 Coadjoint Orbit Quantization and Character Formulas

and $\text{ad}^*(T)$ denotes the infinitesimal coadjoint action of the element $T \in \mathbf{g}$.

The 2-form (5.3) is closed and non-degenerate on the orbit (5.1), and by construction the group G acts on $O_{\Lambda'}$ by symplectic (canonical) transformations with respect to the Kirillov-Kostant 2-form. Its main characteristic is that the Poisson algebra with respect to (5.3) isomorphically represents the group G,

$$\{X_1(\Lambda), X_2(\Lambda)\}_{\omega_\Lambda} = [X_1, X_2](\Lambda) \tag{5.5}$$

where $X_i \in \mathbf{g}$ are regarded as linear functionals on the orbit $O_{\Lambda'}$ with $X_i(\Lambda) \equiv \Lambda(X_i)$. Alekseev, Faddeev and Shatashvili [2, 3] have studied the phase space path integrals for such dynamical systems with Hamiltonians defined on the coadjoint orbit (5.1) (e.g. Cartan generators of $\mathbf{g}$) and have shown that, quite generally, the associated quantum mechanical matrix elements correspond to matrix elements of the Hamiltonian generator of $\mathbf{g}$ in some irreducible representation of the group G. We shall see this feature explicitly later on.

However, for our purposes here, there is a much nicer description of the orbit space (5.1) using its representation as the quotient space $\mathcal{M}_G = G/H_C$ [68]. As a smooth space, $\mathcal{M}_G$ is an example of a complex manifold of complex dimension n (real dimension $2n$), i.e. a manifold which is covered by open sets each homeomorphic to $\mathbb{C}^n$ and for which the coordinate transformations on the overlap of 2 open sets are given by holomorphic functions. Here the complexification of the group G is defined by exponentiating the complexification $\mathbf{g} \otimes \mathbb{C}$ of the finite-dimensional vector space $\mathbf{g}$. Let us quickly review some facts about the differential geometry of complex manifolds. In local coordinates $x = (z^1, \ldots, z^n) \in \mathbb{C}^n$, we can define the tangent space $T_x^{(0,1)}\mathcal{M}_G$ at $x \in \mathcal{M}_G$ as the complex vector space spanned by the $\bar{z}$ derivatives $\{\frac{\partial}{\partial \bar{z}^\mu}\}_{\mu=1}^n$, and analogously $T_x^{(1,0)}\mathcal{M}_G$ is the complex vector space spanned by the z derivatives $\{\frac{\partial}{\partial z^\mu}\}_{\mu=1}^n$. The key feature is that barred and unbarred vectors do not mix under a holomorphic change of coordinates, and therefore it makes sense (globally) to consider tensors with definite numbers k and ℓ of holomorphic and anti-holomorphic indices of either covariant or contravariant type. We refer to these as tensors of type (k, ℓ). The vector space of $(p+q)$-forms of type (p, q) is denoted $\Lambda^{(p,q)}\mathcal{M}_G$ and the exterior algebra of $\mathcal{M}_G$ now refines to

$$\Lambda \mathcal{M}_G = \bigoplus_{p,q=0}^n \Lambda^{(p,q)} \mathcal{M}_G \tag{5.6}$$

The DeRham exterior derivative operator $d \sim \eta^\mu \frac{\partial}{\partial x^\mu} : \Lambda^k \mathcal{M}_G \to \Lambda^{k+1}\mathcal{M}_G$ now decomposes into holomorphic and anti-holomorphic exterior derivative operators as

$$d = \partial + \bar{\partial} \tag{5.7}$$

where $\partial \sim \eta^\mu \frac{\partial}{\partial z^\mu} : \Lambda^{(p,q)}\mathcal{M}_G \to \Lambda^{(p+1,q)}\mathcal{M}_G$ ($\eta^\mu \sim dz^\mu$) and $\bar{\partial} \sim \bar{\eta}^\mu \frac{\partial}{\partial \bar{z}^\mu} : \Lambda^{(p,q)}\mathcal{M}_G \to \Lambda^{(p,q+1)}\mathcal{M}_G$ ($\bar{\eta}^\mu \sim d\bar{z}^\mu$). The anti-holomorphic exterior derivative $\bar{\partial}$ is called the Dolbeault operator, and the nilpotency of d now translates into the set of conditions

$$0 = \partial^2 = \bar{\partial}^2 = \partial\bar{\partial} + \bar{\partial}\partial \tag{5.8}$$

Finally, let us note that a complex manifold always possesses a globally defined rank (1,1) tensor field J (i.e. an endomorphism of the space $T^{(1,0)}\mathcal{M}_G \oplus T^{(0,1)}\mathcal{M}_G$) with $J^2 = -\mathbf{1}$. It can be defined locally by

$$J^\mu_\nu = i\delta^\mu_\nu \quad , \quad J^{\bar{\mu}}_{\bar{\nu}} = -i\delta^{\bar{\mu}}_{\bar{\nu}} \tag{5.9}$$

with all other components vanishing, and it is known as a complex structure.

Given this important property of the coadjoint orbit, we now introduce local complex coordinates $(z^\mu, \bar{z}^{\bar{\mu}})$ on $\mathcal{M}_G$ which are generated by a complex structure J. The orbit (5.1) has the topological features $H^1(\mathcal{M}_G; \mathbb{Z}) = 0$ and $H^2(\mathcal{M}_G; \mathbb{Z}) = H^1(H_C; \mathbb{Z}) = \mathbb{Z}^r$, where $r = \dim H_C$ is the rank of G and $\mathbb{Z}^r$ corresponds to the lattice of roots of H_C [162]. The cohomology classes in $H^2(\mathcal{M}_G; \mathbb{Z})$ are then represented by r closed non-degenerate 2-forms of type (1,1) [68]

$$\omega^{(i)} = \frac{i}{2} g^{(i)}_{\mu\bar{\nu}}(z, \bar{z}) dz^\mu \wedge d\bar{z}^{\bar{\nu}} \tag{5.10}$$

The components $g_{\mu\bar{\nu}}$ of (5.10) define Hermitian matrices, $g^*_{\bar{\nu}\mu} = g_{\mu\bar{\nu}}$, and the non-degeneracy condition implies that they define metrics on $T^{(1,0)}\mathcal{M}_G \oplus T^{(0,1)}\mathcal{M}_G$ by

$$g^{(i)} = g^{(i)}_{\mu\bar{\nu}}(z, \bar{z}) dz^\mu \otimes d\bar{z}^{\bar{\nu}} \tag{5.11}$$

The closure condition on the 2-forms (5.10) can be written in terms of the holomorphic and anti-holomorphic components of the exterior derivative (5.7) as

$$\partial \omega^{(i)} = \bar{\partial} \omega^{(i)} = 0 \tag{5.12}$$

The analogue of the Poincaré lemma for the Dolbeault operator $\bar{\partial}$ is the Dolbeault-Grothendieck lemma. Since the 2-forms $\omega^{(i)}$ in the case at hand are closed under both ∂ and $\bar{\partial}$, the Dolbeault-Grothendieck lemma implies that locally they can be expressed in terms of C^∞-functions $F^{(i)}$ on $\mathcal{M}_G$ as

$$\omega^{(i)} = -i\partial\bar{\partial} F^{(i)} \tag{5.13}$$

or in local coordinates

$$g^{(i)}_{\mu\bar{\nu}}(z, \bar{z}) = \frac{\partial^2 F^{(i)}(z, \bar{z})}{\partial z^\mu \partial \bar{z}^{\bar{\nu}}} \tag{5.14}$$

In general, a complex manifold with a symplectic structure such as (5.10) is called a Kähler manifold. The closed 2-forms (5.10) are then refered to as Kähler classes or Kähler 2-forms, the associated metrics (5.11) are called Kähler metrics, and the locally-defined functions $F^{(i)}$ in (5.14) are called Kähler potentials. For an elementary, comprehensive introduction to complex manifolds and Kähler structures, we refer to [41] and [61]. In the case

5.1 Coadjoint Orbit Quantization and Character Formulas

at hand here, the above construction yields a G-action on $\mathcal{M}_G$ by symplectic (canonical) transformations [68], i.e. holomorphic functions $f(z)$ on $\mathcal{M}_G$ which act on the Kähler potentials by

$$F^{(i)}(z,\bar{z}) \xrightarrow{f} \tilde{F}^{(i)}(z,\bar{z}) = F^{(i)}(z,\bar{z}) + f(z) + \bar{f}(\bar{z}) \qquad (5.15)$$

This follows from the fact that the cotangent bundle of G is

$$T^*G = G \times \mathbf{g}^* \qquad (5.16)$$

so that the natural symplectic action of G on T^*G is $g \cdot (\tilde{g}, \Lambda) = (g \cdot \tilde{g}, \Lambda)$. Consequently, the closed 2-forms $\omega^{(i)}$ define G-invariant integral symplectic structures on $\mathcal{M}_G$. Since $H^2(G) = 0$, the 2-cocycles in (3.39) vanish and this G-action determines group isomorphisms into the Poisson algebras of $\mathcal{M}$. This also follows directly from the property (5.5) of the Kirillov-Kostant 2-form above. Notice that the only non-vanishing components (up to permutation of indices and complex conjugation) of the Riemannian connection and curvature associated with the Kähler metric (5.11) are

$$\Gamma^{\mu}_{\nu\lambda} = g^{\mu\bar{\rho}} \partial_\nu g_{\lambda\bar{\rho}} \quad , \quad R^{\lambda}_{\mu\nu\bar{\rho}} = -\partial_{\bar{\rho}} \Gamma^{\lambda}_{\mu\nu} \qquad (5.17)$$

The Cartan basis of $\mathbf{g}$ is defined by the root space decomposition

$$\mathbf{g} = \mathbf{h} \oplus \left(\bigoplus_{\alpha} \mathbf{g}_\alpha \right) \qquad (5.18)$$

of $\mathbf{g}$, where $\alpha = (\alpha_1, \ldots, \alpha_r)$ are the roots of $\mathbf{g}$ (i.e. the eigenvalues of the Cartan generators in the adjoint representation of G) and $\mathbf{g}_\alpha$ are one-dimensional subspaces of $\mathbf{g}$ [53, 162]. In this basis, the generators have the non-vanishing Lie brackets

$$[H_i, E_\alpha] = \alpha_i E_\alpha \quad , \quad [E_\alpha, E_\beta] = \begin{cases} N_{\alpha\beta} E_{\alpha+\beta} & , \quad \alpha+\beta \neq 0 \\ \sum_{i=1}^{r} \alpha_i H_i & , \quad \beta = -\alpha \end{cases} \qquad (5.19)$$

where α, β are the roots of $\mathbf{g}$, $H_i = H_i^\dagger$, $i = 1, \ldots, r$, are the generators of the Cartan subalgebra $\mathbf{h} \otimes \mathbb{C}$ of $\mathbf{g} \otimes \mathbb{C}$, and $E_\alpha = E_{-\alpha}^\dagger$ are the step operators of $\mathbf{g} \otimes \mathbb{C}$ which, for each α, span $\mathbf{g}_\alpha$ in (5.18) and which act as raising operators by $\alpha > 0$ (relative to some Weyl chamber hyperplane in root space) on the representation states $|\lambda\rangle$ which diagonalize the Cartan generators (the weight states), i.e. $H_i|\lambda\rangle \propto |\lambda\rangle$ and $E_\alpha|\lambda\rangle \propto |\lambda+\alpha\rangle$ for $\alpha > 0$. The unitary irreducible representations of G are characterized by highest weights λ_i, $i = 1, \ldots, r$. For each i, λ_i is an eigenvalue of H_i whose eigenvector is annihilated by all the E_α for $\alpha > 0$. Corresponding to each highest weight vector $\lambda = (\lambda_1, \ldots, \lambda_r)$ we introduce the G-invariant symplectic 2-form

$$\omega^{(\lambda)} = \sum_{i=1}^{r} \lambda_i \omega^{(i)} \qquad (5.20)$$

The symplectic potentials associated with (5.20) are

$$\theta^{(\lambda)} = \sum_{i=1}^{r} \lambda_i \left(\frac{\partial F^{(i)}}{\partial z^\mu} dz^\mu - \frac{\partial F^{(i)}}{\partial \bar{z}^{\bar\mu}} d\bar{z}^{\bar\mu} \right) + dF \qquad (5.21)$$

To construct a topological path integral from this symplectic structure, we need to construct a Hamiltonian satisfying (3.27), i.e. a Hamiltonian which is given by generators of the subalgebra of $\mathbf{g} \otimes \mathbb{C}$ which leave the symplectic potential (5.21) invariant. These are the canonical choices that give well-defined functions on the coadjoint orbit (5.1). As remarked at the end of Section 3.2, there usually exists a choice of function $F(z,\bar{z})$ in (5.21) for which this subalgebra contains the Cartan subalgebra $\mathbf{h} \otimes \mathbb{C}$ of $\mathbf{g} \otimes \mathbb{C}$. Let $H_i^{(\lambda)}$ be the generators of $\mathbf{h} \otimes \mathbb{C}$ in the representation with highest weight vector λ. Then the Hamiltonian

$$H^{(\lambda)} = \sum_{i=1}^{r} h_i H_i^{(\lambda)} \qquad (5.22)$$

satisfies the required conditions and the corresponding path integral will admit the topological form (4.157). Note that this is also consistent with the integrability arguments of the previous Chapters, which showed that the localizable Hamiltonians were those given by the Cartan generators of an isometry group G. Thus the path integral for the above dynamical system determines a cohomological topological quantum field theory which depends only on the second cohomology class of the symplectic 2-form (5.20), i.e. on the representation with highest weight vector $\lambda = (\lambda_1, \ldots, \lambda_r)$.

To apply the equivariant localization formalism to these dynamical systems, we note that since the Kähler metrics $g^{(i)}$ above are G-invariant, the metric

$$g^{(\lambda)} = \sum_{i=1}^{r} \lambda_i g^{(i)} \qquad (5.23)$$

obeys the usual localization criteria. We shall soon see that these group theoretic structures are in fact *implied* by the localization constraints, in that they are the only equivariant Hamiltonian systems associated with homogeneous symplectic manifolds as above. Through numerous examples of such systems and others we shall verify the localization formulas of the last Chapter and discuss the common features that these quantum theories all represent. For now, however, we will just explore what the localization formulas will represent for the propagators tr $e^{iTH^{(\lambda)}}$ assuming that they admit the phase space path integral form of the last Chapter with the above symplectic structure. This will introduce some important group theoretic notions, and later

5.1 Coadjoint Orbit Quantization and Character Formulas

we shall show more precisely how to arrive at these path integral representations and discuss some of the intricacies involved in evaluating the localization formulas.

To apply the Niemi-Tirkkonen localization formula (4.130) to the dynamical system above, we first observe that the tangent and normal bundles of $O_{\Lambda'}$ in $\mathbf{g}^*$ are related by [21]

$$T_{\Lambda'}\mathbf{g}^* = TO_{\Lambda'} \oplus \mathcal{N}O_{\Lambda'} = O_{\Lambda'} \times \mathbf{g}^* \tag{5.24}$$

From the construction of the coadjoint orbit it follows that the normal bundle $\mathcal{N}O_{\Lambda'}$ in $\mathbf{g}^*$ is a trivial bundle with trivial G-action on the fibers, and the product $O_{\Lambda'} \times \mathbf{g}^*$ is a trivial bundle over $\mathbf{g}^*$ with the coadjoint action of G in the fibers. Then using (2.94) and the multiplicativity property (2.99), we can write the G-equivariant $\hat{A}$-genus of the orbit $O_{\Lambda'}$ as

$$\hat{A}_V = \sqrt{\det\left[\frac{\text{ad } X}{\sinh(\text{ad } X)}\right]} \equiv \frac{1}{\sqrt{j(\text{ad } X)}} \tag{5.25}$$

where ad X is the Cartan element $X \in \mathbf{h}$ in the adjoint representation of $\mathbf{g}$. We now choose the radius of the orbit to be the Weyl shift of the weight vector λ, i.e. $\Lambda' = \lambda + \rho$ where

$$\rho = \frac{1}{2}\sum_{\alpha>0}\alpha \tag{5.26}$$

is the Weyl vector (the half-sum of positive roots of G), where λ and α are regarded as linear functions on $\mathbf{g}$ by returning the total value of the weight or root associated to $X \in \mathbf{h}$. Then the localization formula (4.130) is none other than the celebrated Kirillov character formula [87, 140]

$$\text{tr}_\lambda\, e^{iTX} = \frac{1}{\sqrt{j(\text{ad } TX)}}\int_{O_{\lambda+\rho}}\frac{(\omega^{(\lambda)})^n}{n!}\, e^{iTH^{(\lambda)}} \tag{5.27}$$

where tr_λ denotes the trace in the representation with highest weight vector λ and $H^{(\lambda)}$ is the Hamiltonian (5.22) associated with the Cartan element $X \in \mathbf{h}$. If we further apply the finite-dimensional Duistermaat-Heckman theorem to the Fourier transform of the orbit on the right-hand side of (5.27) (i.e. the localization formula (4.135)) we arrive at the famous Harish-Chandra formula [21, 67, 140].

The resulting character formula associated with the Harish-Chandra formula for the Fourier transform of the orbit is the classical Weyl character formula of G [2]–[4], [51, 87, 117, 143, 156, 162]. Let $W(H_C) = N(H_C)/H_C$ be the Weyl group of H_C, where $N(H_C)$ is the normalizer subgroup of H_C, i.e. the subgroup of $g \in G$ with $hgH_C = gH_C, \forall h \in H_C$, so that $N(H_C)$ is the subgroup of fixed points of the left action of H_C on the orbit $\mathcal{M}_G = G/H_C$.

136   5. Equivariant Localization on Simply Connected Phase Spaces

Given $w = nH_C \in W(H_C)$, with $n = e^{iN} \in N(H_C)$, let $X^{(w)} = n^{-1}Xn$ be the respective adjoint representation $e^{iX^{(w)}} = n^{-1}e^{iX}n$. The Weyl character formula can then be written as

$$\mathrm{tr}_\lambda\, e^{iTX} = \sum_{w \in W(H_C)} e^{iT(\lambda+\rho)(X^{(w)})} \prod_{\alpha>0} \frac{1}{2i\sin\frac{T}{2}\alpha(X^{(w)})} \tag{5.28}$$

where $\alpha(X^{(w)})$ are the roots associated to the Cartan elements $X^{(w)}$. We shall see explicitly later on how these character formulas arise from the equivariant localization formulas of the last Chapter, but for now we simply note here the deep group theoretical significance that the localization formulas will represent for the path integral representations of the characters $\mathrm{tr}_\lambda\, e^{iTX}$ in that the equivariant localization formalism reproduces some classical results of group theory. Note that the Weyl character formula writes the character of a Cartan group element as a sum of terms, one for each element of the Weyl group, the group of symmetries of the roots of the Lie algebra $\mathbf{g}$. In the context of the formalism of Chapter 4, the Weyl character formula will follow from the coadjoint orbit path integral over $LO_{\lambda+\rho}$. It was Stone [156] who first related this derivation of the Weyl character formula to the index of a Dirac operator from a supersymmetric path integral and hence to the semi-classical WKB evaluation of the spin partition function, as we did quite generally in Section 4.2 above. The path integral quantization of the coadjoint orbits of semi-simple Lie groups is essential to the quantization of spin systems. One important feature of the above topological field theories is that there is a one-to-one correspondence between the points on the orbits G/H_C and the so-called coherent states associated with the Lie group G in the representation with highest weight vector λ [137]. The above character formulas can therefore be represented in complex polarizations using coherent state path integrals. We shall discuss these and other aspects of the path integral representations of character formulas as we go along in this Chapter.

From the point of view of path integral quantization, the necessity of performing a Weyl shift $\lambda \to \lambda + \rho$ in the above is rather unsatisfactory. This Weyl shift problem has been a point of some controversy in the literature [49]. As we shall see, the Weyl character formula follows *directly* from the WKB formula for the spin partition function [156], and a proper discretization of the trace in (5.27) really does give the path integral over the orbit O_λ [3, 117]. The Weyl shift is in fact an artifact of the regularization procedure [3, 102, 117, 143, 160] discussed in Section 4.2 in evaluating the fluctuation determinant there which led to the Niemi-Tirkkonen localization formula (4.130) and which leads directly to the Kirillov character formula (5.27). As the $\hat{A}$-genus is inherently related to tangent bundles of real manifolds, the problem here essentially is that the regularization discussed in Section 4.2 does not respect the complex structure defined on the orbit. We shall see later on how a coherent state formulation avoids this problem and leads to a correct localization formula without the need to introduce an explicit Weyl shift.

5.2 Isometry Groups of Simply Connected Riemannian Spaces

Given the large class of localizable dynamical systems of the last Section and their novel topological and group theoretical properties, we now turn to an opposite point of view and begin examining what Hamiltonian systems in general fit within the framework of equivariant localization. For this we shall analyse the fundamental isometry condition on the physical theory in a quite general setting, and show that the localizable systems "essentially" all fall into the general framework of the coadjoint orbit quantization of the last Section. Indeed, this will be consistent with the integrability features implied by the equivariant localization criteria.

We consider a simply-connected, connected and orientable Riemannian manifold $(\mathcal{M}, g)$ of dimension d (not necessarily symplectic for now) and with metric g of Euclidean signature, for definiteness. The isometry group $\mathcal{I}(\mathcal{M}, g)$ is the diffeomorphism subgroup of C^∞ coordinate transformations $x \to x'(x)$ which preserve the metric distance on $\mathcal{M}$, i.e. for which $g'_{\mu\nu}(x') = g_{\mu\nu}(x')$. The generators V^a of the connected component of $\mathcal{I}(\mathcal{M}, g)$ form the vector field Lie algebra

$$\mathcal{K}(\mathcal{M}, g) = \{V \in T\mathcal{M} : \mathcal{L}_V g = 0\} \qquad (5.29)$$

and obey the commutation relations (2.47). For a generic simply-connected space, the Lie group $\mathcal{I}(\mathcal{M}, g)$ is locally compact in the compact-open topology induced by $\mathcal{M}$ [68]. In particular, if $\mathcal{M}$ is compact then so is $\mathcal{I}(\mathcal{M}, g)$. When $\dim \mathcal{K}(\mathcal{M}, g) \neq 0$, we shall say that the Riemannian manifold $(\mathcal{M}, g)$ is a symmetric space.

We shall now quickly run through some of the basic facts concerning isometries of simply-connected Riemannian manifolds, all of whose proofs can be found in [42, 43, 68, 159, 164]. First of all, by analysing the possible solutions of the first order linear partial differential equations $\mathcal{L}_V g = 0$, it is possible to show that the number of linearly independent Killing vectors (i.e. generators of (5.29)) is bounded as

$$\dim \mathcal{K}(\mathcal{M}, g) \leq d(d+1)/2 \qquad (5.30)$$

when $\mathcal{M}$ has dimension d, so that the infinitesimal isometries of $(\mathcal{M}, g)$ are therefore characterized by finitely-many linearly independent Killing vectors in $\mathcal{K}(\mathcal{M}, g)$. There are 2 important classes of metric spaces $(\mathcal{M}, g)$ characterized by their possible isometries. We say that a metric space $(\mathcal{M}, g)$ is homogeneous if there exists infinitesimal isometries V that carry any given point $x \in \mathcal{M}$ to any other point in its immediate neighbourhood. $(\mathcal{M}, g)$ is said to be isotropic about a point $x \in \mathcal{M}$ if there exists infinitesimal isometries V that leave the point x fixed, and, in particular, if $(\mathcal{M}, g)$ is isotropic about all of its points then we say that it is isotropic. The homogeneity condition means that the metric g must admit Killing vectors that at any given point of $\mathcal{M}$ take on all possible values (i.e. any point on $\mathcal{M}$ is geometrically like any other point). The isotropy condition means that an isotropic point x_0 of $\mathcal{M}$ is always a fixed point of an $\mathcal{I}(\mathcal{M}, g)$-action on $\mathcal{M}$, $V(x_0) = 0$ for some $V \in \mathcal{K}(\mathcal{M}, g)$, but whose first derivatives take on all possible values, subject only to the Killing equation $\mathcal{L}_V g = 0$.

It follows that a homogeneous metric space always admits $d = \dim \mathcal{M}$ linearly independent Killing vectors (intuitively generating translations in the d directions), and a space that is isotropic about some point admits $d(d-1)/2$ Killing vector fields (intuitively generating rigid rotations about that point). The connection between isotropy and homogeneity of a metric space lies in the fact that any metric space which is isotropic is also homogeneous. The spaces which have the maximal number $d(d+1)/2$ of linearly independent Killing vectors enjoy some very special properties, as we shall soon see. We shall refer to such spaces as maximally symmetric spaces. The above dimension counting shows that a homogeneous metric space which is isotropic about some point is maximally symmetric, and, in particular, any isotropic space is maximally symmetric. The converse is also true, i.e. a maximally symmetric space is homogeneous and isotropic. In these cases, there is only one orbit under the $\mathcal{I}(\mathcal{M}, g)$-action on $\mathcal{M}$, i.e. $\mathcal{M}$ can be represented as the orbit $\mathcal{M} = \mathcal{I}(\mathcal{M}, g) \cdot x$ of any element $x \in \mathcal{M}$, and the space of orbits $\mathcal{M}/\mathcal{I}(\mathcal{M}, g)$ consists of only a single point. In this case we say that the group $\mathcal{I}(\mathcal{M}, g)$ acts transitively on $\mathcal{M}$.

Conversely, if a Lie group G acts transitively on a C^∞-manifold $\mathcal{M}$, then $\mathcal{M}$ is a homogeneous space and the stabilizer $G_x = \{g \in G : g \cdot x = x\}$ of any point $x \in \mathcal{M}$ is a closed subgroup of G. The map $hG_x \to h \cdot x$ defines a homeomorphism $G/G_x \simeq \mathcal{M}$ with the quotient topology on G/G_x induced by the natural (continuous and surjective) projection map $\pi : G \to G/G_x$. On the other hand, if G is locally compact and H is any closed subgroup of G, then there is a natural action of G on G/H defined by $g \cdot \pi(h) = \pi(gh)$, $g, h \in G$, which is transitive and for which H is the stabilizer of the point $\pi(\mathbf{1})$. In other words, homogeneous spaces are essentially coadjoint orbits G/H of Lie groups [68], with $H = G_\Lambda$ the stabilizer group of a point $\Lambda \in \mathbf{g}^*$ under the coadjoint action $\mathcal{O}_\Lambda \equiv \mathrm{Ad}^*(G)\Lambda \subset \mathbf{g}^*$ of G on $\mathbf{g}^*$. A sufficient condition for the coset space $G/H = \{gH : g \in G\}$ to be a symmetric space is that $\mathbf{g}$ admit a reductive decomposition, i.e. an orthogonal decomposition $\mathbf{g} = \mathbf{h} \oplus \mathbf{h}^\perp$ such that $[\mathbf{h}^\perp, \mathbf{h}^\perp] \subset \mathbf{h}$ [68]. Furthermore, it is possible to introduce Kähler structures (the Kirillov-Kostant 2-form introduced in the previous Section) on the group orbits for which G is the associated isometry group. These spaces therefore generalize the maximal coadjoint orbit models of the last Section where H was taken as the Cartan subgroup H_C and which

5.2 Isometry Groups of Simply Connected Riemannian Spaces

can in fact be shown to be maximally symmetric [68]. We shall see explicit examples later on.

We shall now describe the rich features of maximally symmetric spaces. It turns out that these spaces are uniquely characterized by a special curvature constant K. Specifically, $(\mathcal{M}, g)$ is a maximally symmetric Riemannian manifold if and only if there exists a constant $K \in \mathbb{R}$ such that the Riemann curvature tensor of g can be written locally almost everywhere as

$$R_{\lambda\rho\sigma\nu} \equiv g_{\lambda\mu} R^{\mu}_{\rho\sigma\nu} = K(g_{\sigma\rho} g_{\lambda\nu} - g_{\nu\rho} g_{\lambda\sigma}) \tag{5.31}$$

In dimension $d \geq 3$, Schur's lemma [68] states that the existence of such a form for the curvature tensor automatically implies the constancy of K. For $d = 2$, however, this is not the case, and indeed dimension counting shows that the curvature of a Riemann surface always takes the form (5.31). In this case K is called the Gaussian curvature of $(\mathcal{M}, g)$ and it is in general not constant. The above result implies that the Gaussian curvature K of a maximally symmetric simply connected Riemann surface is constant.

The amazing result here is the isometric correspondence between maximally symmetric spaces. Any 2 maximally symmetric spaces $(\mathcal{M}_1, g_1)$ and $(\mathcal{M}_2, g_2)$ of the same dimension and with the same curvature constant K are isometric, i.e. there exists a diffeomorphism $f : \mathcal{M}_1 \to \mathcal{M}_2$ between the 2 manifolds relating their metrics by $g_1(x) = g_2(f(x))$. Thus given any maximally symmetric space we can map it isometrically onto any other one with the same curvature tensor (5.31). We can therefore model maximally symmetric spaces by some "standard" spaces, which we now proceed to describe. Consider a flat $(d+1)$-dimensional space with coordinates (x^{μ}, z) and metric

$$\eta_{d+1} = \frac{1}{|K|} dx_{\mu} \otimes dx^{\mu} + \frac{1}{K} dz \otimes dz \tag{5.32}$$

where K is a real-valued constant. A d-dimensional space can be embedded into this larger space by restricting the variables x^{μ} and z to the surface of a (pseudo-)sphere,

$$\operatorname{sgn}(K) x^2 + z^2 = 1 \tag{5.33}$$

Using (5.33) to solve for $z(x)$ and substituting this into (5.32), the metric induced on the surface by this embedding is then

$$g_K = \begin{cases} \frac{1}{K} \left(dx_{\mu} \otimes dx^{\mu} + \frac{x_{\mu} x_{\nu}}{1 - x^2} dx^{\mu} \otimes dx^{\nu} \right) & \text{for } K > 0 \\ \frac{1}{|K|} \left(dx_{\mu} \otimes dx^{\mu} - \frac{x_{\mu} x_{\nu}}{1 - x^2} dx^{\mu} \otimes dx^{\nu} \right) & \text{for } K < 0 \\ dx_{\mu} \otimes dx^{\mu} & \text{for } K = 0 \end{cases} \tag{5.34}$$

These 3 cases represent, respectively, the standard metrics on the d-sphere S^d of radius $K^{-1/2}$, the hyperbolic Lobaschevsky space $\mathcal{H}^d$ of constant negative curvature K, and Euclidean d-space $\mathbb{R}^d$ with its usual flat metric η_{E^d}.

From the embedding condition (5.33) and the manifest invariances of the embedding space geometry (5.32) it is straightforward to show that the above spaces all admit a $d(d+1)/2$-parameter group of isometries. These consist of $d(d-1)/2$ rigid rotations about the origin and d (quasi-)translations. The first set of isometries always leave some points on the manifold fixed, while the second set translate any point on $\mathcal{M}$ to any other point in its vicinity. The 3 spaces above are therefore the 3 unique (up to isometric equivalence) maximally symmetric spaces in d-dimensions, and any other maximally symmetric space will be isometric to one of these spaces, depending on whether $K = 0$, $K > 0$ or $K < 0$. It is this feature of maximally symmetric spaces that allows the rather complete isometric correspondence which will follow. The Killing vector fields that generate the above stated isometries are, respectively,

$$V_K = \begin{cases} \left(\Omega^\mu_\nu x^\nu + \alpha^\mu \left[1 - \mathrm{sgn}(K)x^2\right]^{1/2}\right) \dfrac{\partial}{\partial x^\mu} & \text{for } K \neq 0 \\ (\Omega^\mu_\nu x^\nu + \alpha^\mu) \dfrac{\partial}{\partial x^\mu} & \text{for } K = 0 \end{cases} \quad (5.35)$$

where $\Omega^\mu_\nu = -\Omega^\nu_\mu$ and α^μ are real-valued parameters. These Killing vectors generate the respective isometry groups

$$\mathcal{I}(S^d) = SO(d+1) \quad , \quad \mathcal{I}(\mathcal{H}^d) = SO(d,1) \quad , \quad \mathcal{I}(\mathbb{R}^d) = E^d \quad (5.36)$$

where E^d denotes the Euclidean group in d-dimensions, i.e. the semi-direct product of the rotation and translation groups in $\mathbb{R}^d$, $SO(d+1)$ is the rotation group of $\mathbb{R}^{d+1}$, and $SO(d,1)$ is the Lorentz group in $(d+1)$-dimensional Minkowski space. From this we see therefore what sort of group actions should be considered within the localization framework for maximally symmetric spaces. Note that the maximal symmetry of the spaces S^d and $\mathcal{H}^d$ are actually implied by that of $\mathbb{R}^d$, because S^d can be regarded as the one-point compactification of $\mathbb{R}^d$, i.e. $S^d = \mathbb{R}^d \cup \{\infty\}$ (also known as stereographic projection), and $\mathcal{H}^d$ can be obtained from S^d by Wick rotating one of its coordinates to purely imaginary values.

The next situation of interest is the case where $(\mathcal{M}, g)$ is not itself maximally symmetric, but contains a smaller, d_H-dimensional maximally symmetric subspace $\mathcal{M}_0$ (e.g. a homogeneous but non-isotropic space). The general theorem that governs the structure of such spaces is as follows. We can distinguish $\mathcal{M}_0$ from $\mathcal{M} - \mathcal{M}_0$ by $d - d_H$ coordinates v^α, and locate points within the subspace $\mathcal{M}_0$ with d_H coordinates u^i. It can then be shown [68, 164] that it is possible to choose the local u-coordinates so that the metric of the entire space $\mathcal{M}$ has the form

$$g = \frac{1}{2}g_{\mu\nu}(x) dx^\mu \otimes dx^\nu = \frac{1}{2}g_{\alpha\beta}(v) dv^\alpha \otimes dv^\beta + \frac{1}{2}f(v)\tilde{g}_{ij}(u) du^i \otimes du^j \quad (5.37)$$

where $g_{\alpha\beta}(v)$ and $f(v)$ are functions of the v-coordinates alone, and $\tilde{g}_{ij}(u)$ is a function of only the u-coordinates that is itself the metric of $\mathcal{M}_0$. As $(\mathcal{M}_0, \tilde{g})$

5.2 Isometry Groups of Simply Connected Riemannian Spaces 141

is a d_H-dimensional maximally symmetric space, it is isometric to one of the 3 standard spaces in d_H-dimensions above and $\tilde{g}$ can be represented in one of the forms given in (5.34) depending on the curvature of the maximally symmetric subspace $\mathcal{M}_0$.

Our final general result concerning Killing vectors on generic d-dimensional simply connected manifolds is for the cases where the isometry group of $(\mathcal{M}, g)$ has the opposite feature of maximal symmetry, i.e. when $\mathcal{I}(\mathcal{M}, g)$ is 1-dimensional. Consider a 1-parameter group of isometries acting on the metric space $(\mathcal{M}, g)$. Let $V = V^\mu(x)\frac{\partial}{\partial x^\mu} \in T\mathcal{M}$ be a generator of $\mathcal{I}(\mathcal{M}, g)$, and let $\chi^\mu(x)$ be differentiable functions on $\mathcal{M}$ such that the change of variables $x'^\mu = \chi^\mu(x)$ has non-trivial Jacobian

$$\det\left[\frac{\partial \chi^\mu}{\partial x^\nu}\right] \neq 0 \tag{5.38}$$

For $\mu = 2, \ldots, d$ we can choose the diffeomorphisms $\chi^\mu(x)$ to in addition be the $d-1$ linearly independent solutions of the first order linear homogeneous partial differential equation

$$V(\chi^\mu) = \mathcal{L}_V \chi^\mu = V^\nu \partial_\nu \chi^\mu = 0 \quad , \quad \mu = 2, \ldots, d \tag{5.39}$$

given by the constant coordinate lines $\chi^\mu(x) = \text{constant}$ embedded into $\mathcal{M}$ from $\mathbb{R}^{d-1}$. The functions $\chi^\mu(x)$ for $\mu = 2, \ldots, d$ also have an invertible Jacobian matrix since then

$$\text{rank}_{2 \leq \mu, \nu \leq d}\left[\frac{\partial \chi^\mu}{\partial x^\nu}\right] = d - 1 \tag{5.40}$$

which owes to the existence of paths under the flow of the isometry group such that

$$\frac{dx^1}{V^1} = \frac{dx^2}{V^2} = \cdots = \frac{dx^d}{V^d} \tag{5.41}$$

as implied by (5.39) and the flow equation (2.46).

If we now choose the function $\chi^1(x)$ so that $\frac{\partial \chi^1}{\partial x^\mu} \neq 0$ for $\mu = 1, \ldots, d$, then the coordinate transformation $x^\mu \to x'^\mu(x) = \chi^\mu(x)$ changes the components of the vector field V to

$$V'^\mu = V^\nu \frac{\partial}{\partial x^\nu} \chi^\mu \quad , \quad \mu = 1, \ldots, d \tag{5.42}$$

It follows from (5.39) that in these new x'-coordinates V therefore has components $V'^1 \neq 0$ and $V'^\mu = 0$ for $\mu = 2, \ldots, d$. Now further change coordinates $x' \to x''$ defined by

$$x''^1 = \int_{x'_0}^{x'} \frac{dx'^1}{V'^1(x')} \quad , \quad x''^\mu = x'^\mu = \chi^\mu \quad \text{for} \quad \mu = 2, \ldots, d \tag{5.43}$$

142   5. Equivariant Localization on Simply Connected Phase Spaces

where x_0 is a fixed basepoint in $\mathcal{M}$. In this way we have shown that, in the case of a 1-parameter isometry group action on $(\mathcal{M}, g)$, there exists a local system of x''-coordinates defined almost everywhere on $\mathcal{M} - \mathcal{M}_V$ in which the Killing vector of the isometry group has components

$$V''^1 = 1 \quad , \quad V''^\mu = 0 \quad \text{for} \quad \mu = 2, \ldots, d \tag{5.44}$$

Furthermore, an application of the Killing equation (2.112) shows that $(\mathcal{M}, g)$ admits a Killing vector if and only if there are local coordinates x'' on $\mathcal{M}$ in which the metric tensor components $g''_{\mu\nu}(x'')$ are independent of the coordinate x'''^1,

$$\frac{\partial g''_{\mu\nu}(x'')}{\partial x'''^1} = 0 \tag{5.45}$$

and then the integral curves of x'''^1 parametrize the paths of the infinitesimal isometry and of the finite total isometry according to (2.46). Moreover, the above derivation also shows that 2 distinct isometries V_1 and V_2 of $(\mathcal{M}, g)$ cannot have the same path, since they can be independently chosen to have the single non-vanishing components $V_1''^1 = V_2''^2 = 0$. These results mean that locally any isometry of g looks like translations in a single coordinate, and this therefore gives the representation of a 1-parameter isometry as an explicit $\mathbb{R}^1$-action on $(\mathcal{M}, g)$ (which is either bounded or is a $U(1)$-action when $\mathcal{M}$ is compact). We shall refer to this system of coordinates as a prefered set of coordinates with respect to a Killing vector field V.

For simplicity, we shall now concentrate on the cases where $\mathcal{M}$ is a simply connected 2-dimensional symplectic manifold with metric g. Notice that the standard Cauchy-Riemann equations of complex analysis imply that a Riemann surface is always a complex manifold. The advantage of this insofar as the localization formalism is concerned is that the Riemann uniformization theorem [74, 111, 145] tells us that $g_{\mu\nu}(x)$ has globally only 1 independent component. This situation is therefore amenable to a detailed analysis of the equivariant localization constraints in terms of the single degree of freedom of the metric g. Defining complex coordinates $z, \bar{z} = x^1 \pm ix^2$, we can represent the metric as

$$g = \lambda(dz + \mu d\bar{z}) \otimes (d\bar{z} + \bar{\mu}dz) \tag{5.46}$$

where

$$\lambda = (\operatorname{tr} g + 2\sqrt{\det g})/4 \quad , \quad \mu = (g_{11} - g_{22} + 2ig_{12})/4\lambda \tag{5.47}$$

Orientation-preserving diffeomorphisms of $\mathcal{M}$ which only change the function $\lambda > 0$ above are called conformal transformations. The function μ determines the complex structures of $\mathcal{M}$, and therefore the set of inequivalent complex structures of $\mathcal{M}$ is in a one-to-one correspondence with the space of conformal equivalence classes of metrics on $\mathcal{M}$. A complex coordinate w is said to be an isothermal coordinate for g if $g = \rho \, dw \otimes d\bar{w}$ for some function $\rho > 0$. Using the tensor transformation law for g, it follows from (5.46) that

5.2 Isometry Groups of Simply Connected Riemannian Spaces 143

an isothermal coordinate w for g exists if and only if the Beltrami partial differential equation

$$\frac{\partial w}{\partial \bar{z}} = \mu \frac{\partial w}{\partial z} \tag{5.48}$$

has a C^∞-solution $w(z,\bar{z})$. Such a solution always exists provided that the function $\mu(z,\bar{z})$ is uniformly bounded as $\|\mu\|_\infty < 1$. A complex structure on $\mathcal{M}$ can therefore be identified with the conformal structure represented by the Riemannian metric g.

The simple-connectivity of a 2-manifold $\mathcal{M}$ implies that via a diffeomorphism and Weyl rescaling $g \to e^\varphi g$ of the coordinates the metric can be put globally[1] into the isothermal form

$$g_{\mu\nu}(x) = e^{\varphi(x)}\delta_{\mu\nu} \quad \text{or} \quad g = e^{\varphi(z,\bar{z})}dz \otimes d\bar{z} \tag{5.49}$$

where $\varphi(x)$ is a globally-defined real-valued function on $\mathcal{M}$ which we shall refer to as the conformal factor of the metric. This means that there is a unique complex structure on the Riemann surface $\mathcal{M}$ which we can define by the standard local complex coordinates $z, \bar{z} = x^1 \pm ix^2$. Notice that these remarks are *not* true if $H^1(\mathcal{M}; \mathbb{Z}) \neq 0$, because then the metric has additional degrees of freedom from moduli parameters (see the torus example in Section 3.5), i.e. (5.49) should be replaced by

$$g = e^\varphi \hat{g}(\tau) \tag{5.50}$$

where τ labels the additional modular degress of freedom of the metric. We shall discuss the case of multiply-connected phase spaces in the next Chapter.

With this complex structure we define $V^{z,\bar{z}} = V^1 \pm iV^2$ for any vector field V, and we set $\partial, \bar{\partial} = \frac{1}{2}(\frac{\partial}{\partial x^1} \mp i\frac{\partial}{\partial x^2})$. The Killing equations (2.112) in these complex coordinates can be written as

$$\bar{\partial}V^z = \partial V^{\bar{z}} = 0 \quad , \quad \partial V^z + \bar{\partial}V^{\bar{z}} + V^z\partial\varphi + V^{\bar{z}}\bar{\partial}\varphi = 0 \tag{5.51}$$

The first set of equations in (5.51) are the Cauchy-Riemann equations and they imply that in these local coordinates the Killing vector field V^z is a holomorphic function on $\mathcal{M}$. The other equation is a source equation for V^z and $V^{\bar{z}}$ that explicitly determines the Killing fields in terms of the single degree of freedom of the metric g (i.e. the conformal factor φ).

The Gaussian curvature scalar $K(x)$ of $(\mathcal{M}, g)$, which is always defined by (5.31) in 2 dimensions, can be written in these isothermal coordinates as

$$K(x) = -\frac{1}{2}e^{-\varphi(x)}\nabla^2\varphi(x) \tag{5.52}$$

---

[1] The fact that this holds *globally* follows from an application of the classical Riemann-Roch theorem, or the more modern Atiyah-Singer index theorem [41, 61, 111].

144   5. Equivariant Localization on Simply Connected Phase Spaces

where $\nabla^2 = g^{\mu\nu}\nabla_\mu\nabla_\nu = \partial\bar\partial$ is the 2-dimensional scalar Laplacian on $\mathcal{M}$ associated with the metric (5.49). This follows from noting that the only non-vanishing connection coefficients of the metric (5.49) are

$$\Gamma^z_{zz} = \partial\varphi \quad , \quad \Gamma^{\bar z}_{\bar z\bar z} = \bar\partial\varphi \tag{5.53}$$

The Gaussian curvature of $(\mathcal{M}, g)$ then uniquely characterizes the isometry group acting on the phase space. If K is constant, then $(\mathcal{M}, g)$ is maximally symmetric with 3 linearly independent Killing vectors. Moreover, in this case $(\mathcal{M}, g)$ is isometric to either the 2-sphere S^2, the Lobaschevsky plane $\mathcal{H}^2$ or the Euclidean plane $\mathbb{R}^2$. We shall soon examine these 3 distinct Riemannian spaces in detail. Notice, however, that if $\mathcal{M} = \Sigma^h$ is a compact Riemann surface of genus h, then the Gauss-Bonnet-Chern theorem (2.103) in the case at hand reads

$$\int_{\Sigma^h} d\mathrm{vol}(g(x))\, K(x) = 4\pi(1-h) \tag{5.54}$$

where $d\mathrm{vol}(g(x)) = d^2x\,\sqrt{\det g(x)}$ is the metric volume form of $(\mathcal{M}, g)$. Thus a maximally symmetric compact Riemann surface of constant negative curvature must have genus $h \geq 2$. It follows, under the simple-connectivity assumption of this Chapter, that when $K = 0$ or $K > 0$ the phase space $\mathcal{M}$ can be either compact or non-compact, but when $K < 0$ it is necessarily non-compact.

The other extremal case is where $(\mathcal{M}, g)$ admits only a 1-parameter group of isometries. From the above general discussion it follows that in this case there exist 2 differentiable functions χ^1 and χ^2 on $\mathcal{M}$ and local coordinates x' on $\mathcal{M}$ such that

$$V^\mu \frac{\partial}{\partial x^\mu}\chi^2(x^1, x^2) = 0 \quad , \quad x'^2 = \chi^2(x^1, x^2) \tag{5.55}$$

and in these coordinates the Killing vector field has components $V'^1 = 1, V'^2 = 0$. Moreover, the characteristic curves of the coordinate $x'^2 = \chi^2$, defined by the initial data surfaces of the partial differential equation in (5.55), can be chosen to be orthogonal to the paths defined by the isometry generator V, i.e. we can choose the initial conditions for the solutions of (5.55) to lie on a non-characteristic surface. This means that in these new coordinates $g'_{12}(x') = 0$. Thus in this case the metric can be written locally as

$$g = g'_{11}dx'^1 \otimes dx'^1 + g'_{22}dx'^2 \otimes dx'^2 \tag{5.56}$$

and from (5.45) it follows that g'_{11} and g'_{22} are functions only of x'^2. The phase space therefore describes a surface of revolution, for example a cylinder or the 'cigar-shaped' geometries that are described in typical black hole theories [164].

The only other case left to consider here is when $(\mathcal{M}, g)$ has a 2-dimensional isometry group. In this case we have 2 independent vector fields

5.2 Isometry Groups of Simply Connected Riemannian Spaces

$V_1 = V_1^\mu \frac{\partial}{\partial x^\mu}$ and $V_2 = V_2^\mu \frac{\partial}{\partial x^\mu}$ which obey the Lie algebra (2.47) with $a, b, c = 1, 2$. There are 2 possibilities for this Lie algebra – either the isometry group is abelian, $f^{abc} = 0$, or it is non-abelian, $f^{abc} \neq 0$ for some a, b, c. Since V_1 and V_2 cannot have the same path in $\mathcal{M}$, we can choose paths for the constant coordinate lines so that $V_1^2 = V_2^1 = 0$. In the abelian case, the commutativity of V_1 and V_2

$$[V_1, V_2] = 0 \tag{5.57}$$

implies that V_1^1 is a function of x^1 alone and V_2^2 is a function only of x^2. As above, we can choose local coordinates almost everywhere on $\mathcal{M}$ in which $V_1^1 = V_2^2 = 1$. In these coordinates, the Killing equations imply that the metric components $g_{\mu\nu}(x)$ are all constant. Thus in this case $(\mathcal{M}, g)$ is isometric to flat Euclidean space, which contradicts the standard maximal symmetry arguments above.

In the non-abelian case, we can choose linear combinations of the isometry generators V_1 and V_2 so that their Lie algebra is

$$[V_1, V_2] = V_1 \tag{5.58}$$

which implies that

$$\partial_1 V_2^2 = 0 \quad , \quad \partial_2 \log V_1^1 = -1/V_2^2 \tag{5.59}$$

and so we can choose local coordinates almost everywhere on $\mathcal{M}$ in which $V_2^2 = 1$ and $V_1^1 = e^{-x^2}$. The Killing equations then become

$$\partial_2 g_{\mu\nu} = \partial_1 g_{11} = 0 \quad , \quad \partial_1 g_{12} = g_{11} \quad , \quad \partial_1 g_{22} = 2g_{12} \tag{5.60}$$

which have solutions

$$g_{11} = \alpha \quad , \quad g_{12} = \alpha x^1 + \beta \quad , \quad g_{22} = \alpha(x^1)^2 + 2\beta x^1 + \gamma \tag{5.61}$$

where α, β and γ are real-valued constants. It is then straightforward to compute the Gaussian curvature of g from the identity

$$K(x) = -R_{1212}(x)/\det g(x) \tag{5.62}$$

which gives $K(x)$ as the constant $K = \alpha/(\beta^2 - \alpha\gamma)$, again contradicting the maximal symmetry theorems quoted above.

Thus a 2-dimensional phase space is *either* maximally symmetric with a 3-dimensional isometry group, or it admits a 1-parameter group of isometries (or, equivalently, has a single 1-dimensional maximally symmetric subspace), because the above arguments show that it clearly cannot have a 2-dimensional isometry group. The fact that there are only 2 distinct classes of isometries in 2 dimensions is another very appealing feature of these cases for the analysis which follows. For the remainder of this Chapter we shall analyse the equivariant Hamiltonian systems which can be studied on the various isometric types of spaces discussed in this Section and discuss the features of the integrable quantum models that arise from the localization formalism. We shall primarily develop these systems in 2 dimensions, and present higher-dimensional examples in Sections 5.7 and 5.8. This will provide a large set of explicit examples of the formalism developed thus far and at the same time clarify some other issues that arise within the formalism of path integral quantization.

5.3 Euclidean Phase Spaces and Holomorphic Quantization

We begin our study of general localizable Hamiltonian systems with the case where the phase space $\mathcal{M}$ is locally flat, i.e. $K = 0$. The conformal factor φ in (5.49) and (5.52) then satisfies the 2-dimensional Laplace equation

$$\nabla^2 \varphi(z, \bar{z}) = \partial \bar{\partial} \varphi(z, \bar{z}) = 0 \tag{5.63}$$

whose general solutions are

$$\varphi(z, \bar{z}) = f(z) + \bar{f}(\bar{z}) \tag{5.64}$$

where $f(z)$ is *any* holomorphic function on $\mathcal{M}$. The Riemannian manifold $(\mathcal{M}, g)$ is isometric to the flat Euclidean space $(\mathbb{R}^2, \eta_{E^2})$ and from the metric tensor transformation law it follows that this coordinate change $z \to w$ taking the metric (5.49) to $dw \otimes d\bar{w}$ satisfies

$$\frac{\partial w}{\partial z}\frac{\partial \bar{w}}{\partial \bar{z}} + \frac{\partial \bar{w}}{\partial z}\frac{\partial w}{\partial \bar{z}} = e^{\varphi(z,\bar{z})} = e^{f(z)} e^{\bar{f}(\bar{z})} \quad , \quad \frac{\partial w}{\partial z}\frac{\partial \bar{w}}{\partial z} = \frac{\partial w}{\partial \bar{z}}\frac{\partial \bar{w}}{\partial \bar{z}} = 0 \tag{5.65}$$

It follows from (5.65) that this isometric transformation is the 2-dimensional conformal transformation $z \to w_f(z)$ (i.e. an analytic rescaling of the standard flat Euclidean metric of the plane) where

$$w_f(z) = \int_{C_z} d\xi \; e^{f(\xi)} \tag{5.66}$$

and $C_z \subset \mathcal{M}$ is a simple curve from some fixed basepoint in $\mathcal{M}$ to z. From the last Section (eq. (5.35)) we know that the Killing vectors of $(\mathbb{R}^2, \eta_{E^2})$ in the complex coordinates $(w, \bar{w})$ take on the general form

$$V_{\mathbb{R}^2}^w = -i\Omega w + \alpha \quad , \quad V_{\mathbb{R}^2}^{\bar{w}} = i\Omega \bar{w} + \bar{\alpha} \tag{5.67}$$

where $\Omega \in \mathbb{R}$ and $\alpha \in \mathbb{C}$ are constants. The Killing vectors (5.67) follow directly from (5.51) with $\varphi = 0$ there, and they generate the groups of 2-dimensional rotations $w \to e^{i\Omega} w$ and translations $w \to w + \alpha$ whose semi-direct product forms the Euclidean group E^2 of the plane.

5.3 Euclidean Phase Spaces and Holomorphic Quantization

In these local complex coordinates on $\mathbb{R}^2$ the Hamiltonian equations $dH = -i_V\omega$ take the form

$$\partial H = \frac{i}{2}\omega(w,\bar{w})V^{\bar{w}} \quad , \quad \bar{\partial} H = -\frac{i}{2}\omega(w,\bar{w})V^w \tag{5.68}$$

where

$$\omega = \frac{i}{2}\omega(w,\bar{w})dw \wedge d\bar{w} \tag{5.69}$$

The symplectic 2-form (5.69) can be explicitly determined here by recalling that the Hamiltonian group action on the phase space is symplectic so that $\mathcal{L}_V\omega = 0$. In local coordinates this means that

$$\partial_\mu(V^\lambda \omega_{\nu\lambda}) - \partial_\nu(V^\lambda \omega_{\mu\lambda}) = 0 \tag{5.70}$$

for each μ and ν. Requiring this symplecticity condition for the full isometry group action of E^2 on $\mathbb{R}^2$, we substitute into (5.70) each of the 3 linearly independent Killing vectors represented by (5.67) (corresponding to $\Omega = 0$, $\alpha^1 = 0$ and $\alpha^2 = 0$ there). The differential equations (5.70) for the function $\omega(w,\bar{w})$ now easily imply that it is constant on $\mathbb{R}^2$ with these substitutions. Thus $\omega(w,\bar{w})$ is the Riemannian volume (and in this case the Darboux) 2-form globally on $\mathbb{R}^2$. Substituting the Darboux value $\omega(w,\bar{w}) = 1$ and the Killing vectors (5.67) into the Hamiltonian equations above and integrating them up to get $H(w,\bar{w})$, we see that the most general equivariant Hamiltonian on a planar phase space $\mathcal{M}$ is

$$H_0(z,\bar{z}) = \Omega w_f(z)\bar{w}_f(\bar{z}) + \bar{\alpha}w_f(z) + \alpha\bar{w}_f(\bar{z}) + C_0 \tag{5.71}$$

where $C_0 \in \mathbb{R}$ is a constant of integration and $w_f(z)$ is the conformal transformation (5.66) from the flat Euclidean space back onto the original phase space.

The fact that the symplectic 2-form here is uniquely determined to be the volume form associated with the phase space geometry is a general feature of any homogeneous symplectic manifold. Indeed, when a Lie group G acts transitively on a symplectic manifold there is a unique G-invariant measure [68], i.e. a unique solution for the $d(d-1)/2$ functions $\omega_{\mu\nu}$ from the $d(d-1) \cdot d(d+1)/4$ differential equations (5.70). Thus $\omega^n/n!$ is necessarily the maximally symmetric volume form of $(\mathcal{M}, g)$ and the phase space is naturally a Kähler manifold, as in Section 5.1. We shall soon see the precise connection between maximally symmetric phase spaces and the coadjoint orbit models of Section 5.1. In the present context, this is one of the underlying distinguishing features between the maximally symmetric and inhomogeneous cases. In the latter case ω is not uniquely determined from the requirement of symplecticity of the isometry group action on $\mathcal{M}$, leading to numerous possibilities for the equivariant Hamiltonian systems. In the case at hand here, the Darboux 2-form on $\mathbb{R}^2$ is the unique 2-form which is invariant under the full Euclidean

148   5. Equivariant Localization on Simply Connected Phase Spaces

group, i.e. invariant under rotations and translations in the plane, and on $\mathcal{M}$ it is the Kähler form associated with the Kähler metric (5.49) and (5.64).

The form (5.71) for the planar equivariant Hamiltonian systems illustrates how the integrable dynamical systems which obey the localization criteria depend on the phase space geometry which needs to be introduced in this formalism. These systems are all, however, holomorphic copies of the same initial dynamical system on $\mathbb{R}^2$ defined by the Darboux Hamiltonian

$$H_0^D(z,\bar{z}) = \Omega z\bar{z} + \bar{\alpha}z + \alpha\bar{z} + C_0 \quad ; \quad z \in \mathbb{C} \qquad (5.72)$$

or identifying $z, \bar{z} = p \pm iq$ with (p,q) canonical momentum and position variables, these dynamical Hamiltonians are of the form

$$H_0^D(p,q) = \Omega(p^2 + q^2) + \alpha_1 p + \alpha_2 q + C_0 \qquad (5.73)$$

Thus the dependence on the phase space Riemannian geometry is trivial in the sense that these systems all lift to families of holomorphic copies of the planar dynamical systems (5.72). This sort of trivial dependence is to be expected since the (classical or quantum) dynamical problem is initially independent of any Riemannian geometry of the phase space. It is also anticipated from the general topological field theory arguments that we presented earlier. Nonetheless, the general functions $H_0(z,\bar{z})$ in (5.71) illustrate how the geometry required for equivariant localization is determined by the different dynamical systems, and vice versa, i.e. the geometries that make these dynamical systems integrable. This probes into what one may consider to be the geometry of the classical or quantum dynamical system, and it illustrates the strong interplay between the Hamiltonian and Riemannian symmetries that are responsible for localization.

Thus essentially the only equivariant Hamiltonian system on a planar symplectic manifold is the displaced harmonic oscillator Hamiltonian

$$H_0^D = \Omega(z+a)(\bar{z}+\bar{a}) = \Omega\left\{(p+a_1)^2 + (q+a_2)^2\right\} + C_0 \qquad (5.74)$$

and in this case we can replace the requirement that H generate a circle action with the requirement that it generate a semi-bounded group action. To compare the localization formulas with some well-known results from elementary quantum mechanics, we note that the Hamiltonian (5.73) can only describe 2 distinct 1-dimensional quantum mechanical models. These are the harmonic oscillator $\frac{1}{2}z\bar{z} = \frac{1}{2}(p^2 + q^2)$ wherein we take $\Omega = \frac{1}{2}$ and $\alpha = 0$ in (5.72) and apply either the WKB or the Niemi-Tirkkonen localization formulas of the last Chapter, and the free particle $\frac{1}{2}p^2$ where we take $\Omega = 0$ and $\alpha = 1/2\sqrt{2}$ in (5.72) and apply the quadratic localization formula (4.150) (or equivalently (4.142)). In fact, these are the original classic examples, which were for a long time the *only* known examples, where the Feynman path integral can be evaluated exactly because then their functional (and classical statistical mechanical) integrals are Gaussian. For the same reasons, these

5.3 Euclidean Phase Spaces and Holomorphic Quantization

are also the basic examples where the WKB approximation is known to be exact [147].

It is straightforward to verify the Niemi-Tirkkonen localization formula (4.130) for the harmonic oscillator. In polar coordinates $z = r\,\mathrm{e}^{i\theta}$ with $r \in \mathbb{R}^+$ and $\theta \in [0, 2\pi]$, we have $\omega_{r\theta} = r$, $(\Omega_V)_{\theta r} = -2r$ and $R = 0$ on flat $\mathbb{R}^2$, so that the integral in (4.130) gives

$$Z_{\text{harm}}(T) \sim \int_0^\infty dr\, \frac{T}{2\sin\frac{T}{2}}\, \mathrm{e}^{-iTr^2/2} = \frac{1}{2i\sin\frac{T}{2}} \qquad (5.75)$$

That this is the correct result can be seen by noting that the energy spectrum determined by the Schrödinger equation for the harmonic oscillator is $E_k = k + \frac{1}{2}$, $k \in \mathbb{Z}^+$ [101], so that

$$\mathrm{tr}\| \,\mathrm{e}^{-iT(\hat{p}^2+\hat{q}^2)/2} \| = \sum_k \mathrm{e}^{-iTE_k} = \sum_{k=0}^\infty \mathrm{e}^{-iT(k+\frac{1}{2})} = \frac{1}{2i\sin\frac{T}{2}} \qquad (5.76)$$

This result also follows from the WKB formula (4.115) after working out the regularized fluctuation determinant as described there. Here the classical trajectories determined by the flows of the vector field $V^z = iz/2$ are the circular orbits $z(t) = z(0)\,\mathrm{e}^{it/2}$. Note that the only way these orbits can be defined on the loop space $L\mathbb{C}$ is to regard $z(t) = z(0)\,\mathrm{e}^{it/2}$ and $\bar{z}(t) = \bar{z}(T)\,\mathrm{e}^{i(T-t)/2}$ as independent complex variables. This means that the functional integral should be evaluated in a holomorphic polarization. We shall return to this point shortly.

Alternatively, we note that for $T \neq 2\pi n$ the only T-periodic critical trajectories of this dynamical system are the critical points $z, \bar{z} = 0$ of the harmonic oscillator Hamiltonian $z\bar{z}$ and (5.76) also follows from (4.135) which gives

$$Z_{\text{harm}}(T) \sim \frac{1}{\mathrm{Pfaff}\begin{pmatrix} 0 & 1 \\ -1 & 0 \end{pmatrix}} \frac{\frac{1}{2i}}{\sin\frac{T}{2}} = \frac{1}{2i\sin\frac{T}{2}} \qquad (5.77)$$

For the discretized values $T = 2\pi n$ any initial condition $z(0) \in \mathbb{C}$ leads to T-periodic orbits, and the moduli space of critical trajectories is non-isolated and coincides with the entire phase space $\mathcal{M} = \mathbb{R}^2$. In that case the degenerate path integral formula (4.122) yields the correct result. These results therefore all agree with the general assertions made at the beginning of Section 4.6 concerning the structure of the moduli space of T-periodic classical trajectories for a Hamiltonian circle action on the phase space.

For the free particle partition function, we have $R = \Omega_V = 0$, and so the $\hat{A}$-genus term in the localization formula (4.150) contributes 1. The ϕ_0-integral in (4.150) is thus a trivial Gaussian one and we find

$$Z_{\text{free}}(T) \sim \int_{-\infty}^\infty dp\,dq \int_{-\infty}^\infty d\phi_0\, \mathrm{e}^{iT\phi_0^2 - iT\phi_0 p/2\sqrt{2}} \sim \int_{-\infty}^\infty dp\,dq\, \mathrm{e}^{-iTp^2/2} \qquad (5.78)$$

which also coincides with the exact propagator $\mathrm{tr}\| \, \mathrm{e}^{-iT\hat{p}^2/2} \|$ in the phase space representation. In this case the Hamiltonian $\frac{1}{2}p^2$ is degenerate on $\mathbb{R}^2$, so that the WKB localization formula is unsuitable for this dynamical system and the result (5.78) follows from the degenerate formula (4.122) by noting that $L\mathcal{M}_S = \mathbb{R}^2$ in this case. Notice also that (5.78) coincides exactly with the classical partition function of this dynamical system as there are no quantum fluctuations.

There is another way to look at the path integral quantization of the Darboux Hamiltonian system (5.72) which ties in with some of the general ideas of Section 5.1 above. The Heisenberg-Weyl algebra $\mathbf{g}_{HW}$ [101] is the algebra generated by the usual harmonic oscillator creation and annihilation operators

$$\hat{a}, \hat{a}^\dagger = \frac{1}{\sqrt{2}}(\hat{p} \pm i\hat{q}) \qquad (5.79)$$

in the canonical quantum theory associated with the phase space $\mathbb{R}^2$ and the operator algebra (4.1). The Lie algebra $\mathbf{g}_{HW}$ is generated by the operators $\hat{a}^\dagger$, $\hat{a}$ and $\hat{N} \equiv \hat{a}^\dagger \hat{a} = \frac{1}{2}(\hat{p}^2 + \hat{q}^2 - 1)$ with the commutation relations

$$[\hat{a}^\dagger, \hat{a}] = 1 \qquad (5.80)$$

The (infinite-dimensional) Hilbert space which defines a representation of these operators is spanned by the bosonic number basis $|n\rangle$, $n \in \mathbb{Z}^+$, which form the complete orthonormal system of eigenstates of the number operator $\hat{N}$ with eigenvalue n,

$$\hat{N}|n\rangle = \hat{a}^\dagger \hat{a}|n\rangle = n|n\rangle \qquad (5.81)$$

and on which $\hat{a}^\dagger$ and $\hat{a}$ act as raising and lowering operators, respectively,

$$\hat{a}^\dagger|n\rangle = \sqrt{n+1}|n+1\rangle \quad , \quad \hat{a}|n\rangle = \sqrt{n}|n-1\rangle \qquad (5.82)$$

We now define the canonical coherent states [45, 101, 137] associated with this representation of the Heisenberg-Weyl group G_{HW} as

$$|z) \equiv \mathrm{e}^{z\hat{a}^\dagger}|0\rangle = \sum_{n=0}^{\infty} \frac{z^n}{\sqrt{n!}}|n\rangle \quad ; \quad z \in \mathbb{C} \qquad (5.83)$$

These states are normalized as

$$(z|z) = \mathrm{e}^{z\bar{z}} \qquad (5.84)$$

with $(z| \equiv |z)^\dagger$, and they obey the completeness relation

5.3 Euclidean Phase Spaces and Holomorphic Quantization

$$\int \frac{d^2z}{2\pi} |z))((z| = \int \frac{d^2z}{2\pi} \; e^{-z\bar{z}} \sum_{n,m} \frac{z^n \bar{z}^m}{\sqrt{n!m!}} |n\rangle\langle m|$$

$$= \frac{1}{2\pi} \int_0^\infty dr \; r \; e^{-r^2} \sum_{n,m} \frac{r^{n+m}}{\sqrt{n!m!}} \int_0^{2\pi} d\theta \; e^{i(n-m)\theta} |n\rangle\langle m|$$

$$= \frac{1}{2\pi} \int_0^\infty dr \; r \; e^{-r^2} \sum_{n,m} \frac{r^{n+m}}{\sqrt{n!m!}} \cdot 2\pi \delta_{nm} |n\rangle\langle m| = \sum_{n=0}^\infty |n\rangle\langle n| = 1 \tag{5.85}$$

where we have as usual written $z = r \; e^{i\theta}$ and

$$|z)) \equiv |z)/\sqrt{(z|z)} = e^{-z\bar{z}/2}|z) \tag{5.86}$$

are the normalized coherent states.

The normalized matrix elements of the algebra generators in these states are

$$((z|\hat{a}^\dagger \hat{a}|z)) = z\bar{z} \; , \quad ((z|\hat{a}|z)) = \bar{z} \; , \quad ((z|\hat{a}^\dagger|z)) = z \tag{5.87}$$

Thus the 3 independent terms in the Darboux Hamiltonian (5.72) are none other than the normalized canonical coherent state matrix elements of the Heisenberg-Weyl group generators. These 3 observables represent the Poisson Lie group action of the Euclidean group E^2 on the coadjoint orbit $G_{HW}/H_C = G_{HW}/U(1) = \mathbb{C}^1$ with the Darboux Poisson bracket

$$\{z, \bar{z}\}_{\omega_D} = 1 \tag{5.88}$$

which is the Poisson algebra representation of the Heisenberg-Weyl algebra (5.80). This correspondence with the coset space G_{HW}/H_C and the general framework of Section 5.1 is not entirely surprising, since homogeneous symplectic manifolds are in *general* essentially coadjoint orbits of Lie groups [68], i.e. they can be represented as the quotient of their isometry groups by a maximal torus according to the general discussion of the last Section. The integrable Hamiltonian systems in this case are functionals of Cartan elements of $\mathbf{g}_{HW}$ (e.g. the harmonic oscillator $\hat{a}^\dagger \hat{a}$ or the free particle $(\hat{a} + \hat{a}^\dagger)^2$).

The canonical coherent states (5.83) are those quantum states which minimize the Heisenberg uncertainty principle $\Delta q \cdot \Delta p \geq \frac{1}{2}$ [101], because they diagonalize the annihilation operator $\hat{a}$, $\hat{a}|z) = z|z)$, and they can be generalized to arbitrary Lie groups [137], as we shall soon see. The Darboux 2-form

$$\omega_D = \frac{i}{2} dz \wedge d\bar{z} \tag{5.89}$$

is defined globally on $\mathbb{C}$ and, since $\mathbb{R}^2$ is contractable and hence $H^2(\mathbb{R}^2; \mathbb{Z}) = 0$, it can be be generated *globally* by the symplectic potential

$$\theta_D = -\frac{i}{2}(\bar{z}dz - zd\bar{z}) \tag{5.90}$$

152   5. Equivariant Localization on Simply Connected Phase Spaces

The canonical 1-form (5.90) and the flat Kähler metric associated with (5.89) on $\mathbb{R}^2$ can be written in terms of the coherent states (5.83) as

$$\theta_D = \frac{i}{2} (\!(z|d|z)\!) \tag{5.91}$$

$$g_D = dz \otimes d\bar{z} = \| \, d|z)\!) \, \| \otimes \| \, d|z)\!) \, \| - (\!(z|d|z)\!) \otimes (\!(z|d|z)\!)^* \tag{5.92}$$

and the globally-defined Kähler potential associated with (5.90) is

$$F_{\mathbb{R}^2}(z,\bar{z}) = z\bar{z} \tag{5.93}$$

The path integral here then coincides with the standard coherent state path integral

$$\operatorname{tr} e^{-iT\hat{\mathcal{H}}} = \int \frac{d^2z}{2\pi} (\!(z| \, e^{-iT\hat{\mathcal{H}}} |z)\!)$$

$$= \int_{L\mathbb{R}^2} \prod_{t \in [0,T]} \frac{dz(t) \, d\bar{z}(t)}{2\pi} \, \exp\left\{ i \int_0^T dt \; (z\dot{\bar{z}} - \bar{z}\dot{z} - H(z,\bar{z})) \right\} \tag{5.94}$$

where

$$H(z,\bar{z}) = (\!(z|\hat{\mathcal{H}}|z)\!) \tag{5.95}$$

is the coherent state matrix element of some operator $\hat{\mathcal{H}} = \hat{\mathcal{H}}(\hat{a}, \hat{a}^\dagger)$ on the underlying representation space of the Heisenberg-Weyl algebra. The derivation of (5.94) is identical to that in Section 4.1 except that now we use the modified completeness relation (5.85) for the coherent state representation. This manner of describing the quantum dynamics goes under the names of holomorphic, coherent state or Kähler polarization. One of its nice features in general is that it provides a natural identification of the path integral and loop space Liouville measures. We recall from (4.21) that in the former measure there is one unpaired momentum in general and, besides the periodic boundary conditions, there is a formal analog between the measures in (4.21) and (4.25) only if the initial configuration of the propagator depends on the position variables q and the final configuration on the momentum variables p, or vice versa. In the holomorphic polarization above, however, the initial configuration depends on the z variables, the final one on the $\bar{z}$ variables, and the path integral measure is the formal $N \to \infty$ limit of $\prod_{i=1}^N dz_i \, d\bar{z}_i/2\pi$. Since the number of z and $\bar{z}$ integrations are the same, we obtain the desired formal identifications. Besides providing one with a formal analog between the path integral localization formulas and the Duistermaat-Heckman theorem and its generalizations, this enables one to also ensure that the loop space supersymmetry encountered in Section 4.4, which is intimately connected with the definition of the path integral measure (as are the boundary conditions for the propagator), is consistent with the imposed boundary conditions.

Thus on a planar phase space essentially the only equivariant Hamiltonian systems are harmonic oscillators, generalized as in (5.71) to the inclusion of a generic flat geometry so that the remaining Hamiltonian systems are merely holomorphic copies of these displaced oscillators defined by the analytic coordinate transformation (5.66). These systems generate a topological quantum theory of the sort discussed in Sections 4.10 and 5.1, with the Darboux Hamiltonian (5.72) related to the symplectic potential (5.90) by the usual topological condition $H_0^D = i_{V_{\mathbb{R}^2}} \theta_D$ reflecting the fact that (5.90) is invariant under the action of the rotation group of the plane. It is not, however, invariant under the translation group action, so that the translation generators do not determine a Witten-type topological field theory like the harmonic oscillator Hamiltonians do. This means that there are no E^2-invariant symplectic potentials on the plane, i.e. it is impossible to find a function F in (5.21) that gives an invariant potential simultaneously for all 3 of the independent generators in (5.72). The harmonic oscillator nature of these systems is consistent with their global integrability properties. The holomorphic polarization of the quantum theory associates the canonical quantum theory above with the topological coadjoint orbit quantum theory of Section 5.1 and the coherent state path integral (5.94) yields character formulas for the isometry group of the phase space. This will be the general characteristic feature of *all* localizable systems we shall find. In the case at hand, the character formulas are associated with the Cartan elements of the Heisenberg-Weyl group.

5.4 Coherent States on Homogeneous Kähler Manifolds and Holomorphic Localization Formulas

Before carrying on with our geometric determination of the localizable dynamical systems and their path integral representations, we pause to briefly discuss how the holomorphic quantization introduced above on the coadjoint orbit $\mathbb{R}^2$ can be generalized to the action of an arbitrary semi-simple Lie group G [17, 90, 137]. This representation of the quantum dynamics proves to be the most fruitful on homogeneous spaces G/H_C, and later on we shall generalize this construction to apply to non-homogeneous phase spaces and even non-symmetric multiply connected phase spaces. As the coherent states are those which are closest to "classical" states, in that they are the most tightly peaked ones about their locations, they are the best quantum states in which to study the semi-classical localizations for quantum systems. We shall also see that they are related to the geometric quantization of dynamical systems [172].

Given any irreducible unitary representation $D(G)$ of the group G and some normalized state $|0\rangle$ in the representation space, we define the (normalized) state $|g\rangle$ by

$$|g\rangle = D(g)|0\rangle \tag{5.96}$$

If Dg denotes Haar measure of G, then Schur's lemma [162] and the completeness of the representation $D(G)$ implies the completeness relation

$$\frac{\dim D(G)}{\mathrm{vol}(G)} \int_G Dg \, |g\rangle\langle g| = 1 \qquad (5.97)$$

Following the derivation of Section 4.1, it follows that the partition function associated with an operator $\hat{\mathcal{H}}$ acting on the representation space of $D(G)$ can be represented by the path integral

$$\mathrm{tr}_{D(G)} \, e^{-iT\hat{\mathcal{H}}} = \int_G Dg \, \langle g| \, e^{-iT\hat{\mathcal{H}}} |g\rangle$$

$$= \int_{LG} \prod_{t\in[0,T]} \frac{\dim D(G)}{\mathrm{vol}(G)} Dg(t) \, \exp\left\{ i \oint_{\gamma(g)} \langle g|d|g\rangle - i \int_0^T dt \, \langle g|\hat{\mathcal{H}}|g\rangle \right\} \qquad (5.98)$$

However, if we take $|0\rangle$ to be a simultaneous eigenstate of the generators of $H_C \subset G$ (i.e. a weight state), then the 'coherent' states $|g\rangle$ associated with any one coset of G/H_C are all phase multiples of one another. Thus the set of coherent states form a principal H_C-bundle $L \to \mathcal{M}_G$ over G/H_C and the coherent state path integral (5.98) is in fact taken over paths in the homogeneous space G/H_C. This geometrical method for constructing irreducible representations of semi-simple Lie groups as sections of a line bundle $L \to \mathcal{M}_G$ associated to the principal fiber bundle $G \to G/H_C$ is known as the Borel-Weil-Bott method [142]. The holomorphic sections of this complex line bundle (the coherent states) form a basis for the irreducible representation.

What is most interesting about the character representation (5.98) is that it is closely related to the Kähler geometry of the homogeneous space G/H_C. To see this, we first define the Borel subgroups $B_\pm$ of G which are the exponentiations of the subalgebras $\mathcal{B}_\pm$ spanned by $H_i \in \mathbf{h} \otimes \mathbb{C}$ and E_α for $\alpha > 0$ and $\alpha < 0$, respectively (see (5.18)). The complexification of the coadjoint orbit $\mathcal{M}_G$ is then provided by the isomorphism $G/H_C \simeq G^c/B_\pm$, where G^c is the complexification of G [162]. Almost any $g \in G$ can be factored as a Gauss decomposition

$$g = \zeta_+ h \zeta_- \qquad (5.99)$$

where $h \in H_C^c$ and

$$\zeta_+ = e^{\sum_{\alpha>0} z^\alpha E_\alpha} \quad , \quad \zeta_- = e^{\sum_{\alpha<0} z^\alpha E_\alpha} \qquad (5.100)$$

Here $z^\alpha \in \mathbb{C}$, and if we now apply the representation operator $D(g)$ to a lowest weight state, then ζ_- acts as the identity and the set of physically distinct states are in a one-to-one correspondence with the coset space G^c/B_+. Since

5.4 Holomorphic Localization Formulas

$\mathcal{B}_+$ is a closed subalgebra of $\mathbf{g} \otimes \mathbb{C}$, the parameters $z^\alpha \in \mathbb{C}$ define a complex structure on $\mathcal{M}_G$. In this way, we can now write down coherent state path integral representations of the character formulas of Section 5.1 above. The choice of $|0\rangle \equiv |\lambda\rangle$ above as a lowest weight state ensures that the coherent states

$$|z\rangle \equiv \zeta_+|\lambda\rangle \qquad (5.101)$$

are holomorphic. Note that their coherency follows from the fact that $E_{-\alpha}|z\rangle = z^\alpha|z\rangle$ for $\alpha > 0$.

The coherent state representation can be used to provided the one-to-one identification of the given representation space with the coadjoint orbit of Section 5.1. This is provided by the injective mapping on $\mathbf{g} \to \mathbb{R}$ defined by $X \to \langle g|X|g\rangle$. The Kirillov-Kostant 2-form on the coadjoint orbit is then mapped onto the flag manifold $\mathcal{M}_G$ as

$$\omega_\lambda = \langle \lambda | D(g^{-1}dg) \wedge D(g^{-1}dg)^\dagger | \lambda \rangle \qquad (5.102)$$

and likewise for the canonical G-invariant orbit metric

$$g_\lambda = \langle \lambda | D(g^{-1}dg) \otimes D(g^{-1}dg)^\dagger | \lambda \rangle \qquad (5.103)$$

in terms of the Cartan-Maurer 1-form $g^{-1}dg$ in the representation $D(G)$. (5.102) coincides with the appropriate symplectic potential θ_λ associated with the kinetic term $\langle g|\dot g\rangle = \langle g|\frac{d}{dt}|g\rangle$ in (5.98). Furthermore, the matrix element $\langle g|\hat{\mathcal{H}}|g\rangle$ in (5.98) for $\hat{\mathcal{H}} \in D(H_C)$ generates the Hamiltonian flow $(g,t) \to e^{t\hat{\mathcal{H}}} g$ for the action of the direct product $LS^1 \otimes H_C$ on LG [138]. This follows from the identity $i_V(g^{-1}dg) = g^{-1}\hat{\mathcal{H}}g$, so that the action in (5.98) admits the usual topological form $i_{\dot g + V}\theta_\lambda$. The Kähler potentials are thus given by the normalizations

$$e^{F^{(h)}(z,\bar z)} = \langle z|z\rangle \equiv \langle \lambda|g|\lambda\rangle = \langle \lambda|h|\lambda\rangle \qquad (5.104)$$

with the potentials $F^{(i)}(z,\bar z)$, $i = 1,\ldots,r$, each associated with the Cartan generator H_i in (5.104) (compare eqs. (5.84) and (5.93)). From this it follows that the associated Kähler metrics $g^{(i)}$ and symplectic potentials $\theta^{(i)}$ can be represented as coherent state matrix elements as in (5.91) and (5.92). In this way the kinetic term in the coherent state path integral (5.98) coincides with the usual one induced by the symplectic Kähler structures of the homogeneous space $\mathcal{M}_G$ and the path integral measure becomes the loop space Liouville measure.

A particularly important aspect of the holomorphic quantization scheme above is that it resolves some ambiguities in the localization formulas when compared with the classical group character formulas [102]. For this, we consider the usual symplectic line bundle $L^{(\lambda)} \to \mathcal{M}_G$ associated with the principal G-bundle $G \to \mathcal{M}_G$ with connection 1-form $\theta^{(\lambda)}$ given in (5.21). On this line bundle we then define the twisted (covariant) Dolbeault derivative

$$\bar{\partial}_\lambda = \bar{\partial} + \theta_{\bar{z}}^{(\lambda)} \qquad (5.105)$$

The Riemann-Roch-Hirzebruch index theorem (the analog of the Atiyah-Singer index theorem for the twisted elliptic Dolbeault exterior derivative) [21, 102, 156] relates the analytical index of (5.105) to the topological invariant

$$\text{index}(\bar{\partial}_\lambda) = \int_{\mathcal{O}_\lambda} \text{ch}(\omega^{(\lambda)}) \wedge \text{td}(R^{(\lambda)}) \qquad (5.106)$$

Furthermore, the index of the twisted Dolbeault complex is

$$\text{index}(\bar{\partial}_\lambda) = \sum_{k=0}^{n} (-1)^k \dim H^{(0,k)}(\mathcal{M}_G; L^{(\lambda)}) \qquad (5.107)$$

where we recall that $\Lambda^{(p,q)}(\mathcal{M}_G; L^{(\lambda)}) \equiv \Lambda^{(p,q)}\mathcal{M}_G \otimes L^{(\lambda)}$. When the line bundle $L^{(\lambda)} \to \mathcal{M}_G$ is trivial, the index (5.107) is called the arithmetic genus of the complex manifold $\mathcal{M}_G$ and the expression (5.106) for it in terms of the Todd class is called the Riemann-Roch theorem [41, 111]. This is the complex analog of the Gauss-Bonnet theorem for the Euler characteristic (the index of the DeRham complex) and it can also be generalized to higher-dimensional complex vector bundles.

The crucial point here is that the twisted Dolbeault operator (5.105) annihilates the normalized coherent states $|z\rangle\rangle = |z\rangle/\langle z|z\rangle$, so that $|z\rangle\rangle \in H^{(0,0)}(\mathcal{M}_G; L^{(\lambda)})$. An application of the Lichnerowicz vanishing theorem [4, 21] shows that all other cohomology groups $H^{(0,q)}(\mathcal{M}_G; L^{(\lambda)})$ for $q > 0$ are trivial, so that the dimension of the representation $\mathcal{R}_\lambda$ with highest weight vector λ coincides with $\dim H^{(0,0)}(\mathcal{M}_G; L^{(\lambda)})$ which is just the index of the twisted Dolbeault complex,

$$\dim \mathcal{R}_\lambda = \text{tr}_\lambda \mathbf{1} = \text{index}(\bar{\partial}_\lambda) \qquad (5.108)$$

Since [162]

$$\dim \mathcal{R}_\lambda = \prod_{\alpha > 0} \frac{\alpha(\lambda + \rho)}{\alpha(\rho)} \qquad (5.109)$$

where $\alpha(\tilde{\lambda}) \equiv \sum_{i=1}^{r} \alpha_i \tilde{\lambda}_i$, we see that the holomorphic polarization above naturally incorporates the Weyl shift $\lambda \to \lambda + \rho$ of the highest weight vector λ. Furthermore, in this case the Dirac operator $\bar{\partial}_\lambda^\dagger$ has no zero modes and according to the arguments used in Section 4.10 above the Lefschetz number

$$\text{index}_{H^{(\lambda)}}(\bar{\partial}_\lambda; T) = \lim_{\beta \to \infty} \text{tr}\| e^{iTH^{(\lambda)}}(e^{-\beta \bar{\partial}_\lambda^\dagger \bar{\partial}_\lambda} - e^{-\beta \bar{\partial}_\lambda \bar{\partial}_\lambda^\dagger})\| \qquad (5.110)$$

coincides with the character of the zero mode representation defined by $\bar{\partial}_\lambda$, i.e.

$$\text{index}_{H^{(\lambda)}}(\bar{\partial}_\lambda; T) = \text{str}_\lambda\, e^{iTH^{(\lambda)}} \qquad (5.111)$$

5.4 Holomorphic Localization Formulas

The character of the representation $\mathcal{R}_\lambda$ is therefore the equivariant index of the twisted Dolbeault complex.

This identification with the holomorphic properties of the complex manifold $\mathcal{M}_G$ motivates another way to regularize the fluctuation determinants discussed in Section 4.2. The proper way in this case to carry out the regularization procedure is to attach different signs to the factors of b in Section 4.2 (see (4.67)) corresponding to the holomorphic and antiholomorphic sectors in the regularization (4.65),(4.67) [102]. This restores the holomorphic properties of the path integral wherein the skew-eigenvalues $(\lambda_j, -\lambda_j)$ of the block $R_V^{(j)}$ correspond, respectively, to the holomorphic and anti-holomorphic components of the equivariant curvature 2-form. The above argument implies that the correct way to treat the complex tangent bundle here is to then restrict to the holomorphic component of this curvature. In doing this, the fluctuation determinant in (4.129) is not under a square root in this complex case, because now it arises from Berezin integration over *complex* Grassmann variables, i.e. the localization symmetries in a Kähler polarization are determined now by a larger $N = 1$ "hidden" supersymmetry. Now the determinant (4.65) for the case at hand is replaced with

$$\mathrm{det}'_{\mathrm{hol}} \| \partial_t - R_{V(\lambda)}^{(j)} \| = \prod_{k \neq 0} \left(\frac{2\pi i k}{T} - \lambda_j \right) \tag{5.112}$$

where the determinant in (5.112) is taken only over those eigenvalues corresponding to the holomorphic indices. To regularize (5.112) properly we now have to take into account that the Dirac operator $\partial_t - R_{V(\lambda)}^{(j)}$ has an infinite number of negative eigenvalues, i.e. its spectral asymmetry must be regularized using the eta-invariant (4.63). Thus we take the fluctuation regularization factor of Section 4.2 to be $b = \frac{\pi}{2}$, which explicitly takes into account the spectral asymmetry determined by the Atiyah-Patodi-Singer eta-invariant and maintains the original symmetry of the path integral under the "large gauge transformations" $R_{V(\lambda)} \to R_{V(\lambda)} + 2\pi n/T$, $n \in \mathbb{Z}$. The evaluation of the fluctuation determinant in (4.129) now leads instead to the equivariant Todd class (2.98) of the complex tangent bundle,

$$\frac{1}{\mathrm{det}'_{\mathrm{hol}} \| \partial_t - R_{V(\lambda)}^{(j)} \|} = \prod_{j=1}^{n} \frac{\lambda_j/2}{\sinh(T\lambda_j/2)} \, \mathrm{e}^{iT\lambda_j/2} = \frac{1}{T^n} \, \mathrm{td}_{V(\lambda)}(TR^{(\lambda)}) \tag{5.113}$$

Then, using the usual Niemi-Tirkkonen localization prescription, we arrive at the Kirillov character formula *without* an explicit Weyl shift,

$$\mathrm{tr}_\lambda \, \mathrm{e}^{iTX} \sim \int_{\mathcal{O}_\lambda} \mathrm{ch}_{V(\lambda)}(-iT\omega^{(\lambda)}) \wedge \mathrm{td}_{V(\lambda)}(TR^{(\lambda)}) \tag{5.114}$$

158   5. Equivariant Localization on Simply Connected Phase Spaces

which can be derived as well from the coadjoint orbit path integral over LO_λ (as opposed to $LO_{\lambda+\rho}$ as in (5.27))[2]. The corresponding localization onto the critical points of the Hamiltonian $H^{(\lambda)}$ (c.f. Section 4.7), or equivalently the Weyl group, is

$$\operatorname{tr}_\lambda e^{iTX} \sim \sum_{z \in \mathcal{M}_{V(\lambda)}^+} \frac{e^{-iTH^{(\lambda)}(z,\bar{z})}}{\det_{\mathrm{hol}}^+ dV^{(\lambda)}(z,\bar{z})} \operatorname{td}(dV^{(\lambda)}(z,\bar{z})) \qquad (5.115)$$

where we have identified the Pfaffian in the real polarization with the determinant over the holomorphic eigenvalues of the matrix $dV^{(\lambda)}$. We recall from Section 4.10 that the Atiyah-Singer index contribution to (5.27) evaluates the spectral asymmetry of the zero mode representation of **g** determined by the pertinent Dirac operator, while the Lefschetz number coincides with the character of that representation of the spin complex. From the formulas (5.114),(5.110), however, we see that the character of a Lie group G is a Lefschetz number related to the G-index theorem of the holomorphic Dolbeault complex, rather than to the Atiyah-Singer index theorem of the spin complex. The localizations onto the (equivariant) Todd classes hold quite generally for any phase space path integral in a holomorphic representation. One just replaces the $\hat{A}$-genus factors everywhere in Chapter 4 with the corresponding Todd classes.

5.5 Spherical Phase Spaces and Quantization of Spin Systems

We are now ready to continue with our general isometric classification and hence work out some more explicit examples. The next case we consider is when the phase space $\mathcal{M}$ has a positive constant Gaussian curvature $K > 0$. In this case the conformal factor solves the Liouville field equation

$$\nabla^2 \varphi(z,\bar{z}) = -2K\, e^{\varphi(z,\bar{z})} \qquad (5.116)$$

which is a completely integrable system [35] whose general solutions are

$$\varphi(z,\bar{z}) = \log\left[\frac{\partial f(z) \bar{\partial}\bar{f}(\bar{z})}{\left(\frac{K}{4} + f(z)\bar{f}(\bar{z})\right)^2}\right] \qquad (5.117)$$

By the essential uniqueness of maximally symmetric spaces, we know that in this case $(\mathcal{M},g)$ is isometric to the sphere S^2 of radius $K^{-1/2}$ with its

---

[2] The analog of the Weyl shift ambiguity for the coadjoint orbit models of the last Section is associated with the appearence of the correct zero-point energy $E_0 = \frac{1}{2}$ for the harmonic oscillator Hamiltonian $\hat{a}^\dagger \hat{a}$ in certain ordering prescriptions. See [102] for details.

5.5 Spherical Phase Spaces and Quantization of Spin Systems

standard round metric given in (5.34). From the transformation law of the metric tensor g and (5.117) it is straightforward to work out the explicit diffeomorphism $(z, \bar{z}) \to (w(z, \bar{z}), \bar{w}(z, \bar{z}))$ which accomplishes this isometric correspondence.

First of all, we rewrite the spherical metric in (5.34) in complex coordinates $w, \bar{w} = x^1 \pm ix^2$, with x^μ the spherical coordinates defined in (5.34), to get

$$g_{S^2} = \frac{1}{4K} \left[\frac{\bar{w}^2}{1 - w\bar{w}} dw \otimes dw + \frac{w^2}{1 - w\bar{w}} d\bar{w} \otimes d\bar{w} \right.$$
$$\left. + 2 \left(2 + \frac{w\bar{w}}{1 - w\bar{w}} \right) dw \otimes d\bar{w} \right] \tag{5.118}$$

where $w\bar{w} \leq 1$. If we view the unit sphere as centered in the $x'y'$-plane in $\mathbb{R}^3$ and symmetrically about the z'-axis, then we can map S^2 onto the complex plane via the standard stereographic projection from the south pole $z' = -1$,

$$w = \frac{2w'}{1 + w'\bar{w}'} \quad , \quad z' = \sqrt{1 - w\bar{w}} = \frac{1 - w'\bar{w}'}{1 + w'\bar{w}'} \tag{5.119}$$

This gives a diffeomorphism of S^2 with the compactified plane $\mathbb{C} \cup \{\infty\}$. From (5.118), the metric tensor transformation law and (5.117) we find after some algebra that the coordinate transformation above must satisfy

$$\frac{1}{(1 + w'\bar{w}')^2} \left(\frac{\partial w'}{\partial z} \frac{\partial \bar{w}'}{\partial \bar{z}} + \frac{\partial w'}{\partial \bar{z}} \frac{\partial \bar{w}'}{\partial z} \right) = K\, e^{\varphi(z, \bar{z})} = K \frac{\partial f(z) \bar{\partial} \bar{f}(\bar{z})}{\left(\frac{K}{4} + f(z)\bar{f}(\bar{z}) \right)^2} \tag{5.120}$$

From (5.120) and (5.119) it then follows that the desired coordinate transformation from $(\mathcal{M}, g)$ to S^2 with the standard round metric (5.118) is given by

$$w(z, \bar{z}) = \frac{4K^{-1/2} f(z)}{1 + 4K^{-1} f(z)\bar{f}(\bar{z})} \tag{5.121}$$

The mapping (5.121) is just a generalized stereographic projection from the south pole of S^2 where $f(z)$ maps $(\mathcal{M}, g)$ onto the entire complex plane with the usual Kähler geometry of S^2 defined by the coordinates in (5.119),

$$g_{S^2} = 4\partial\bar{\partial} F_{S^2}(z, \bar{z}) dz \otimes d\bar{z} = \frac{4}{(1 + z\bar{z})^2} dz \otimes d\bar{z}$$
$$\omega_{S^2} = 2i\partial\bar{\partial} F_{S^2}(z, \bar{z}) dz \wedge d\bar{z} = \frac{2i}{(1 + z\bar{z})^2} dz \wedge d\bar{z} \tag{5.122}$$

where the associated Kähler potential is

$$F_{S^2}(z, \bar{z}) = \log(1 + z\bar{z}) \tag{5.123}$$

Notice that the diffeomorphism (5.121) obeys $w\bar{w} \leq 1$, as required for $(w, \bar{w}) \in S^2$, and that the Kähler metric g_{S^2} in (5.122) coincides with the original phase space geometry (5.49) when $f(z) = \frac{1}{2} K^{1/2} z$ in (5.117) above.

160   5. Equivariant Localization on Simply Connected Phase Spaces

From the general considerations of Section 5.2 above we know that the Killing vectors of the metric (5.118) are

$$V_{S^2}^w = -i\Omega w + \alpha(1 - w\bar{w})^{1/2} \quad , \quad V_{S^2}^{\bar{w}} = i\Omega\bar{w} + \bar{\alpha}(1 - w\bar{w})^{1/2} \quad (5.124)$$

The Killing vectors (5.124) generate the rigid rotations $w \to e^{i\Omega}w$ of the sphere and the quasi-translations $w \to w + \alpha(1 - w\bar{w})^{1/2}$ (i.e. translations along the geodesical great circles of S^2), and they together generate the Lie group $SO(3)$. Requiring the symplecticity condition (5.70) again under the full $SO(3)$ group action generated by (5.124) on the symplectic 2-form (5.69), we find after some algebra that the equations (5.70) are uniquely solved by

$$\omega_{S^2}(w, \bar{w}) = 1/K(1 - w\bar{w})^{1/2} \quad (5.125)$$

This symplectic 2-form is again the volume form associated with (5.118). It is a non-trivial element of $H^2(S^2; \mathbb{Z}) = \mathbb{Z}$ and it coincides with the Kähler classes in (5.122) in the stereographic coordinates (5.119). We now substitute (5.124) and (5.125) into the Hamiltonian equations (5.68), which are easily solved on S^2 in the w-coordinates above, and then apply the generalized stereographic projection (5.121) to get the most general equivariant Hamiltonian on a spherical phase space as

$$H_+(z, \bar{z}) = \frac{\Omega\left(\frac{K}{4} - f(z)\bar{f}(\bar{z})\right)}{\frac{K}{4} + f(z)\bar{f}(\bar{z})} + \frac{\alpha\bar{f}(\bar{z}) + \bar{\alpha}f(z)}{\frac{K}{4} + f(z)\bar{f}(\bar{z})} + C_0 \quad (5.126)$$

Thus, again the Riemannian geometry of the phase space $\mathcal{M}$ is realized (or even determined) by the equivariant Hamiltonian systems which can be defined on $\mathcal{M}$. The transformation to Darboux coordinates on $\mathcal{M}$, defined as usual as those coordinates $(v, \bar{v})$ in which the symplectic 2-form is locally $\omega_{S^2} = \frac{i}{2}dv \wedge d\bar{v}$, can be found from the fact that ω_{S^2} is the (Kähler) volume form associated with (5.117) and applying the tensor transformation law (2.12) for ω. After some algebra we find that the local Darboux coordinates on $\mathcal{M}$ are defined by the diffeomorphism $(z, \bar{z}) \to (v(z, \bar{z}), \bar{v}(z, \bar{z}))$, where the function

$$v(z, \bar{z}) = \frac{f(z)}{\left(\frac{K}{4} + f(z)\bar{f}(\bar{z})\right)^{1/2}} \quad (5.127)$$

maps $\mathcal{M}$ onto the unit disc

$$D^2 = \{z \in \mathbb{C} : z\bar{z} \leq 1\} \quad (5.128)$$

which is the Darboux phase space associated with a general spherical phase space geometry. Thus, applying the transformation (5.127) to (5.126), we see that the general Darboux Hamiltonians in the present case are

$$H_+^D(z, \bar{z}) = \Omega z\bar{z} + (\bar{\alpha}z + \alpha\bar{z})(1 - z\bar{z})^{1/2} + C_0 \quad ; \quad z \in D^2 \quad (5.129)$$

which correspond to the quasi-displaced harmonic oscillators

5.5 Spherical Phase Spaces and Quantization of Spin Systems

$$H_+^D(z,\bar{z}) = \Omega \left[z + a(1-z\bar{z})^{1/2}\right] \left[\bar{z} + \bar{a}(1-z\bar{z})^{1/2}\right] \tag{5.130}$$

with compactified position and momentum ranges. Thus here the criterion of a (compact) circle action *cannot* be removed, in contrast to the case of the planar geometries of Section 5.3 where the Darboux phase space was the entire complex plane $\mathbb{C}$. Notice that all translations in the planar case become quasi-translations in the spherical case, which is a measure of the presence of a curved Riemannian geometry on $\mathcal{M}$.

The mapping onto Darboux coordinates above shows that once again all the general spherical Hamiltonians are holomorphic copies of each other, as they all define the same Darboux dynamics. We shall therefore focus our attention to the quantum dynamics defined on the phase space S^2 (i.e. $f(z) = K^{1/2}z/2$ above), and for simplicity we normalize the coordinates so that now $K = 1$, i.e. S^2 has unit radius. First of all, we write the 3 independent observables appearing in (5.126) above as

$$J_3^{(j)}(z,\bar{z}) = -j\frac{1-z\bar{z}}{1+z\bar{z}} \;,\quad J_+^{(j)}(z,\bar{z}) = 2j\frac{\bar{z}}{1+z\bar{z}} \;,\quad J_-^{(j)}(z,\bar{z}) = 2j\frac{z}{1+z\bar{z}} \tag{5.131}$$

where the parameter j will be specified below. Using (5.122) we define the Kähler 2-form

$$\omega^{(j)} = j\omega_{S^2} \tag{5.132}$$

and working out the associated Poisson algebra of the functions (5.131)

$$\left\{J_3^{(j)}, J_\pm^{(j)}\right\}_{\omega^{(j)}} = \pm J_\pm^{(j)} \;,\quad \left\{J_+^{(j)}, J_-^{(j)}\right\}_{\omega^{(j)}} = 2J_3^{(j)} \tag{5.133}$$

shows that they realize the $SU(2)$ (angular momentum) Lie algebra [162]. The functions (5.131) therefore generate the Poisson-Lie group action of the S^2 isometry group $SO(3)$ on the coadjoint orbit

$$G/H_C = SU(2)/U(1) \simeq S^3/S^1 = S^2 \tag{5.134}$$

and we obtain the usual coadjoint orbit topological quantum theory by choosing the Hamiltonian to be an element of the Cartan subalgebra $\mathbf{u}(1)$ of $\mathbf{su}(2)$. The homogeneous space $SU(2) \to SU(2)/U(1) = S^2$ is often called the magnetic monopole bundle. For the basic representation space $W = \mathbb{C}$ of $U(1)$, the associated vector bundle $(SU(2) \times \mathbb{C})/U(1)$ is the usual symplectic line bundle over S^2. The Borel-Weil-Bott wavefunctions in the presence of a magnetic monopole take values in this bundle.

Notice that, comparing (5.131) with the stereographic coordinates (5.119), we see that these observables just describe the Larmor precession of a classical spin vector of unit length $J = \pm 1$. The coadjoint orbit path integral associated with the observables (5.131) will therefore describe the quantum dynamics of a classical spin system, e.g. the system with Hamiltonian $H = J_3$ describes the Pauli magnetic moment interaction between a spin J and a uniform magnetic field directed along the z-axis. Thus in this case S^2 is actually naturally the *configuration* space for a spin system [156], which has on it a natural symplectic structure and so the corresponding path integral can be regarded as one for the Lagrangian formulation of the theory, rather than the Hamiltonian one [139, 156]. This is also immediate from noting that the stereographic complex coordinates above can be written as

$$z = e^{-i\phi}\tan(\theta/2) \qquad (5.135)$$

in terms of the usual spherical polar coordinates (θ, ϕ), so that the observable J_3 in (5.131) coincides with the height function (2.1) of S^2 with $a = 1$ (up to an additive constant), the Kähler geometry above becomes the standard round geometry of S^2, and the kinetic term in the action is

$$\theta_\mu(x)\dot{x}^\mu = \cos\theta\dot{\phi} \qquad (5.136)$$

after an integration by parts over t. The classical partition function, evaluated in Section 2.1, yields the usual Langevin formula for the classical statistical mechanics of a spin system. Alternatively, the motion of the precessing spin can be reduced to that of a charged particle around a monopole which is isomorphic to the problem for the motion of an excited diatomic molecule where the electrons are in a state with angular momentum j about the axis joining the nuclei (i.e. a rigid rotator with fixed angular momentum j about its axis) [156]. It is the balance between the the Lorentz force on the particle, due to the fictitious magnetic field $\omega = d\theta$ of the monopole located at the center of the sphere, and the potential force on the moment, due to the real magnetic field, that leads to the characteristic Larmor precession about the direction of the field.

To construct a topological Hamiltonian along the lines of the theory of Section 5.1, we consider an irreducible spin-j representation of $SU(2)$, where $j = \frac{1}{2}, 1, \frac{3}{2}, 2, \ldots$ [162]. The state space for this representation with heighest weight j is spanned by the complete set of orthonormal basis states $|j, m\rangle$, where m are the magnetic quantum numbers with the range $m = -j, -j+1, \ldots, j-1, j$. The $SU(2)$ generators act on these states as

$$\hat{J}_3|j, m\rangle = m|j, m\rangle \quad , \quad \hat{J}_\pm|j, m\rangle = \sqrt{(j \mp m)(j \pm m + 1)}\,|j, m\pm1\rangle \quad (5.137)$$

Following the last Section, we define the $SU(2)$ coherent states by successive applications of the raising operator $\hat{J}_+$ to the lowest weight (vacuum) state $|j, -j\rangle$ [45, 137],

$$|z\rangle = e^{-ij\rho}\,e^{z\hat{J}_+}|j, -j\rangle = e^{-ij\rho}\sum_{m=-j}^{j}\binom{2j}{j+m}^{1/2} z^{j+m}|j, m\rangle \quad ; \quad z \in \mathbb{C} \quad (5.138)$$

where for $n, m \in \mathbb{Z}^+$ with $n \geq m$ the binomial coefficient is defined by

$$\binom{n}{m} = \frac{n!}{m!(n-m)!} \qquad (5.139)$$

5.5 Spherical Phase Spaces and Quantization of Spin Systems

and where the function $\rho(z,\bar{z})$ is an arbitrary phase which as we shall see is related to the function $F(z,\bar{z})$ in (5.21). It is easily verified that then the $SU(2)$ generators (5.131) are the normalized matrix elements of the operators $\hat{J}_3, \hat{J}_\pm$ in the coherent states (5.138), respectively[3].

The coherent states (5.138) are normalized as

$$(z_2|z_1) = (1 + z_1\bar{z}_2)^{2j}\, e^{ij[\rho(z_2,\bar{z}_2)-\rho(z_1,\bar{z}_1)]} \qquad (5.140)$$

where we have used the binomial theorem

$$(x+y)^n = \sum_{k=0}^{n} \binom{n}{k} x^k y^{n-k} \qquad (5.141)$$

They obey the completeness relation

$$\int d\mu^{(j)}(z,\bar{z})\, |z))((z| = \mathbf{1}^{(j)} \qquad (5.142)$$

where $\mathbf{1}^{(j)}$ is the identity operator in the spin-j representation of $SU(2)$ and the coherent state measure is

$$d\mu^{(j)}(z,\bar{z}) = \frac{i}{2\pi}\frac{2j+1}{(1+z\bar{z})^2} dz \wedge d\bar{z} \qquad (5.143)$$

which coincides with the symplectic 2-form of the spin system above. The identity (5.142) follows from a calculation analogous to that in (5.85). Note that, as explained in the last Section, the Kähler structure is generated through the identity $(z|z) = e^{2jF_{S^2}(z,\bar{z})}$.

We want to evaluate the propagator

$$\mathcal{K}(z_2,z_1;T) = ((z_2|\, e^{-iT\hat{\mathcal{H}}}|z_1)) \qquad (5.144)$$

for some $SU(2)$ operator $\hat{\mathcal{H}}$ given the one-to-one correspondence between the points on the coadjoint orbit $SU(2)/U(1) = S^2 \simeq \mathbb{C} \cup \{\infty\}$ and the $SU(2)$ coherent states (5.138). Dividing the time interval in (5.144) up into N segments and letting $N \to \infty$, following the analogous steps as in Section 4.1 using the completeness relation (5.142) we arrive at the coherent state path integral

$$\mathcal{K}(z_2,z_1;T) = \mathcal{N} \int_{L\mathbb{R}^2} \prod_{t \in [0,T]} dz(t)\, d\bar{z}(t)\, \sqrt{\det \|\Omega^{(j)}\|}$$

$$\times \exp\left\{ j\log(1+z_2\bar{z}_2) + j\log(1+z_1\bar{z}_1) \right.$$

$$\left. +i\int_0^T dt\, \left[\frac{ij}{1+z\bar{z}}(\bar{z}\dot{z} - z\dot{\bar{z}}) - ij\left(\frac{\partial\rho}{\partial z}\dot{z} + \frac{\partial\rho}{\partial \bar{z}}\dot{\bar{z}}\right) - H(z,\bar{z})\right] \right\}$$

$$(5.145)$$

---

[3] The dimension $\dim \mathcal{R}_j = 2j+1$ of the spin-j representation of $SU(2)$ can also be derived from the index theorems of the last Section (see (5.108)).

where
$$\mathcal{N} = \lim_{N\to\infty} \prod_{k=1}^{N-1} \frac{2j+1}{2j\pi} \tag{5.146}$$

is a normalization constant and $H(z,\bar{z})$ denotes the matrix elements (5.95) in the coherent states (5.138). Here we see once again the formal equivalence of the path integral and Liouville measures defined by the Kähler polarization above. In particular, the local symplectic potential generating the Kähler structures (5.132) are

$$\theta^{(j)} = \frac{ij}{1+z\bar{z}}(\bar{z}dz - zd\bar{z}) - ijd\rho \tag{5.147}$$

and they coincide with the standard coherent state canonical 1-forms (5.91). Similarly, the Kähler structure (5.122) can be represented in the standard coherent state form (5.92).

The Wess-Zumino-Witten quantization condition (4.161) applied to $\omega^{(j)}$ implies that j must be a half-integer, since $\int_{S^2} \omega_{S^2} = 4\pi$, corresponding to the unitary irreducible representations of $G = SU(2)$ [162]. This is the topological (or Dirac) quantization of spin. The quantization of the magnetic quantum numbers m above then follows from an application of the semi-classical Bohr-Sommerfeld quantization condition [101] for the spin system. To construct a topological quantum theory (or equivalently an integrable quantum system) as described in Sections 4.10 and 5.1, we need to choose the phase function $\rho(z,\bar{z})$ in the above so that $i_V \theta^{(j)} = H$. This problem was analysed in detail by Niemi and Pasanen [123] who showed that it is impossible to satisfy this integrability requirement simultaneously for all 3 of the generators in (5.131). Again, this means that there are no $SU(2)$-invariant symplectic potentials on the sphere S^2. However, such 1-forms do exist on the cylindrical representation of $SU(2)$ [123], i.e. the complex plane with the origin removed, which is conformally equivalent to the Kähler representation of S^2 above under the transformation $z = e^{s^1 + is^2}$ which maps $(s^1, s^2) \in \mathbb{R} \times S^1$ to $z \in \mathbb{C} - \{0\}$. In this latter representation, the Hamiltonian in (5.145) can be taken to be an arbitrary linear combination of the $SU(2)$ generators, and the coherent state path integral (5.145) determines a topological quantum field theory with $\rho = 0$ in (5.147). This is not true, however, in the Kähler representation above, but we do find, for example, that the symplectic invariance condition can be fulfilled by choosing the basis $H(z,\bar{z}) = J_3^{(j)}(z,\bar{z})$ of the Cartan subalgebra $\mathbf{u}(1)$ and $\rho(z,\bar{z}) = \frac{1}{2}\log(z/\bar{z})$. The ensuing topological path integral (5.145) then describes the quantization of spin.

To evaluate this spin partition function, we set $\rho = 0$ above. Although the ensuing quantum theory now does not have the topological form in terms of a BRST-exact action, it still maintains the Schwarz-type topological form described in Section 4.10, since the Hamiltonian then satisfies (3.45) with $C = j$ and the function K in (3.46) is

5.5 Spherical Phase Spaces and Quantization of Spin Systems 165

$$K(z,\bar{z}) = \frac{i}{2}\log\left(\frac{z}{\bar{z}}\right) \tag{5.148}$$

so that (5.145) is a topological path integral of the form (4.165), i.e. the quantum theory determines a Schwarz-type topological field theory, as opposed to a Witten-type one as above. We first analyse the WKB localization formula (4.115) for the coadjoint orbit path integral (5.145). We note first of all that the boundary conditions in (5.145) are $z(0) = z_1$ and $\bar{z}(T) = \bar{z}_2$. In particular, the final value $z(T)$ and the initial value $\bar{z}(0)$ are not specified, and the boundary terms in (5.145) ensure that with these boundary conditions there is no boundary contribution to the pertinent classical equations of motion

$$\dot{z} + iz = 0 \quad , \quad \dot{\bar{z}} - i\bar{z} = 0 \tag{5.149}$$

In general, if $z(t)$ and $\bar{z}(t)$ are complex conjugates of each other, then there are no classical trajectories that connect $z(0) = z_1$ with $\bar{z}(T) = \bar{z}_2$ on the sphere S^2. But if we view the path integral (5.145) instead as a matrix element between 2 configurations in *different* polarizations, then there is always the following solution to the equations of motion (5.149) with the required boundary conditions for arbitrary z_1 and $\bar{z}_2$,

$$z(t) = z_1\, e^{-it} \quad , \quad \bar{z}(t) = \bar{z}_2\, e^{-i(T-t)} \tag{5.150}$$

The solution (5.150) is complex, and hence $z(t)$ and $\bar{z}(t)$ must be regarded as independent variables. This is one of the characteristic features behind the holomorphic quantization formalism that makes it suitable to describe topological field theories. The trajectories (5.150) are therefore regarded as describing a *complex* saddle-point of the path integral [44, 85, 143]. We shall see other forms of this feature later on.

Substituting the solutions (5.150) into the WKB formula (4.115) we find the propagator[4]

$$\mathcal{K}(z_2, z_1; T) = \frac{(1 + z_1\bar{z}_2\, e^{-iT})^{2j}\, e^{-ijT}}{(1 + z_1\bar{z}_1)^j(1 + z_2\bar{z}_2)^j} \tag{5.151}$$

The exact propagator from a direct calculation is

$$(\!(z_2|\, e^{-iT\hat{J}_3}|z_1)\!) = \frac{1}{(1 + z_1\bar{z}_1)^j(1 + z_2\bar{z}_2)^j}\sum_{m=-j}^{j}\binom{2j}{j+m}(z_1\bar{z}_2\, e^{-iT})^{j+m}\, e^{ijT} \tag{5.152}$$

which coincides with (5.151) upon application of the binomial theorem (5.141). In particular, setting $z_1 = z_2 = z$ and integrating over $z \in \mathbb{C}$ using the coherent state measure (5.143), we find the partition function

---

[4] In this case the fluctuation determinant in (4.115) is regulated using the generic non-periodic boundary conditions discussed above.

166   5. Equivariant Localization on Simply Connected Phase Spaces

$$Z_{SU(2)}(T) = \int d\mu^{(j)}(z,\bar{z})\,(\langle z|\,\mathrm{e}^{-iT\hat{J}_3}|z\rangle)$$

$$= \int_0^\infty dr\, r\, \frac{(2j+1)(1+r^2\,\mathrm{e}^{-iT})^{2j}\,\mathrm{e}^{-ijT}}{(1+r^2)^{2j}} = \frac{\sin\left(\frac{T}{2}(2j+1)\right)}{\sin\frac{T}{2}}$$

(5.153)

which also coincides with the exact result

$$\mathrm{tr}_j\,\mathrm{e}^{-iT\hat{J}_3} = \sum_{m=-j}^{j} \mathrm{e}^{-iTm} = \frac{\mathrm{e}^{iTj}}{1-\mathrm{e}^{-iTj}} + \frac{\mathrm{e}^{-iTj}}{1-\mathrm{e}^{iTj}} \qquad (5.154)$$

The right-hand side of (5.154) is precisely what one anticipates from the Weyl character formula (5.28). The roots of $SU(2)$ are $\alpha = \pm 1$ [162], and the Cartan subalgebra is $\mathbf{u}(1)$ consisting of the single element $\hat{J}_3$. The Weyl group is $W = \mathbb{Z}_2$ and it has 2 elements, the identity map and the reflection map $\hat{J}_3 \to -\hat{J}_3$. Thus the formula (5.154) is simply the Weyl character formula (5.28) for the spin-j representation of $SU(2)$.

Within the framework of the Duistermaat-Heckman theorem, the terms summed in (5.154) are each associated with one of the poles of the sphere S^2, i.e. with the critical points of the height function on S^2. Indeed, since this Hamiltonian is a perfect Morse function with even Morse indices, we expect that the Weyl character formula above coincides with the pertinent stronger version (4.135) of the localization formulas. Because of the Kähler structure (5.122) on S^2 (see (5.53)), the Riemann moment map has the non-vanishing components

$$(\mu_{V^{(j)}})^z_z = -(\mu_{V^{(j)}})^{\bar{z}}_{\bar{z}} = iJ_3^{(j)}(z,\bar{z})/j \qquad (5.155)$$

and consequently the Dirac $\hat{A}$-genus is

$$\hat{A}(T\Omega_{V^{(j)}}) = \frac{T}{2j}\frac{J_3^{(j)}}{\sin\left(\frac{T}{2j}J_3^{(j)}\right)} \qquad (5.156)$$

Substituting these into the localization formula (4.135) yields precisely the Weyl character formula (5.154). This localization onto the critical points of the Hamiltonian, as for the harmonic oscillator example of Section 5.3, agrees with the general arguments at the beginning of Section 4.6. Substituting the stereographic projection map (5.135) into the classical equations of motion (5.149) gives

$$\dot{\theta}\sin\theta = 0 \quad , \quad \dot{\phi}+1 = 0 \qquad (5.157)$$

For $T \neq 2\pi n$, $n \in \mathbb{Z}$, the only T-periodic critical trajectories coincide with the critical points of the Hamiltonian $j(1-\cos\theta)$, i.e. $\theta = 0, \pi$, and in this case the critical point set of the action is isolated and non-degenerate. However, for $T = 2\pi n$, $n \in \mathbb{Z}$, we find T-periodic classical solutions for any initial value of θ and ϕ in (5.157) and the critical point set of the classical action coincides with the original phase space S^2. Thus the moduli space of classical solutions in this case is $L\mathcal{M}_S = S^2$, and the localization onto this moduli space is now easily verified from (4.122) to give the correct anticipated result above. From the discussion of Section 4.10, it also follows that the sum of the terms in (5.154) describes exactly the properly normalized period group of the symplectic 2-form $\omega^{(j)}$ on the sphere [85], i.e. the integer-valued surface integrals of $\omega^{(j)}$ as in (4.161). We shall see in the next Chapter that quantizations of the propagation time T as above lead to interesting quantum theories in certain other instances of the localization framework.

It is an instructive exercise to work out the Niemi-Tirkkonen localization formula (4.130) for the above dynamical system. For this we note that, again because of the Kähler geometry of S^2, the Riemann curvature 2-form has the non-vanishing components

$$R_z^z = -R_{\bar{z}}^{\bar{z}} = -i\omega^{(j)}/j \tag{5.158}$$

and so combined with (5.155) we see that the equivariant $\hat{A}$-genus here is

$$\hat{A}_{V^{(j)}}(TR) = \frac{T}{2j} \frac{J_3^{(j)} - \omega^{(j)}}{\sin\left(\frac{T}{2j}(J_3^{(j)} - \omega^{(j)})\right)} \tag{5.159}$$

The equivariant extension of $\omega^{(j)}$ is

$$J_3^{(j)} - \omega^{(j)} = j\left(\frac{1-z\bar{z}}{1+z\bar{z}} - \frac{2i}{(1+z\bar{z})^2}\eta\bar{\eta}\right) = j\left(\frac{1-z\bar{z}-\eta\bar{\eta}}{1+z\bar{z}+\eta\bar{\eta}}\right) \tag{5.160}$$

where we have redefined the Grassmann variables $\eta^\mu \to \sqrt{i} \cdot \eta^\mu$. The Niemi-Tirkkonen localization formula (4.130) can then be written as

$$Z_{SU(2)}(T) \sim \frac{i}{\pi T} \int_{\mathbb{R}^2 \otimes \Lambda^1 \mathbb{R}^2} dz\, d\bar{z}\, d\eta\, d\bar{\eta}\, L(z\bar{z} + \eta\bar{\eta}) \tag{5.161}$$

where

$$L(y) = \frac{\frac{T}{2}\frac{1-y}{1+y}}{\sin\left(\frac{T}{2}\left(\frac{1-y}{1+y}\right)\right)} \exp\left[-ijT\left(\frac{1-y}{1+y}\right)\right] \tag{5.162}$$

Using the Parisi-Sourlas integration formula [136]

$$\frac{1}{\pi}\int_{\mathbb{R}^2\otimes\Lambda^1\mathbb{R}^2} d^2x\, d\eta\, d\bar{\eta}\, L(x^2 + \eta\bar{\eta}) = \int_0^\infty du\, \frac{dL(u)}{du} = L(\infty) - L(0) \tag{5.163}$$

we obtain from (5.161) the partition function

$$Z_{SU(2)}(T) \sim \sin(Tj)/\sin(T/2) \tag{5.164}$$

168   5. Equivariant Localization on Simply Connected Phase Spaces

Introducing the Weyl shift $j \to j + \frac{1}{2}$ in (5.164) then yields the correct Weyl character formula (5.153) for $SU(2)$ [5]. Note that (5.163) shows explicitly how the localization in (5.161) comes directly from the extrema of the height function at $z = \infty$ and $z = 0$.

As a final application for the above dynamical system, we examine the quadratic localization formula (4.150). Now the (degenerate) Hamiltonian is

$$\mathcal{F}(J_3^{(j)}) = (J_3^{(j)})^2 = j^2 \left(\frac{1 - z\bar{z}}{1 + z\bar{z}}\right)^2 \tag{5.165}$$

Following the same steps as above, the localization formula (4.150) can be written as

$$Z_{SU(2)}(T|(J_3^{(j)})^2) \sim \frac{i}{\sqrt{4\pi i T}} \int_{-\infty}^{\infty} \frac{d\phi_0}{\phi_0} \int_{\mathbb{R}^2 \otimes \Lambda^1 \mathbb{R}^2} dz\, d\bar{z}\, d\eta\, d\bar{\eta}\; L(\phi_0, z\bar{z} + \eta\bar{\eta}) \tag{5.166}$$

where

$$L(\phi_0, y) = \frac{\frac{T\phi_0}{2} \frac{1-y}{1+y}}{\sin\left(\frac{T\phi_0}{2}\left(\frac{1-y}{1+y}\right)\right)} \exp\left[\frac{iT}{4}\phi_0^2 - ijT\phi_0 \left(\frac{1-y}{1+y}\right)\right] \tag{5.167}$$

and we have redefined $\eta^\mu \to \sqrt{i/\phi_0} \cdot \eta^\mu$. Using the Parisi-Sourlas integration formula (5.163) again and introducing the Weyl shift $j \to j + \frac{1}{2}$, we find

$$Z_{SU(2)}(T|(J_3^{(j)})^2) \sim \sqrt{\frac{T}{4\pi i}} \int_{-\infty}^{\infty} d\phi_0\; e^{iT\phi_0^2/4}\; \frac{\sin[(j + \frac{1}{2})T\phi_0]}{\sin(T\phi_0/2)}$$

$$= \sum_{m=-j}^{j} \sqrt{\frac{T}{4\pi i}} \int_{-\infty}^{\infty} d\phi_0\; e^{-iTm\phi_0}\; e^{iT\phi_0^2/4} = \sum_{m=-j}^{j} e^{-iTm^2} \tag{5.168}$$

which is again the correct character $\operatorname{tr}_j e^{-iT\hat{J}_3^2}$.

Thus on a spherical phase space geometry the equivariant Hamiltonian systems provide a rich example of the topological quantum field theories discussed in Section 4.10, and they are the natural framework for the study of the quantum properties of classical spin systems. The character formula path integrals above describe the quantization of the harmonic oscillator on the sphere, and therefore the only integrable quantum system, up to holomorphic equivalence (i.e. modification by the general geometry of the phase space), that exists within the equivariant localization framework on a general spherical geometry is the harmonic oscillator defined on the reduced compact phase space D^2.

---

[5] Of course, we could alternatively obtain the Weyl character formula using instead the G-index localization formula (5.114) without having to perform this Weyl shift [102].

5.6 Hyperbolic Phase Spaces

The situation for the case where the phase space is endowed with a Riemannian geometry of constant negative Gaussian curvature $K < 0$ parallels that of the last Section, and we only therefore briefly discuss the essential differences [159]. The phase space $\mathcal{M}$ is now necessarily a non-compact manifold, and we can map it onto the maximally symmetric space $\mathcal{H}^2$, the Lobaschevsky plane (or pseudo-sphere) of constant negative curvature, with its standard curved hyperbolic metric $g_{\mathcal{H}^2}$ [42, 43, 68]. The Killing vectors of this metric have the general form

$$V^w_{\mathcal{H}^2} = -i\Omega w + \alpha(1 + w\bar{w})^{1/2} \quad , \quad V^{\bar{w}}_{\mathcal{H}^2} = i\Omega\bar{w} + \bar{\alpha}(1 + w\bar{w})^{1/2} \quad (5.169)$$

and they generate the isometry group $SO(2,1)$. The rest of the analysis at the beginning of the last Section now carries through analogously to the case at hand here, where we replace the K factors everywhere by $-|K|$ and the $K^{1/2}$ factors by $|K|^{1/2}$.

In particular, with these changes the generalized stereographic coordinate transformation (5.121) is the same except that now the holomorphic function $f(z)$ there maps the phase space onto the Poincaré disk of radius $\frac{1}{2}|K|^{1/2}$, i.e. the disk D^2 with the Poincaré metric

$$g_{\mathcal{H}^2} = \frac{4}{(1-z\bar{z})^2} dz \otimes d\bar{z} \quad (5.170)$$

which defines a Kähler geometry on the disk for which the associated symplectic 2-form is the unique invariant volume form under the transitive $SO(2,1)$-action. The Poincaré disk is the stereographic projection image for the Lobaschevsky plane when we regard it through its embedding in $\mathbb{R}^3$ as the pseudo-sphere, so that we can represent it by pseudo-spherical coordinates $(\tau, \phi) \in \mathbb{R} \times [0, 2\pi]$ as $x^1 = \sinh\tau\cos\phi$, $x^2 = \sinh\tau\sin\phi$ and $z = \cosh\tau$. The stereographic projection is again taken from the projection center $z' = -1$, and the boundary of the Poincaré disc corresponds to points at infinity of the hyperboloid $\mathcal{H}^2$. The pseudo-sphere itself is represented by the interior of the disk. The explicit transformation in terms of pseudo-spherical coordinates is

$$z = \frac{w'}{1+z'} = e^{-i\phi}\tanh(\tau/2) \quad (5.171)$$

We also note here that the Poincaré disc is conformally equivalent to the upper half plane $\mathbb{C}^+$ via the Cayley transform $\xi \to z = (\xi - i)/(\xi + i)$ which takes $\xi \in \mathbb{C}^+$ onto the Poincaré disk, and the Poincaré metric (5.170) on the (Poincaré) upper-half plane is

$$g_{\mathcal{H}^2} = \text{Im}(\xi)^{-2} d\xi \otimes d\bar{\xi} \quad (5.172)$$

The path integral over such hyperbolic geometries arises in string theory and studies of quantum chaos [31].

170   5. Equivariant Localization on Simply Connected Phase Spaces

The most general localizable Hamiltonian in a hyperbolic phase space geometry is therefore

$$H_-(z,\bar{z}) = \frac{\Omega\left(\frac{|K|}{4} + f(z)\bar{f}(\bar{z})\right)}{\frac{|K|}{4} - f(z)\bar{f}(\bar{z})} + \frac{\alpha \bar{f}(\bar{z}) + \bar{\alpha} f(z)}{\frac{|K|}{4} - f(z)\bar{f}(\bar{z})} + C_0 \qquad (5.173)$$

The transformation to Darboux coordinates on $\mathcal{M}$ is now accomplished by the diffeomorphism

$$v(z,\bar{z}) = \frac{f(z)}{\left(\frac{|K|}{4} - f(z)\bar{f}(\bar{z})\right)^{1/2}} \qquad (5.174)$$

which maps $\mathcal{M}$ onto the complement of the unit disk $\mathbb{C} - \text{int}(D^2)$ in $\mathbb{R}^2$. The general Darboux Hamiltonians are therefore

$$H_-^D(z,\bar{z}) = \Omega z\bar{z} + (\bar{\alpha}z + \alpha\bar{z})(1 + z\bar{z})^{1/2} \quad ; \quad z \in \mathbb{C} - \text{int}(D^2) \qquad (5.175)$$

We note that here there are 2 inequivalent Hamiltonians, corresponding to a choice of "spacelike" and "timelike" Killing vectors, but the generic hyperbolic Hamiltonians are again all holomorphic copies of one another, again reducing to a quasi-displaced harmonic oscillator. However, given that the Darboux phase space is now non-compact, we can again weaken the requirement of a global circle action on the phase space to a semi-bounded group action.

Considering therefore the quantum problem defined on the Poincaré disk of unit radius, we write the 3 independent observables in (5.173) as

$$S_3^{(k)}(z,\bar{z}) = k\frac{1 + z\bar{z}}{1 - z\bar{z}} \quad , \quad S_+^{(k)}(z,\bar{z}) = 2k\frac{\bar{z}}{1 - z\bar{z}} \quad , \quad S_-^{(k)}(z,\bar{z}) = 2k\frac{z}{1 - z\bar{z}} \qquad (5.176)$$

Defining the Kähler 2-form $\omega^{(k)} = k\omega_{\mathcal{H}^2}$, we see that the associated Poisson algebra of these observables is just the $SU(1,1)$ Lie algebra

$$\left\{S_3^{(k)}, S_\pm^{(k)}\right\}_{\omega^{(k)}} = \pm S_\pm^{(k)} \quad , \quad \left\{S_+^{(k)}, S_-^{(k)}\right\}_{\omega^{(k)}} = -2S_3^{(k)} \qquad (5.177)$$

The Hamiltonians in (5.173) are therefore functions on the coadjoint orbit

$$SU(1,1)/U(1) \simeq \mathcal{H}^2 \qquad (5.178)$$

of the non-compact Lie group $SU(1,1)$, and the generators (5.176) are the normalized matrix elements of the $SU(1,1)$ generators in the $SU(1,1)$ coherent states

$$|z\rangle = e^{z\hat{S}_+}|k,0\rangle = \sum_{n=0}^\infty \binom{2k+n+1}{n}^{1/2} z^n |k,n\rangle \quad ; \quad z \in \text{int}(D^2) \qquad (5.179)$$

for the discrete irreducible representation of $SU(1,1)$ characterized by $k = 1, \frac{3}{2}, 2, \frac{5}{2}, \ldots$ [137]. The representation spaces are now infinite-dimensional because of the non-compactness of the group manifold of $SU(1,1)$, and the representation states $|k, n\rangle$ defined here are the eigenstates of the generator $\hat{S}_3$ with eigenvalues

$$\hat{S}_3|k, n\rangle = (k + n)|k, n\rangle \tag{5.180}$$

The coherent states (5.179) have the normalization

$$(z_2|z_1) = (1 - z_1\bar{z}_2)^{-2k} \tag{5.181}$$

where we have used the binomial series expansion

$$\frac{1}{(1-x)^n} = \sum_{m=0}^{\infty} \binom{m+n-1}{m} x^m \tag{5.182}$$

which is valid for $n \in \mathbb{Z}^+$ and $|x| < 1$.

Again, the integrable Hamiltonian systems are obtained by taking $H = S_3^{(k)}$, which is the height function on $\mathcal{H}^2$, and the corresponding coherent state path integral describes the quantization of the harmonic oscillator on the open infinite space $\mathcal{H}^2$ (and up to holomorphic equivalence these are the only integrable systems on a general hyperbolic phase space). It is straightforward to analyse the localization formulas for the coherent state path integral just as in the last Section. For instance, the WKB localization formula for the coadjoint orbit path integral

$$Z_{SU(1,1)}(T) = \int_{L\mathcal{H}^2} [d\cosh\tau]\,[d\phi]\, \exp\left\{i\int_0^T dt\, \left(k\cosh\tau\dot{\phi} - k(1+\cosh\tau)\right)\right\} \tag{5.183}$$

can be shown to coincide with the exact Weyl character formula for $SU(1,1)$ [51, 143]

$$Z_{SU(1,1)}(T) = \mathrm{tr}_k\, e^{-iT\hat{S}_3} = \sum_{n=0}^{\infty} e^{-iT(k+n)} = 2i\frac{e^{-iT(k-\frac{1}{2})}}{\sin\frac{T}{2}} \tag{5.184}$$

5.7 Localization of Generalized Spin Models and Hamiltonian Reduction

The explicit examples we have given thus far of the localization formalism in both the classical and quantum cases have, for simplicity, focused on dynamical systems with 2-dimensional phase spaces. Our main examples have been the harmonic oscillator, where the localization is trivial because the Hamiltonian is a quadratic function, and the spin partition function, where the exactness of the stationary-phase approximation is a consequence of the conspiracy between the phase space volume and energy which makes this dynamical system resemble a harmonic oscillator. In Chapter 8 we shall present some true field theoretical applications of equivariant localization, but in this Section and the next we wish to overview the results which concern the exactness of the localization formulas for some higher-dimensional coadjoint orbit models which can be considered as generalizations of the spin models of this Chapter (and the previous ones) to larger Lie groups. We have already established quite generally that these are always examples of localizable dynamical systems, and here we shall explicitly examine their features in some special instances.

The generalization of the classical partition function for $SU(2)$ is what is commonly refered to as the Itzykson-Zuber integral [76]

$$I[X, Y; T] = \int_{U(N)} DU \; e^{iT \, \text{tr}(UXU^\dagger Y)} \tag{5.185}$$

where

$$DU \equiv \prod_{i,j=1}^{N} dU_{ij} \; \delta\left(\sum_{k=1}^{N} U_{ik} U_{jk}^* - \delta_{ij}\right) \tag{5.186}$$

is Haar measure on the group $U(N)$ of $N \times N$ unitary matrices $U^\dagger = U^{-1}$ (or on $SU(N) = U(N)/(U(1) \times \mathbb{Z}_2)$). Here $(X, Y) = (X^\dagger, Y^\dagger)$ are Hermitian $N \times N$ matrices (i.e. elements of the $U(N)$ Lie algebra) and which can be therefore diagonalized with eigenvalues $x_i, y_i \in \mathbb{R}$ by unitary transformations $(X, Y) \to (V^\dagger XV, W^\dagger YW)$. By the invariance of the Haar measure in (5.185) under the left-right action $U \to WUV^\dagger$ of $U(N)$, we can thus assume without loss of generality that the matrices X and Y in (5.185) are diagonal so that

$$I[X, Y; T] = \int_{U(N)} DU \; \exp\left(iT \sum_{i,j=1}^{N} x_i y_j |U_{ij}|^2\right) \tag{5.187}$$

The Itzykson-Zuber integral is a fundamental object that appears in matrix models of string theories, low dimensional quantum gravity and higher-dimensional lattice gauge theories [36, 92, 106].

The integration over unitary matrices in (5.185) can be carried out using the Duistermaat-Heckman theorem via the following observation. If we define the Hermitian matrix

$$\bar{\Lambda} \equiv UYU^\dagger \tag{5.188}$$

then we can explicitly compute the Jacobian for the change of integration variables $U \to \bar{\Lambda}$ in (5.185) to get [36, 106]

$$D\bar{\Lambda} \equiv \prod_{i=1}^{N} d\bar{\Lambda}_{ii} \prod_{1 \leq j < k \leq N} d\,\text{Re}\,\bar{\Lambda}_{jk}\,d\,\text{Im}\,\bar{\Lambda}_{jk} = \Delta[y]^2 \; DU \tag{5.189}$$

where
$$\Delta[y] \equiv \det_{i,j}\left[y_i^{j-1}\right] = \prod_{1 \leq i < j \leq N}(y_i - y_j) \tag{5.190}$$
is the Vandermonde determinant. The Itzykson-Zuber integral can therefore be written as
$$I[X,Y;T] = \frac{1}{\Delta[y]^2}\int_{\mathcal{O}_Y} D\bar{A}\ e^{iT\ \text{tr}(X\bar{A})} \tag{5.191}$$
Notice that the diagonal components of U do not act on a diagonal matrix Y under unitary transformation. The integration in (5.191) is therefore over the coadjoint orbit of Y under the action of the unitary group, and as such it is an integral over the symmetric space G/H_C where $H_C = U(1)^N$ is the Cartan subgroup of $G = U(N)$. This can also be noted directly from the definition (5.185) in which the integrand is unchanged if U is multiplied on the right by a diagonal matrix so that the integration is really over the coset space obtained by quotienting $U(N)$ by the subgroup of diagonal unitary matrices (this extra integration then produces a factor $[\text{vol}(U(1))]^N = (2\pi)^N$ in front of the coadjoint orbit integral). This coset space has (even) dimension
$$\dim U(N) - \dim U(1)^N = N^2 - N \tag{5.192}$$
and $D\bar{A}$ is the standard symplectic measure on the coadjoint orbit. The integral (5.191) was explicitly evaluated in [3] using the so-called Gelfand-Tseytlin parametrization of the orbit.

To apply the Duistermaat-Heckman integration formula to the integral (5.187), we note that the extrema of the Hamiltonian
$$H_I[U] = \sum_{i,j=1}^{N} x_i y_j |U_{ij}|^2 \tag{5.193}$$
as a function of $U \in U(N)$, given by differentiating (5.193) with respect to U_{ij} and setting it equal to zero, satisfy the stationary conditions
$$[X, UYU^\dagger] = 0 \tag{5.194}$$
It is easy to see that the solutions of (5.194) for X,Y diagonal are of the form $U = U_d \cdot P$ where U_d is a diagonal matrix and P is a permutation matrix that permutes the diagonal entries of a matrix when acting by unitary conjugation. Diagonal matrices do not contribute to the Hamiltonian (5.193) and thus the sum over extrema in the Duistermaat-Heckman formula is over permutations $P \in S_N$. To evaluate the pertinent fluctuation determinants, we write $U = P \cdot e^{iL}$ in (5.193) where L is an infinitesimal Hermitian matrix. Then Taylor expanding (5.193) to quadratic order in L we find
$$H_I[U = P \cdot e^{iL}] = \sum_{i=1}^{N} x_i y_{P(i)} + \frac{1}{2}\sum_{i,j=1}^{N} |L_{ij}|^2(x_i - x_j)(y_{P(i)} - y_{P(j)}) + \ldots \tag{5.195}$$

174   5. Equivariant Localization on Simply Connected Phase Spaces

and the Duistermaat-Heckman formula yields

$$I[X,Y;T] = \sum_{P \in S_N} \frac{e^{iT \sum_i x_i y_{P(i)}}}{N!}$$

$$\times \int_{-\infty}^{\infty} \prod_{i,j} dL_{ij} \; e^{(iT/2) \sum_{i,j} |L_{ij}|^2 (x_i - x_j)(y_{P(i)} - y_{P(j)})} \quad (5.196)$$

$$= \left(\frac{2\pi i}{T}\right)^{N(N-1)/2} \frac{1}{N!} \sum_{P \in S_N} \mathrm{sgn}(P) \frac{e^{iT \sum_i x_i y_{P(i)}}}{\Delta[x]\Delta[y]}$$

$$= \left(\frac{2\pi i}{T}\right)^{N(N-1)/2} \frac{1}{N!} \frac{\det_{i,j}\left[e^{iT x_i y_j}\right]}{\Delta[x]\Delta[y]}$$

where we have used

$$\prod_{i<j}(y_{P(i)} - y_{P(j)}) = \mathrm{sgn}(P) \prod_{i<j}(y_i - y_j) = \mathrm{sgn}(P) \Delta[y] \quad (5.197)$$

The sign of the permutation P in (5.197) arises from the eta-invariant of the Hessian of (5.193).

The localization formula (5.196), which is a special case of the Harish-Chandra formula, was discovered by Itzykson and Zuber [76] in the context of 2-matrix models which describe conformal matter coupled to 2-dimensional quantum gravity [36] (e.g. the Ising model on a random surface). It was originally derived using heat kernel and $U(N)$ group character expansion methods [76], and orthogonal polynomial techniques for the corresponding 2-matrix model [36]. In matrix models, saddle-point approximations are always employed in the large-N limit where the models describe the relevant continuum physical theories. Notice that the classical spin partition function (2.3) is a special case of the above result where $N = 2$ and X and Y are both proportional to the $SU(2)$ Pauli spin matrix

$$\sigma^3 = \begin{pmatrix} 1 & 0 \\ 0 & -1 \end{pmatrix} \quad (5.198)$$

In the (defining or vector) spin-$\frac{1}{2}$ representation, we can represent an arbitrary matrix $D \in SU(2)$ (obtained from a 3-dimensional rotation matrix in $SO(3)$) as

$$D = \begin{pmatrix} -i \; e^{\frac{i}{2}(\psi - \phi)} \sin\frac{\theta}{2} & i \; e^{\frac{i}{2}(\psi + \phi)} \cos\frac{\theta}{2} \\ i \; e^{-\frac{i}{2}(\psi + \phi)} \cos\frac{\theta}{2} & i \; e^{-\frac{i}{2}(\psi - \phi)} \sin\frac{\theta}{2} \end{pmatrix} \quad (5.199)$$

where θ, ϕ and $\psi \in [0, 4\pi]$ are the usual Euler angles. The projection map of the principal fiber bundle $SU(2) \to SU(2)/U(1)$ is then the Hopf map $(\theta, \phi, \psi) \in S^3 \to (\theta, \phi) \in S^2$ which effectively sets $\psi = 0$ in (5.199). Substituting these identifications into the Itzykson-Zuber integral (5.185), we recover (2.2) and (2.3) now follows from (5.196) (for $a = 0$ in Section 2.1).

5.7 Localization of Generalized Spin Models and Hamiltonian Reduction

There are various extensions of the above generalized classical spin model which are relevant in matrix model theories [92]. First of all, we can consider a unitary matrix integral of the form

$$I[A;T] = \int_{U(N)} DU \, \exp\left(iT \sum_{i,j=1}^{N} A_{ij}|U_{ij}|^2\right) \qquad (5.200)$$

The stationary conditions for the Hamiltonian in (5.200) are

$$\sum_{j=1}^{N} U_{ij}(A_{ji} - A_{jk})U^\dagger_{jk} = 0 \quad , \quad i,k = 1,\ldots,N \qquad (5.201)$$

whose solutions are again the permutation matrices $U = P$. The saddle-point expansion of (5.200) thus yields (c.f. eq. (3.53))

$$I[A;T] \sim \sum_{P \in S_N} \frac{e^{iT \sum_k A_{k,P(k)}}}{\prod_{i<j}(A_{i,P(i)} + A_{j,P(j)} - A_{i,P(j)} - A_{j,P(i)})} (1 + \mathcal{O}(1/T \cdot A)) \qquad (5.202)$$

In general, the higher-order corrections in (5.202) to the lowest-order stationary phase approximation do not vanish because the Hamiltonian in (5.200) is not defined on any coadjoint orbit in general but on the entire group manifold which is not even a symplectic space. The corrections do, however, vanish in some interesting exceptions where (5.200) is a coadjoint orbit integral. One case is that discussed above, namely when A_{ij} is of rank 1, so that $A_{ij} = x_i y_j$. Another interesting case is when $N = 2$, so that (5.200) is slight modification of the spin partition function. For the case of $SU(2)$ one can check explicitly that the leading order term in (5.201) is the exact result for the integral so that

$$I_{SU(2)}[A;T] \sim \frac{e^{iT(A_{11}+A_{22})} - e^{iT(A_{12}+A_{21})}}{A_{11} + A_{22} - A_{12} - A_{21}} \qquad (5.203)$$

The unitary matrix integrals of the form (5.200) are generating functions for the correlation functions in the Itzykson-Zuber model (5.185),

$$I[X,Y;T]^{k_1 l_1 \cdots k_p l_p}_{i_1 j_1 \cdots i_p j_p} = \int_{U(N)} DU \, U_{i_1 j_1} \cdots U_{i_p j_p}(U^\dagger)^{k_1 l_1} \cdots (U^\dagger)^{k_p l_p}$$
$$\times e^{iT \, \text{tr}(UXU^\dagger Y)} \qquad (5.204)$$

which are the only non-vanishing correlators because of the $U(1)$ phase invariance $U_{ij} \to e^{i\theta}U_{ij}$ of the integral (5.185). The evaluation of the unitary matrix integrals (5.204) is very important for matrix models of induced gauge theories and string theories [92] and has been a difficult problem that has received much attention in the last few years. For instance, using Gelfand-Tseytlin coordinates on the group manifold, Shatashvili [154] has shown that

176   5. Equivariant Localization on Simply Connected Phase Spaces

some of the correlators (5.204) are explicitly given by very complicated formulas, such as

$$
I[X,Y;T]^{k_1,1,k_2,1\cdots k_p,1}_{1,j_1,1,j_2,\cdots 1,j_p}
$$
$$
= \frac{\delta_{j_1 k_1}\delta_{j_2 k_2}\cdots\delta_{j_p k_p}}{(iT)^{(N-2)(N-1)}} \frac{\prod_{\ell=1}^{N-1}\ell!}{\Delta[x]\Delta(y_2,\ldots,y_N)} \int_{-\infty}^{\infty} \prod_{k=1}^{N-1} d\lambda_k \prod_{l=1}^{n} \frac{\prod_{q=1}^{N-1}(\lambda_q - x_{j_l})}{\prod_{q\neq j_l}(x_q - x_{j_l})}
$$
$$
\times \exp\left\{iTy_1\left(\sum_{k=1}^N x_k - \sum_{k=1}^{N-1}\lambda_k\right)\right\} \det_{1\leq i,j-1\leq N-1}\left[e^{iT\lambda_i y_j}\right]
$$
(5.205)

where the delta-functions in (5.205) arise from the $U(N)$ gauge invariance $U \to VUV^\dagger$ of the Itzykson-Zuber integral (5.185). From the point of view of equivariant localization theory, these types of integrals fall into the category of the problem of developing a description of the corrections to the Duistermaat-Heckman formula in a universal way. This problem will be discussed in Chapter 7.

There is also a more geometric generalization of the spin partition function in terms of higher-dimensional generalizations of the Kähler structure of S^2. These examples will also introduce another interpretation of the localization symmetries which is directly tied to the integrability properties of these spin systems. Consider the complex N-dimensional projective space $\mathbb{C}P^N$, defined as the space of all complex lines through the origin in $\mathbb{C}^{N+1}$. A homogeneous Kähler structure on the bundle $\mathbb{C}P^N \simeq S^{2N+1}/S^1$ is obtained by the symplectic reduction of that from S^{2N+1}, i.e. the restriction of the unique Kähler structure of S^{2N+1} to $\mathbb{C}P^N$ (note that $\mathbb{C}P^1 = S^2$). The embedding of S^{2N+1} in $\mathbb{C}^{N+1}$ is defined by the constraint

$$
P(z,\bar{z}) \equiv \sum_{\mu=1}^{N+1} z^\mu \bar{z}^\mu - 1 = 0 \tag{5.206}
$$

and the symplectic structure on the maximally symmetric space $\mathbb{C}^{N+1}$ is as usual defined by the Darboux 2-form

$$
\omega_D^{(N+1)} = 2iJ \sum_{\mu=1}^{N+1} d\bar{z}^\mu \wedge dz^\mu \tag{5.207}
$$

The standard complex structure on $\mathbb{C}P^N$ is defined by the complex coordinates $\xi^{\mu-1} = z^\mu/z^1$ for $\mu = 2,\ldots,N+1$, where $z^1 = \bar{z}^1 \neq 0$ solves the constraint (5.206). Solving for z^1 and substituting it along with $z^\mu = \xi^\mu z^1$ into (5.207), the descendent symplectic structure on $\mathbb{C}P^N$ from $\mathbb{C}^{N+1}$ is

$$
\omega_{N+1}^{(J)} = 2iJ \left[\sum_{\mu=1}^N \frac{d\bar{\xi}^\mu \wedge d\xi^\mu}{1+\sum_{\lambda=1}^N \xi^\lambda \bar{\xi}^\lambda} - \sum_{\mu,\nu=1}^N \frac{\xi^\mu \bar{\xi}^\nu d\bar{\xi}^\mu \wedge d\xi^\nu}{(1+\sum_{\lambda=1}^N \xi^\lambda \bar{\xi}^\lambda)^2}\right] \tag{5.208}
$$

5.7 Localization of Generalized Spin Models and Hamiltonian Reduction 177

The Kähler metric associated with (5.208) is usually refered to as the Fubini-Study metric [41] of $\mathbb{C}P^N$, and the associated Kähler potential is

$$F^{(J)}_{N+1} = J\log\left(1 + \sum_{\mu=1}^{N} \xi^\mu \bar{\xi}^{\bar{\mu}}\right) \quad (5.209)$$

Since the symplectic 2-forms (5.208) define non-trivial elements of the cohomology group $H^2(\mathbb{C}P^N;\mathbb{Z})$, for $N > 1$ the spaces $\mathbb{C}P^N$ are not homeomorphic to any of the maximally symmetric spaces discussed in Section 5.2 above and this symplectic manifold leads to our first example of localization on a homogeneous space which is not maximally symmetric. This space for $N > 1$ has N independent isometries which are the rotations in each of the N 2-planes (see below). In the next Section we shall see how to explicitly relate the space $\mathbb{C}P^N$ to a (non-maximal) coadjoint orbit of $SU(N+1)$ so that, physically, this example describes the dynamics of a particle with internal $SU(N+1)$ isospin degrees of freedom in an external magnetic field. There we shall also encounter other examples of this situation, i.e. where the phase space is not maximally symmetric but has a maximally symmetric subspace. For now, we shall concentrate on the properties of the $SU(N+1)$ classical spin partition function associated with an integrable model.

An action of the torus $T^{N+1} = (S^1)^{N+1}$ on $\mathbb{C}^{N+1}$ is given by $z^\mu \to e^{i\theta^\mu}z^\mu$, $\theta^\mu \in [0, 2\pi)$, which is generated by the Darboux Hamiltonian

$$H^{(N+1)}_D(z,\bar{z};\theta) = 2J\sum_{\mu=1}^{N+1} \theta^\mu z^\mu \bar{z}^{\bar{\mu}} \quad (5.210)$$

that describes the dynamics of $N+1$ independent simple harmonic oscillators which each correspond to a conserved charge of this integrable system. The classical partition function is of course given by a trivial Gaussian integration

$$Z^{(N+1)}_D(T) = \int_{\mathbb{R}^{2N+2}} \prod_{\mu=1}^{N+1} \frac{2J\,dz^\mu\,d\bar{z}^{\bar{\mu}}}{\pi} \; e^{iTH^{(N+1)}_D(z,\bar{z};\theta)} = \frac{1}{(-iT)^{N+1}\prod_{\mu=1}^{N+1}\theta^\mu}$$

(5.211)

which coincides with the Duistermaat-Heckman formula as always because the only fixed point of the torus action generated by (5.210) is at $z^\mu = 0$. More interesting though is the Hamiltonian T^N-action on $\mathbb{C}P^N$, $\xi^\mu \to e^{i\tilde\theta^\mu}\xi^\mu$. For each μ, the Hamiltonian function which generates the circle action on the ξ^μ-plane in $\mathbb{C}P^N$ is the conserved charge

$$I_\mu(\xi,\bar{\xi};\tilde\theta) = \frac{2J\tilde\theta^\mu \xi^\mu \bar{\xi}^{\bar{\mu}}}{1 + \sum_{\nu=1}^{N} \xi^\nu \bar{\xi}^{\bar{\nu}}} \quad (5.212)$$

and an integrable Hamiltonian can be constructed as the sum of the N height-functions (5.212),

$$\tilde{H}_N(\xi,\bar{\xi};\tilde{\theta}) = \sum_{\mu=1}^{N} I_\mu(\xi,\bar{\xi};\tilde{\theta}) \qquad (5.213)$$

The conserved charges (5.212) are the action variables of the above integrable model and the associated angle variables $\phi^\mu \in [0, 2\pi]$ are the usual polar angles of the ξ^μ-planes (as for the $\mathbb{C}^{N+1}$ example above). The Liouville-Arnold integrability of this dynamical system can be made more explicit by considering the generalized stereographic projection onto $\mathbb{C}P^N$ in terms of the spherical coordinates $(\theta^\mu, \phi^\mu) \in S^{2N+1}$ (where $0 \leq \theta^\mu \leq \pi$),

$$\xi^1 = \tan(\theta^1/2)\cos(\theta^2/2)\,\mathrm{e}^{-i\phi^1} \quad , \quad \xi^2 = \tan(\theta^1/2)\sin(\theta^2/2)\cos(\theta^3/2)\,\mathrm{e}^{-i\phi^2}$$

$$\cdots\cdots$$

$$\xi^{N-1} = \tan(\theta^1/2)\sin(\theta^2/2)\cdots\sin(\theta^{N-1}/2)\cos(\theta^N/2)\,\mathrm{e}^{-i\phi^{N-1}}$$
$$\xi^N = \tan(\theta^1/2)\sin(\theta^2/2)\cdots\sin(\theta^{N-1}/2)\sin(\theta^N/2)\,\mathrm{e}^{-i\phi^N}$$
$$(5.214)$$

The action variables (5.212) in these spherical coordinates are then

$$I_\mu = 2J\tilde{\theta}^\mu \sin^2(\theta^1/2)\cdots\sin^2(\theta^\mu/2)\cos^2(\theta^{\mu+1}/2) \quad , \quad \mu < N$$
$$I_N = 2J\tilde{\theta}^N \sin^2(\theta^1/2)\cdots\sin^2(\theta^{N-1}/2)\sin^2(\theta^N/2)$$
$$(5.215)$$

and it is straightforward to see that the symplectic 2-form (5.208) takes the usual integrable form $dI_\mu \wedge d\phi^\mu$ (for the oscillator above $I_\mu \sim z^\mu \bar{z}^\mu$ as usual). The explicit torus action on $\mathbb{C}P^N$ generated by the Hamiltonian vector field associated with (5.213) is thus

$$\tilde{V}_N = \sum_{\mu=1}^{N} i\tilde{\theta}^\mu \left(\xi^\mu \frac{\partial}{\partial \xi^\mu} - \bar{\xi}^{\bar{\mu}} \frac{\partial}{\partial \bar{\xi}^{\bar{\mu}}}\right) = \sum_{\mu=1}^{N} \tilde{\theta}^\mu \frac{\partial}{\partial \phi^\mu} \qquad (5.216)$$

This integrable system on the phase space $\mathbb{C}P^N$ is therefore once again isomorphic to a linear combination of independent harmonic oscillators as above, generalizing the dynamics of our standard spin example.

The relation between such generalized spin systems and systems of harmonic oscillators, and hence the exactness of the localization formulas, can be understood from a slightly different perspective than the usual localization symmetries provided by the underlying equivariant cohomological or supersymmetric structures of the dynamical system. To see this, we evaluate the classical partition function explicitly for the above dynamical system by extending the integration over $\mathbb{C}P^N$ to the whole of $\mathbb{C}^{N+1}$ via a Lagrange multiplier which enforces the constraint (5.206). This leads to

5.7 Localization of Generalized Spin Models and Hamiltonian Reduction   179

$$Z_N(T) = \int_{S^{2N+1}/S^1} \frac{\prod_{\mu=1}^N 2J\, d\bar{\xi}^\mu\, d\xi^\mu}{\pi^N(1+\sum_{\nu=1}^N \xi^\nu \bar{\xi}^\nu)^{N+1}}\, e^{iT\tilde{H}_N(\xi,\bar{\xi};\tilde{\theta})}$$

$$= \int_{-\infty}^\infty \frac{d\lambda}{2\pi} \int_{\mathbb{R}^{2N+2}} \prod_{\mu=1}^{N+1} \frac{d\bar{z}^\mu\, dz^\mu}{\pi} \tag{5.217}$$

$$\times \exp\left\{-TH_D^{(N+1)}(z,\bar{z};\theta) + i\lambda\left(\sum_{\mu=1}^{N+1} z^\mu \bar{z}^\mu - 1\right)\right\}$$

where the angular parameters in (5.217) are related by $\tilde{\theta}^\mu = \theta^\mu - \theta^{N+1}$. The Gaussian integral over $\mathbb{C}^{N+1}$ in (5.217) can be performed to yield

$$Z_N(T) = \int_{-\infty}^\infty d\lambda\ e^{-i\lambda} \prod_{\mu=1}^{N+1} \frac{i}{T\theta^\mu + \lambda} \tag{5.218}$$

The remaining integration in (5.218) can be carried out by continuing the integration over a large contour in the complex plane and using the residue theorem to pick up the $N+1$ simple poles of the integrand at $\lambda = -T\theta^\mu$, each of which have residue 1. Thus the classical partition function is

$$Z_N(T) = \sum_{\mu=1}^{N+1} e^{iT\theta^\mu} \prod_{\nu\neq\mu} \frac{i}{T(\theta^\nu - \theta^\mu)} \tag{5.219}$$

It is readily verified that (5.219) coincides the Duistermaat-Heckman integration formula for the Hamiltonian (5.213) which has precisely $N+1$ stationary points [52]. (5.219) therefore represents the T^N-equivariant cohomology of the manifold $\mathbb{C}P^N$.

The explicit evaluation above illustrates an interesting feature of the localization in these cases, as was first pointed out in [44, 52]. This is seen in (5.217) – the dynamical system describing the $SU(N+1)$ isospin is the Hamiltonian reduction of a larger dynamical system which is described by a quadratic Hamiltonian on the Darboux symplectic manifold $\mathbb{R}^{2N+2}$ and for which the localization is therefore trivial. The reduction constraint function (5.206) commutes with the Darboux oscillator Hamiltonian (5.210), i.e.

$$\left\{P(z,\bar{z}), H_D^{(N+1)}\right\}_{\omega_D^{(N+1)}} = 0 \tag{5.220}$$

so that the constraint function $P(z,\bar{z})$ determines a first class constraint on the Darboux dynamical system and is therefore a symmetry (or conserved charge) of the classical dynamics. This commutativity property is in fact the crucial mechanism for the correspondence between the integrations in (5.217), as then the dynamics on $\mathbb{C}^{N+1}$ is restricted to the symplectic subspace determined by the constant values of the conserved charges $P(z,\bar{z})$ (according to the reduction theorem [7] – see (4.144)). Another important feature of the reduction mechanism here is the action of $SU(N+1)$ on $\mathbb{C}^{N+1}$ defined by the usual matrix-vector multiplication by matrices in the defining (vector) representation of $SU(N+1)$. This defines a symplectic Hamiltonian action on the Darboux system above [44] which leaves invariant the sphere S^{2N+1} and hence restricts the dynamics to this submanifold of $\mathbb{R}^{2N+2}$. From the $U(1)$ invariance of the harmonic oscillator Hamiltonian on $\mathbb{C}^{N+1}$ above, this restricts further to the coset space $\mathbb{C}P^N = S^{2N+1}/U(1)$ and the conserved charges under the mapping by the symplectic constraint function $P(z,\bar{z})$ on $\mathbb{C}^{N+1}$ become the components of the projection map of the generalized Hopf bundle $U(1) \to S^{2N+1} \to \mathbb{C}P^N$ [44]. It is easily seen that the critical points of the quadratic Hamiltonian in (5.217) in the space of the λ and z variables coincide exactly with the poles in (5.218) so that (5.219) coincides precisely with the Duistermaat-Heckman formula for this Darboux system. The localization of the isospin partition function is thus the reduction of that from $\mathbb{R}^{2N+2}$.

We recall that such a reduction method in equivariant localization is also that which is used to derive the Callias-Bott index theorems from higher-dimensional Atiyah-Singer index theorems [70] (see comments at the end of Section 4.2). The above reduction procedure is a standard method of integrating dynamical systems by associating to a given Hamiltonian system a related one on a lower dimensional symplectic manifold [1]. In this scheme one reduces the rank of the differential equations of motion or the number of degrees of freedom using invariant relations and constants of the motion onto a coadjoint orbit as above. Furthermore, in the context of generic integrable models, there is the conjecture that every integrable system is the Hamiltonian reduction of a larger linear dynamical system by first class constraints [176]. If this conjecture were true, then the localization of generic integrable systems could be cast in another formalism giving an even stronger connection between the Duistermaat-Heckman theorem and the integrability properties of a dynamical system. Conversely, as the integration formulas we have encountered also always correspond to a sort of Hamiltonian reduction of the original dynamical model, they yield realizations of this reduction conjecture as well as the quantum mechanical conjecture mentioned earlier about the exactness of the semi-classical approximation for the description of the quantum dynamics of generic integrable systems.

5.8 Quantization of Isospin Systems

Let us now quickly describe the quantum generalizations of the results of the last Section. First, we note that, for a general group G, a point particle in an irreducible representation $\mathcal{R}$ of the internal symmetry group G has its time-dependent isospin vectors $\mathcal{R}(t)$ living in a fixed orbit of G in the adjoint representation [12]. As described earlier, this fixed coadjoint orbit determines a unitary irreducible representation of G and if we let Λ' denote a (time-independent) fixed fiducial point on this orbit, then the internal isospin vectors can be written as

$$\mathcal{R}(t) = \mathrm{Ad}^*(\Lambda')g(t) = g(t)\Lambda'g(t)^{-1} \tag{5.221}$$

with $g(t) \in G$ time-dependent group elements. In the absence of any "external" motion of the charged particle, the isospin vector is conserved, $\dot{\mathcal{R}}(t) = 0$. Thus the point particle action describing its motion in a generalized external magnetic field $B = B^a \mathcal{R}(X^a)$ is [12]

$$S[g, \Lambda'] = \int_0^T dt \left(\mathrm{tr}(\Lambda' g^{-1} \dot{g}) + \mathrm{tr}(\mathcal{R} \cdot B) \right) \tag{5.222}$$

When Λ' is taken to be a Cartan element, this action generalizes the spin action before which described the dynamics of a particle moving in both monopole and external fields (Hamiltonianly reduced from a free particle action on $\mathbb{R}^4$) [12, 44]. Thus again we see that the kinetic term in (5.222) is of first order in time derivatives and consequently the action (5.222) is already cast in phase space. The phase space variables are the isospin vectors $\mathcal{R}(X^a)$ and the Poisson algebra of them is just the Lie algebra of the internal symmetry group G [12]. Thus these generalized dynamical systems all fall into the class of those systems whose configuration and phase spaces coincide so that their path integral interpretations on the respective spaces are the same.

To describe the quantum dynamics of these models, we start with the quantization of the classical $SU(N+1)$ isospin model of the last Section over the phase space $\mathbb{C}P^N$. Since [143]

$$SU(N+1) \simeq S^{2N+1} \times SU(N) \simeq S^{2N+1} \times S^{2N-1} \times \cdots \times S^3 \tag{5.223}$$

it follows that this phase space is the coadjoint orbit $O_{\{N,1\}} = SU(N+1)/(SU(N) \times U(1))$ of $SU(N+1)$ [68], and an integrable Hamiltonian on it will be a combination of the N Cartan elements of $SU(N+1)$ which leave the maximal torus $H_C \sim (S^1)^N \subset O_{\{N,1\}}$ invariant. Although the orbit space is not a flag manifold as before, it is still possible to generalize the construction of the coherent states in Section 5.4. They are constructed by taking the highest weight vector $|0\rangle = \sum_\ell n_\ell \lambda_\ell^{(f)}$ where n_ℓ is a non-negative integer, $\lambda_\ell^{(f)}$ is a highest weight vector of the fundamental representation, and the sum over ℓ is taken in such a way that the maximum stabilizer group of $|0\rangle$ is $SU(N) \times U(1)$ (so that the loop space path integral will be canonically associated with $O_{\{N,1\}} = O_{\Lambda'} \simeq \mathrm{Ad}^*(SU(N+1))\Lambda' \simeq SU(N+1)/SU(N+1)_{\Lambda'}$) and so that $|0\rangle$ is appropriate to the geometry of the given coadjoint orbit (e.g. the normalization of the coherent states generates the Kähler potentials of the orbits in the manner described in Section 5.4). The existence of such a weight state is in general always guaranteed by the Borel-Weil-Bott theorem [68, 142].

We now define the coherent states associated with the irreducible $[2J]$ representation of the $SU(N+1)$ group, where $2J \in \mathbb{Z}^+$, $2J \leq N+1$ [53, 87] (i.e. the representation with fundamental highest weight J corresponding to a Young tableau representation by a single column of $2J$ boxes). In this representation, the highest weight vector is denoted as $|J; N+1\rangle$, there are precisely N simple roots $\alpha = 1, 2, \ldots, N$ (i.e. roots which cannot be decomposed into the sum of 2 other positive roots) with $E_\alpha^{(N+1)}|J; N+1\rangle = 0$, and the Cartan generators in the Gell-Mann basis have the (diagonal) matrix elements

$$[H_m]_{ij} = \frac{1}{\sqrt{2m(m+1)}} \left(\sum_{k=1}^m \delta_{ik}\delta_{jk} - m\delta_{i,m+1}\delta_{j,m+1} \right) \quad , \quad m = 1, \ldots, N \tag{5.224}$$

where we have normalized the generators so that $\mathrm{tr}(X^a X^b) = \frac{1}{2}\delta^{ab}$. The dimension of this representation is

$$\dim\,[2J] = \binom{N+1}{2J} = \frac{(N+1)!}{(2J)!(N+1-2J)!} \tag{5.225}$$

The generalized coherent states associated with this representation of $SU(N+1)$ are then [137]

$$|\xi\rangle \equiv e^{\sum_{\alpha=1}^N \xi^\alpha E_{-\alpha}^{(N+1)}} |J; N+1\rangle \quad , \quad \xi = (\xi^1, \ldots, \xi^N) \in \mathbb{C}P^N$$

$$= \sum_{n_1+\ldots+n_{N+1}=2J} \sqrt{\frac{(2J)!}{n_1! \cdots n_{N+1}!}} (\xi^1)^{n_1} \cdots (\xi^N)^{n_N} |\{n_k\}_{k=1}^{N+1}; N+1\rangle \tag{5.226}$$

These coherent states have the normalizations

$$(\xi_2|\xi_1) = \left(1 + \sum_{\alpha=1}^N \xi_1^\alpha \bar{\xi}_2^{\bar{\alpha}}\right)^{2J} \tag{5.227}$$

and they obey the completeness relation

$$\hat{P}_J \equiv \int d\mu^{(J)}(\xi, \bar{\xi})\, |\xi\rangle\!\rangle\langle\!\langle\xi| = 1^{(J)} \tag{5.228}$$

where

$$d\mu^{(J)}(\xi, \bar{\xi}) = \frac{(2J+N)!}{(2J)!} \frac{\prod_{\alpha=1}^N d\bar{\xi}^{\bar{\alpha}} \wedge d\xi^\alpha}{\pi^N (1 + \sum_{\mu=1}^N \xi^\mu \bar{\xi}^{\bar{\mu}})^{N+1}} \tag{5.229}$$

is the associated coherent state measure.

Consider the quantum propagator

5.8 Quantization of Isospin Systems

$$\mathcal{K}^{(J)}(\xi_2,\xi_1;T) = \langle\!\langle \xi_2 | \, e^{-iT\hat{\mathcal{H}}} | \xi_1 \rangle\!\rangle \tag{5.230}$$

defined in terms of the invariant $SU(N+1)$ operator that is a linear combination of the Cartan generators (5.224)

$$\hat{\mathcal{H}} = 2J \sum_{m=1}^{N} \tilde{\theta}^m H_m + 2JC_0 \tag{5.231}$$

which leads as usual to a topological and integrable quantum theory. Then the standard calculation yields the coherent state path integral

$$\mathcal{K}^{(J)}(\xi_2,\xi_1;T)$$
$$= \int_{L(S^{2N+1}/S^1)} [d\mu^{(J)}(\xi,\bar{\xi})] \, \exp\Bigg\{ 2J \log\Bigg(1 + \sum_{\alpha=1}^{N} \xi_2^{\alpha} \bar{\xi}_2^{\bar{\alpha}}\Bigg)$$
$$+ 2J \log\Bigg(1 + \sum_{\alpha=1}^{N} \xi_1^{\alpha} \bar{\xi}_1^{\bar{\alpha}}\Bigg) + i \int_0^T dt \left[\sum_{\alpha=1}^{N} \frac{2iJ(\bar{\xi}^{\bar{\alpha}} \dot{\xi}^{\alpha} - \xi^{\alpha} \dot{\bar{\xi}}^{\bar{\alpha}})}{1 + \sum_{\mu} \xi^{\mu} \bar{\xi}^{\bar{\mu}}} - H(\xi,\bar{\xi}) \right] \Bigg\} \tag{5.232}$$

where the boundary conditions in the path integral are $\xi(0) = \xi_1$ and $\bar{\xi}(T) = \bar{\xi}_2$, and

$$H(\xi,\bar{\xi}) \equiv \langle\!\langle \xi | \hat{\mathcal{H}} | \xi \rangle\!\rangle = 2JC_0 + 2J \sum_{\alpha=1}^{N} \frac{\tilde{\theta}^{\alpha} \xi^{\alpha} \bar{\xi}^{\bar{\alpha}}}{1 + \sum_{\mu} \xi^{\mu} \bar{\xi}^{\bar{\mu}}} \tag{5.233}$$

is the generalized height function (5.213) on $\mathbb{C}P^N$. The WKB approximation for the coherent state path integral has been discussed extensively in [52, 131]. In fact, all of the localization formulas of Chapter 4 can be explicitly verified as the above dynamical system is just a multi-dimensional generalization of the Kähler polarization for the spin propagator in Section 5.5 above. The classical equations of motion are

$$\dot{\xi}^{\alpha} + 2iJ\tilde{\theta}^{\alpha}\xi^{\alpha} = 0 \quad , \quad \dot{\bar{\xi}}^{\bar{\alpha}} - 2iJ\tilde{\theta}^{\alpha}\bar{\xi}^{\bar{\alpha}} = 0 \tag{5.234}$$

whose solutions in the stereographic coordinates (5.214) are the conditionally periodic motions

$$\theta^{\alpha}(t) = \theta^{\alpha}(0) \quad , \quad \phi^{\alpha}(t) = \phi^{\alpha}(0) + 2J\tilde{\theta}^{\alpha} t \tag{5.235}$$

generalizing the results of Section 5.5.

In any case, we arrive at the propagator

$$\mathcal{K}^{(J)}(\xi_2,\xi_1;T) = \frac{\left(1 + \sum_{\alpha=1}^{N} \xi_1^{\alpha} \bar{\xi}_2^{\bar{\alpha}} \, e^{-2iJ\tilde{\theta}^{\alpha} T}\right)^{2J}}{(1 + \sum_{\mu} \xi_1^{\mu} \bar{\xi}_1^{\bar{\mu}})^J (1 + \sum_{\mu} \xi_2^{\mu} \bar{\xi}_2^{\bar{\mu}})^J} \, e^{2iJC_0 T} \tag{5.236}$$

184   5. Equivariant Localization on Simply Connected Phase Spaces

and the associated quantum partition function

$$Z_{SU(N+1)}(T) = \int d\mu^{(J)}(\xi,\bar{\xi}) \; ((\xi| \, e^{-i\hat{\mathcal{H}}T} |\xi))$$
$$= \sum_{\alpha=1}^{N+1} e^{-2iJ\tilde{\theta}^{\alpha}T} \prod_{\beta \neq \alpha} \frac{1}{1 - e^{-2iJ(\tilde{\theta}^{\beta}-\tilde{\theta}^{\alpha})T}} \qquad (5.237)$$

where $\tilde{\theta}^{N+1} \equiv C_0$. These generalize the previous results for $SU(2)$ and, in particular, (5.237) coincides with the anticipated Weyl character formula for the $[2J]$ representation of $G = SU(N+1)$. What is also interesting here is that the Hamiltonian reduction mechanism of the last Section has a quantum counterpart which can be used to interpret the quantum localization of this dynamical system. To see this, we recall from the last Section that the Hamiltonian (5.233) is the reduction of an $(N+1)$-dimensional simple harmonic oscillator Hamiltonian, which in the quantum case is the Hermitian operator

$$\hat{\mathcal{H}}^{(D)} = 2J \sum_{\alpha=1}^{N+1} \tilde{\theta}^{\alpha} \hat{a}_{\alpha}^{\dagger} \hat{a}_{\alpha} \qquad (5.238)$$

acting on a multi-dimensional Heisenberg-Weyl algebra $\bigoplus_{\alpha=1}^{N+1} \mathbf{g}_{HW}$, where $\hat{a}_{\alpha}^{\dagger}, \hat{a}_{\alpha}$ are $N+1$ mutually commuting copies of the raising and lowering operators (5.79). Relating the coordinates on $\mathbb{C}^{N+1}$ and $\mathbb{C}P^N$ as usual via the constraint function (5.206), it is straightforward to show that the projection operator in (5.228) is [52]

$$\hat{P}_J = \int_0^{2\pi} \frac{d\lambda}{2\pi} \; e^{i\lambda(\sum_{\alpha} \hat{a}_{\alpha}^{\dagger} \hat{a}_{\alpha} - 2J)} = \sum_{n_1+\ldots+n_{N+1}=2J} |n_1,\ldots,n_{N+1}\rangle\langle n_1,\ldots,n_{N+1}|$$

(5.239)

where

$$|n_1,\ldots,n_{N+1}\rangle \equiv \frac{1}{\sqrt{n_1!\cdots n_{N+1}!}} (\hat{a}_1^{\dagger})^{n_1} \cdots (\hat{a}_{N+1}^{\dagger})^{n_{N+1}} |0\rangle \qquad (5.240)$$

are the states of the orthonormal number basis for $\bigoplus_{\alpha=1}^{N+1} \mathbf{g}_{HW}$. This identifies the weight states of the $SU(N+1)$ representation above as $|\{n_k\}_{k=1}^{N+1}; N+1\rangle = |n_1,\ldots,n_{N+1}\rangle$ in terms of the bosonic Fock space states (5.240). This method of constructing coherent states is known as the Schwinger boson formalism [49]–[52].

The trace formula (5.237) can now be represented in terms of the canonical coherent states

$$|z)) \equiv e^{\sum_{\alpha=1}^{N+1} \hat{a}_{\alpha}^{\dagger} z^{\alpha}} |0\rangle / e^{\frac{1}{2} \sum_{\mu} z^{\mu} \bar{z}^{\mu}} \quad , \quad z = (z^1,\ldots,z^{N+1}) \in \mathbb{C}^{N+1} \quad (5.241)$$

as

$$Z_{SU(N+1)}(T) = \int \frac{\prod_{\alpha=1}^{N+1} d\bar{z}^{\alpha} \, dz^{\alpha}}{\pi^{N+1}} \left(\!\left(z | \hat{P}_J \, e^{-i\hat{\mathcal{H}}^{(D)}T} | z\right)\!\right) \tag{5.242}$$

Then using the important property

$$\left[\hat{\mathcal{H}}^{(D)}, \hat{P}_J\right] = 0 \tag{5.243}$$

it is straightforward to show that (5.242) becomes [52]

$$Z_{SU(N+1)}(T)$$
$$= \int_0^{2\pi} \frac{d\lambda}{2\pi} \, e^{-2iJ\lambda} \int_{L\mathbb{R}^{2N+2}} \prod_{t\in[0,T]} \frac{\prod_\mu d\bar{z}^\mu(t) \, dz^\mu(t)}{\pi^{N+1}} \tag{5.244}$$
$$\times \exp\left\{-\int_0^T dt \sum_{\alpha=1}^{N+1} \left[e^{-2iJT\tilde{\theta}^\alpha + i\lambda}\left(\bar{z}^{\bar{\alpha}}\dot{z}^\alpha - z^\alpha \dot{\bar{z}}^{\bar{\alpha}}\right) - z^\alpha \bar{z}^{\bar{\alpha}}\right]\right\}$$

where we have also used the resolution of unity for the canonical coherent states (5.241). The Gaussian functional integration in (5.244) is now trivial to carry out and we find

$$Z_{SU(N+1)}(T) = \int_0^{2\pi} \frac{d\lambda}{2\pi} \, e^{-2iJ\lambda} \prod_{\alpha=1}^{N+1} \frac{1}{1 - e^{-2iJ\tilde{\theta}^\alpha T + i\lambda}}$$
$$= \oint_{S^1} \frac{dw}{2\pi} \, w^{2J+N} \prod_{\alpha=1}^{N+1} \frac{1}{w - e^{-2iJ\tilde{\theta}^\alpha T}} \tag{5.245}$$

where we have transformed the angular integration in (5.245) into a contour integral over the unit circle S^1. Carrying out the contour integration picks up $N+1$ simple poles each of residue 1 and leads to Weyl character formula (5.237).

Thus the classical Hamiltonian reduction mechanism discussed in the last Section also implies localization at the quantum level, and again the terms summed over in the WKB formula represent the singular points of the symplectic reduction of $\mathbb{C}^{N+1}$ to $\mathbb{C}P^N$. In the quantum case the first class constraint algebra (5.243) is determined by the projection operator $\hat{P}_J$ which coincides with the identity operator on the group representation space when restricted to the $SU(N+1)$ coherent states as above. This Hamiltonian reduction is always just a manifestation of the fact that these localizable dynamical systems are always in some way just a set of harmonic oscillators, because of the large "hidden" supersymmetry in these problems. The above analysis can also be straightforwardly generalized to the non-compact hyperbolic space $D^{N,1}$ (the complex open $(N+1)$-dimensional Poincaré ball [68]) with the associated $SU(N,1)$ coherent states, generalizing the results of Section

5.6 for $SU(1,1)$ [52]. In particular, the localization formulas all lead to the Weyl character formula for $SU(N,1)$

$$Z_{SU(N,1)}(T) = e^{-iC_0KT} \prod_{\alpha=1}^{N} \frac{1}{1 - e^{-i\tilde{\theta}^{\alpha}T}} \tag{5.246}$$

The $\mathbb{C}P^N$ model above is a special case of the more general Kähler space called a Grassmann manifold. Geometrically, this is defined as the $k \cdot (N-k)$ complex-dimensional space $\text{Gr}(N,k)$ of k-planes through the origin of $\mathbb{C}^N$ (note that $\text{Gr}(N,1) = \mathbb{C}P^{N-1}$). Algebraically, it is the space of $N \times N$ Hermitian matrices obeying a quadratic constraint,

$$\text{Gr}(N,k) = \{P : P^{\dagger} = P, P^2 = P, \text{ tr } P = k\} \tag{5.247}$$

from which it can be shown to be isomorphic to the $U(N)$ coadjoint orbit [68, 143]

$$\text{Gr}(N,k) \simeq U(N)/(U(k) \times U(N-k)) \tag{5.248}$$

with the transitive $U(N)$ (coadjoint) action $P \to UPU^{\dagger}$ on $\text{Gr}(N,k)$. Note that the quadratic constraint in (5.247) implies that the operators P have eigenvalues 0 or 1. They can therefore be interpreted as fermionic occupation number operators and the trace condition in (5.247) can be interpreted as the total number of fermions in a given state. The Grassmann manifolds are therefore intimately related to an underlying free fermion theory [143]. The symplectic structure on (5.247) is the associated Kirillov-Kostant 2-form

$$\omega^{(N,k)} = i \text{ tr}(PdP \wedge dP) \tag{5.249}$$

whose explicit form can be written using the local coordinatization provided by the diffeomorphism (5.223) and $U(N) \simeq SU(N) \times S^1$ [49]. Similarly, the non-compact hyperbolic analog of the Grassmann manifold is known as the Siegel disc $D^{N,k}$ [68]. It is defined algebraically as

$$D^{N,k} = \{\tilde{P} : \tilde{P}^{\dagger} = \tilde{P}, \tilde{P}^{\dagger}\eta_{(N,k)}P = \eta_{(N,k)}\} \tag{5.250}$$

where $\eta_{(N,k)}$ is the flat Minkowski metric with diagonal elements consisting of $N-k$ entries of -1 and k entries of $+1$. It is isomorphic to the coadjoint orbit

$$D^{N,k} \simeq U(N,k)/(U(N) \times U(k)) \tag{5.251}$$

of the non-compact Lie group $U(N,k)$, and its Kähler structure is defined by

$$\tilde{\omega}^{(N,k)} = \text{tr}\left(\tilde{P}\eta_{(N,k)}d\tilde{P} \wedge \eta_{(N,k)}d\tilde{P}\eta_{(N,k)}\right) \tag{5.252}$$

The Grassmann and Siegel spaces above are the representative spaces for homogeneous manifolds [68]. The coherent state quantization and semi-classical exactness of these dynamical systems has been discussed extensively

5.8 Quantization of Isospin Systems

in [49, 143]. The integrable Hamiltonians on (5.247) are parametrized by a constant $N \times N$ Hermitian matrix X and are defined by

$$H_X(P) = \mathrm{tr}(XP) \qquad (5.253)$$

so that the partition function generalizes the Itzykson-Zuber integral (5.185). Their Poisson algebra is

$$\{H_X, H_Y\}_{\omega^{(N,k)}} = H_{[X,Y]} \qquad (5.254)$$

so that the associated partition functions define topological theories. The Duistermaat-Heckman formula

$$Z_{\mathrm{cl}}^{\mathrm{Gr}(N,k)}(T) = \left(\frac{2\pi i}{T}\right)^{k(N-k)} \sum_{1 \leq i_1 < \ldots < i_k \leq N} \frac{e^{iT \sum_{l=1}^{k} h_{i_l}}}{\prod_{l=1}^{k} \prod_{j \neq i_l}(h_j - h_{i_l})}, \qquad (5.255)$$

with $h_i \in \mathbb{R}$ the eigenvalues of the Hermitian matrix X in (5.253) parametrizing the Hamiltonian, has been verified for this dynamical system and shown to be associated with a Hamiltonian reduction mechanism as described above for the case of $\mathbb{C}P^N$. The construction of coherent states via the algebraic or Schwinger boson formalisms parallels that above for the $\mathbb{C}P^N$ coherent states, although in the present case it is somewhat more involved. We shall not go into the technical details here, but refer to [49] where it was also shown that the WKB localization formula

$$Z_{U(N)}^{\mathrm{Gr}(N,k)}(T) \sim \sum_{1 \leq i_1 < \ldots < i_k \leq N} \frac{e^{-ikT \sum_{l=1}^{k} h_{i_l}}}{\prod_{l=1}^{k} \prod_{j \neq i_l}(1 - e^{-iT(h_j - h_{i_l})})} \qquad (5.256)$$

$$= \frac{\det_{i,j}\left[e^{-iTh_i n_j}\right]}{\Delta\left[e^{-iTh_i}\right]}$$

yields the anticipated Weyl character formula in the representation defined by the coset space (5.248). Here $k \in \mathbb{Z}^+$ characterizes the highest weight of the pertinent $U(N)$ representation, the non-negative integers $n_i > n_{i+1}$ are the components of the vector $\lambda + \rho$ (defined as $n_i = N - i + b_i$ where b_i is the number of boxes in the i-th row of the associated Young tableau representation), and here we have taken the basis of the Cartan subalgebra in which $(H_i)_{jk} = \delta_{ij}\delta_{jk}$. Similarly, the integrable Hamiltonians on the hyperbolic space (5.250) are $H_{\tilde{X}}(\tilde{P}) = -\mathrm{tr}(\eta_{(N,k)}\tilde{X}\eta_{(N,k)}\tilde{P})$ with $\tilde{X}$ a constant real-valued diagonal matrix.

These constructions of coherent states can also be generalized to the coadjoint orbit spaces [66, 86, 130]

$$O_{\{n_1, n_2, \ldots, n_\ell\}} \equiv SU(N)/(SU(n_1) \times \cdots \times SU(n_\ell) \times U(1)^{\ell-1}) \qquad (5.257)$$

of dimension $N^2 - \sum_{i=1}^{\ell} n_i^2$, where $\sum_{i=1}^{\ell} n_i = N$ and the dimension of the $SU(N)$ subgroup $SU(n_1) \times \ldots \times SU(n_\ell) \times U(1)^{\ell-1}$ is $N - 1$. The complex structure on the orbit (5.257) is defined by the isomorphism $O_{\{n_1,n_2,\ldots,n_\ell\}} \simeq SL(N,\mathbb{C})/P_{\{n_1,n_2,\ldots,n_\ell\}}$, where $SL(N,\mathbb{C})$ is the complexification of $SU(N)$ and $P_{\{n_1,n_2,\ldots,n_\ell\}}$ is a parabolic subgroup of $SL(N,\mathbb{C})$ which is the subgroup of block upper triangular matrices in the $(n_1+\ldots+n_\ell) \times (n_1+\ldots+n_\ell)$ block decomposition of elements of $SL(N,\mathbb{C})$. Note that for $n_1 = \ldots = n_N = 1$ the coadjoint orbit (5.257) is the $SU(N)$ flag manifold (the maximal orbit) with $P_{\{1,1,\ldots,1\}}$ the Borel subgroup B_N, while for $n_1 = N-1$ and $n_2 = 1$ non-zero only, (5.257) coincides with the $\mathbb{C}P^N$ manifold discussed above (the minimal orbit). The isospin degrees of freedom on the coadjoint orbit (5.257) are defined as [49, 130] (c.f. eq. (5.221))

$$Q = \mathrm{Ad}^*(X)g = gXg^{-1} \quad , \quad g \in SU(N) \tag{5.258}$$

where X is a diagonal matrix with entries $x_i \in \mathbb{R}$ satisfying $\sum_{i=1}^{N} x_i = 0$ and $x_1 = x_2 = \ldots = x_{n_1} > x_{n_1+1} = x_{n_1+2} = \ldots = x_{n_2} > \ldots = x_{n_\ell-1} = x_{n_\ell}$.

The complexification of any element $g \in SU(N)$ can be parametrized by N column vectors $Z_i \in \mathbb{C}^N$, $i = 1,\ldots,N$, with

$$Z_i^\dagger Z_j = \delta_{ij} \quad , \quad \det[Z_1,\ldots,Z_N] = 1 \tag{5.259}$$

The canonical 1-form on the orbit is then

$$\theta^{\{n_i\}} = \mathrm{tr}\left(Xg^{-1}dg\right) = i\sum_{i=1}^{N-1} J_i Z_i^\dagger dZ_i \tag{5.260}$$

where $J_i = x_1 + \ldots + 2x_i + \ldots + x_{N-1} \geq 0$ and we have used the determinant constraint in (5.259). The associated Kähler 2-form is

$$\omega^{\{n_i\}} = d\theta^{\{n_i\}} = -i\sum_{i=1}^{N-1} J_i dZ_i \wedge dZ_i^\dagger \tag{5.261}$$

and the isospin element (5.258) can be expressed as

$$Q = i\sum_{i=1}^{N-1} \left(J_i Z_i Z_i^\dagger - \frac{J_i}{N}\mathbf{1}\right) \tag{5.262}$$

Defining the isospin generators $Q^a = \mathrm{tr}(QX^a)$ with the normalization of the $SU(N)$ generators defined as above for the $\mathbb{C}P^N$ case, it is straightforward to see that these functions generate the $SU(N)$ Lie algebra in the Poisson bracket generated by the symplectic structure (5.261) once the first set of orthonormal constraints in (5.259) are substituted into the above relations. These functions generalize the angular momentum generators in the $SU(2)$ case (where the only coadjoint orbit is the maximal one). The constraint functions in (5.259) are easily seen to be first class constraint functions [49, 130] so that the coadjoint orbit is determined again as the symplectic

5.8 Quantization of Isospin Systems

reduction of a larger space. These constraint functions generate the subgroup $SU(n_1) \times \ldots \times SU(n_\ell) \times U(1)^{\ell-1}$ of $SU(N)$. The localization properties of the topological Hamiltonians generated by the orbit functions (5.262) can therefore be interpreted again in terms of generalized sorts of harmonic oscillator dynamical systems, the characteristic feature of the equivariant localization mechanism. The coherent states in these generic cases are defined analogously to those over $\mathbb{C}P^N$ above, with the restrictions discussed at the beginning of this Section.

The structure of the integrable models defined on these coset spaces has been discussed in [66, 86]. The Poisson-Lie relations of the $SU(N)$ symmetry of each orbit (5.257) lead to a maximal number $N^2 - N$ of mutually commuting functions, which is equal to half the (real) dimension of the maximal coadjoint orbit. In particular, the dynamical properties of the flag manifold $SU(N)/U(1)^{N-1}$ have been related to ideas in the theory of non-commutative integrability. The coherent state quantization of this flag manifold in the case of $SU(3)$ has also been worked out in detail in [66, 86] (see also [79, 100]). In that case, the Kähler potential is [139]

$$F_{SU(3)}(z,\bar{z}) = \log\left[(1 + z_1\bar{z}_1 + z_2\bar{z}_2)^p(1 + z_3\bar{z}_3 + (z_2 - z_1z_3)(\bar{z}_2 - \bar{z}_1\bar{z}_3))^q\right]$$
(5.263)

where p and q are integers. Using the usual integrable Hamiltonians Q^3 and Q^8 defined as above, it is straightforward to construct a coherent state path integral representation for the quantum dynamics of this localizable system and verify the localization formulas of Chapter 4 above. For instance, the semi-classical approximation has been verified for the dynamical system with topological Hamiltonian function $H = \sum_{i=\pm} \tilde{\theta}_i Q_i$, where $Q_\pm = Q^3 \pm \sqrt{3}Q^8$ generate the symplectic circle actions $(z_1, z_2, z_3) \to (\,e^{i\theta_1}z_1,\, e^{i\theta_1}z_2, z_3)$ and $(z_1, z_2, z_3) \to (\,e^{i\theta_2}z_1, z_2,\, e^{-i\theta_2}z_3)$, respectively [66, 86] (so that H generates the symplectic action of the 2-torus group $T^2 = S^1 \times S^1$ on $SU(3)/U(1)^2$). The WKB localization formula for the quantum propagator in the coherent state representation for this integrable spin model can then be worked out to be

$$\mathcal{K}_{SU(3)}(\bar{z}', z; T) = (1 + \bar{z}'_1 z_1\, e^{i(\tilde{\theta}_+ + \tilde{\theta}_-)T} + \bar{z}'_2 z_2\, e^{i\tilde{\theta}_+ T})^p$$
$$\times (1 + \bar{z}'_3 z_3\, e^{-i\tilde{\theta}_- T} + (\bar{z}'_2 - \bar{z}'_1\bar{z}'_3)(z_2 - z_1 z_3)\, e^{i\tilde{\theta}_+ T})^q$$
(5.264)

which can be shown to coincide with the exact result from a direct calculation.

Finally, we point out that one can also apply the nonabelian localization formalisms of Sections 3.8 and 4.9 to these generalized spin models regarding the associated partition functions as defined on the coordinate manifolds. As such, they can be applied to problems such as geodesic motion on group manifolds, and in particular we can reproduce the results of Picken [139] in the Hamiltonian framework [161]. More precisely, we note that in these cases there is the natural G-invariant metric

190   5. Equivariant Localization on Simply Connected Phase Spaces

$$g = \text{tr}(\Xi \otimes \Xi^\dagger) \tag{5.265}$$

defined on the group manifold of G, where $\Xi = h^{-1}dh$ is the adjoint representation of the Cartan-Maurer 1-form which takes values in the Lie algebra $\mathbf{g}$ of G. For free geodesic motion on G, the Hamiltonian operator in the Schrödinger polarization is given by the Laplace-Beltrami operator

$$\hat{H}^{(0)} = -\frac{1}{2}\nabla_g^2 \tag{5.266}$$

with respect to the metric (5.265). The invariant measure on G is taken to be the Riemannian volume form associated with (5.265), and it is possible to show that the space of trajectories over G can be made into a Kähler manifold by absorbing the Riemannian volume form into an effective action in the usual way [139].

The generic classical trajectories generated by the Hamiltonian (5.266) are straight lines in the Weyl alcove $\mathcal{A} = H_C/W(H_C)$ of the Lie group G because of the decomposition [162]

$$g_1 g_2^{-1} = vhv^{-1} \quad \forall g_1, g_2 \in G \tag{5.267}$$

where $h \in H_C$ and the elements v can be parametrized by the coset space G/H_C. The geodesic distance between g_1 and g_2 defined by (5.265) is independent of v. Then the sum over extrema in the semi-classical approximation is given by a sum over the lattice $\mathbb{Z}^r$ in r-dimensional Euclidean root space generated by the simple roots of $\mathbf{g}$. The time-evolution kernel is therefore a function of the r-dimensional vector $\boldsymbol{a} = (a_1, \ldots, a_r) \in \mathcal{A}$ and the localization formulas for it can be shown to yield the anticipated result [38, 139, 146, 161]

$$\begin{aligned}\mathcal{K}(\boldsymbol{a};T) \sim &\left(\frac{1}{2\pi i T}\right)^{\dim G/2} \sum_{\boldsymbol{k}=(k_1,\ldots,k_r)\in\mathbb{Z}^r} e^{i\sum_{j=1}^r (a_j+2\pi k_j)^2/2T} \\ &\times \prod_{\alpha>0} \frac{\alpha(\boldsymbol{a}+2\pi\boldsymbol{k})}{2\sin\frac{1}{2}\alpha(\boldsymbol{a}+2\pi\boldsymbol{k})}\end{aligned} \tag{5.268}$$

where $\alpha(\boldsymbol{a}+2\pi\boldsymbol{k}) = \sum_{j=1}^r \alpha_j(a_j+2\pi k_j)$. (5.268) is the configuration space analog of the Weyl character formula (5.28). Applying the Poisson resummation formula [96] leads to a spectral expansion of the quantum propagator which is a series over the unitary irreducible representations of G given by

$$\mathcal{K}(\boldsymbol{a};T) = \sum_{\lambda\in\mathbb{Z}^r} \dim \mathcal{R}_\lambda \, (\text{tr}_\lambda \boldsymbol{a}) \, e^{-ic(\lambda)T} \tag{5.269}$$

where

$$c(\lambda) = \sum_{i=1}^r \left((\lambda_i+\rho_i)^2 - \rho_i^2\right) \tag{5.270}$$

are the eigenvalues of the quadratic Casimir operator $\sum_a (X^a)^2$ in the representation with highest weight vector λ.

Similar considerations also apply to n-spheres $S^n \simeq SO(n+1)/SO(n)$ [96, 100] and their hyperbolic counterparts $\mathcal{H}^n \simeq SO(n-1,1)/SO(n)$ obtained by the usual analytical continuation of S^n [31]. The case of S^2 we saw was associated with the Dirac monopole, that of $S^3 \simeq SU(2)$ describes the emergence of spin, and S^4 corresponds to the BPST instanton-antiinstanton pair with 2 chiral spins [100]. Notice that these localizations also apply to the basic integrable models which are well-known to be equivalent to the group geodesic motion problems above, such as 2-dimensional Yang-Mills theory, supersymmetric quantum mechanics and Calegoro-Moser type theories. These describe the quantum mechanics of integrable models related to Hamiltonian reduction of free field theories [44, 56] and will be discussed again in Chapter 8. Note that these free theory reductions again illustrate the isomorphism between the localizable models and (trivial) harmonic oscillator type theories.

5.9 Quantization on Non-Homogeneous Phase Spaces

Thus far in this Chapter we have examined the localizable dynamical systems on both those phase spaces which are maximally symmetric and those with multi-dimensional maximally-symmetric subspaces. This exhausts all spaces with constant curvature and which are symmetric, and we outlined both the physical and group theoretical features of these dynamical systems. In this final Section of this Chapter we consider the final remaining possible class of Riemannian geometries on the phase space $\mathcal{M}$, i.e. those with a Gaussian curvature $K(x)$ which is a non-constant function of the coordinates on $\mathcal{M}$, so that $\dim \mathcal{K}(\mathcal{M}, g) = 1$. For simplicity we restrict attention again to 2-dimensional phase spaces. The geometries which admit only a single Killing vector are far more numerous than the maximally symmetric or homogeneous ones and it is here that one could hope to obtain more non-trivial applications of the localization formulas. Another nice feature of these spaces is that the corresponding Hamiltonian Poisson algebra will be abelian, so that the Hamiltonians so obtained will automatically be Cartan elements, in contrast to the previous cases where the Lie algebra $\mathcal{K}(\mathcal{M}, g)$ was non-abelian. Thus the abelian localization formulas of the last Chapter can be applied straightforwardly, and the resulting propagators will yield character formulas for the isometry group elements defined in terms of a topological field theory type path integral describing the properties of integrable quantum systems corresponding to Cartan element Hamiltonians.

Given a 1-parameter isometry group $G^{(1)}$ acting on $(\mathcal{M}, g)$, we begin by introducing a set of prefered coordinates (x'^1, x'^2) defined in terms of 2 differentiable functions χ^1 and χ^2 as described in Section 5.2, so that in these coordinates the Killing vector V has components $V'^1 = 1, V'^2 = 0$. For now, the function χ^1 is any non-constant function on $\mathcal{M}$, but we shall soon see how, once a given isometry of the dynamical system is identified, it can be fixed to suit the given problem. For a Hamiltonian system $(\mathcal{M}, \omega, H)$ which generates the flows of the given isometry in the usual way via Hamilton's equations, the defining condition (5.55) for the coordinate function χ^2 now reads

$$\{H, \chi^2\}_\omega = \mathcal{L}_V \chi^2 = 0 \quad (5.271)$$

which is assumed to hold away from the critical point set of H (i.e. the zeroes of V) almost everywhere on $\mathcal{M}$. This means that χ^2 is a conserved charge of the given dynamical system, i.e. a $G^{(1)}$-invariant function of action variables. In higher dimensions there would be many such possibilities for the conserved charges depending on the integrability properties of the system. However, in 2-dimensions this requirement fixes the action variable to be simply a functional of the Hamiltonian H,

$$\chi^2 = \mathcal{F}(H) \quad (5.272)$$

and so even in the non-homogeneous cases we see the intimate connection here between the equivariant localization formalism and the integrability of a (classical or quantum) dynamical system. We note that this only fixes the requirement (5.271) that the coordinate transformation function be constant along the integral curves of the Killing vector field V. The isometry condition (5.70) on the symplectic 2-form now only implies that, in the new x'-coordinates, $\omega_{\mu\nu}(x')$ is independent of x'^1 (just as for the metric). The Hamiltonian equations with $V'^1 = 1, V'^2 = 0$ must be solved consistently now using (5.272) and an associated symplectic structure. Notice that this construction is explicitly independent of the other coordinate transformation function χ^1 used in the construction of the prefered coordinates for V (c.f. Section 5.2).

Thus for a general metric (5.49) that admits a sole isometry, the general "admissible" Hamiltonians within the framework of equivariant localization are given by the functionals in (5.272) determined by the transformation $x \to x'$ to coordinates in which the (circle or translation) action of the corresponding Killing vector is explicit. The rich structure now arises because the integrability condition $\mathcal{L}_V \omega = 0$ for the Hamiltonian equations does not uniquely determine the symplectic 2-form ω, as it did in the case of a homogeneous symmetric geometry. The above construction could therefore be started with any *given* symplectic 2-form obeying this requirement, with the hope of being able to analyse quite general classes of Hamiltonian systems. This has the possibility of largely expanding the known examples of quantum systems where the Feynman path integral could be evaluated exactly, in contrast to the homogeneous cases where we saw that there was only a small number of few-parameter Hamiltonians which fit the localization framework. However, it has been argued that the set of Hamiltonian systems in general for which the localization criteria apply is still rather small [40, 159]. For instance, we could from the onset take ω to be the Darboux 2-form on $\mathcal{M} = \mathbb{R}^2$ and hope to

5.9 Quantization on Non-Homogeneous Phase Spaces

obtain localizable examples of 1-dimensional quantum mechanical problems with static potentials. These are defined by the Darboux Hamiltonians

$$H_{QM}(p,q) = \frac{1}{2}p^2 + U(q) \tag{5.273}$$

where $U(q)$ is some potential which is a C^∞-function of the position $q \in \mathbb{R}^1$. It was Dykstra, Lykken and Raiten [40] who first pointed out that the formalism in Chapter 4 above, which naively seems like it would imply the exact solvability of any phase space path integral, does not work for arbitrary potentials $U(q)$.

To see this, we consider a generic potential $U(q)$ which is bounded from below. By adding an irrelevant constant to the Hamiltonian (5.273) if necessary, we can assume that $U(q) \geq 0$ without loss of generality. We introduce a "harmonic" coordinate $y \in \mathbb{R}$ and polar coordinates $(r, \theta) \in \mathbb{R}^+ \times S^1$ by

$$p = r\sin\theta \quad , \quad U(q) = \frac{1}{2}y^2 = \frac{1}{2}r^2\cos^2\theta \tag{5.274}$$

where we further assume that $U(q)$ is a monotone function. In these coordinates the Hamiltonian (5.273) takes the usual integrable harmonic oscillator form $H = \frac{1}{2}r^2$, so that the function χ^2 above defines the radial coordinate r in (5.274) and $\mathcal{F}(H) = \sqrt{2H}$ in (5.272). The Hamiltonian vector field in these polar coordinates has the single non-vanishing component

$$V^\theta = -\frac{dy}{dq} \tag{5.275}$$

The metric tensor (5.49) will have in general have 3 components g_{rr}, $g_{\theta\theta}$ and $g_{\theta r}$ under the coordinate transformation (5.274), and the Killing equations (2.112) become

$$V^\theta \partial_\theta g_{\theta\theta} + 2g_{\theta\theta}\partial_\theta V^\theta = 0 \quad , \quad \partial_\theta(g_{r\theta}V^\theta) + g_{\theta\theta}\partial_r V^\theta = 0$$
$$V^\theta \partial_\theta g_{rr} + 2g_{r\theta}\partial_r V^\theta = 0 \tag{5.276}$$

The 3 equations in (5.276) can be solved in succession by integrating them and the general solution has the form

$$g_{\theta\theta} = \frac{f(r)}{(V^\theta)^2} \quad , \quad g_{r\theta} = \frac{f(r)}{V^\theta}\int_{\theta_0}^\theta d\theta' \, \partial_r\left(\frac{1}{V^{\theta'}}\right) + \frac{h(r)}{V^\theta} \quad , \quad g_{rr} = \frac{(V^\theta)^2}{f(r)}g_{r\theta}^2 + k(r) \tag{5.277}$$

where $f(r)$, $h(r)$ and $k(r)$ are arbitrary C^∞-functions that are independent of the angular coordinate θ.

Note that, as expected, there is no unique solution for the conformal factor φ in (5.49), only the requirement that it be radially symmetric (i.e. independent of θ). However, the equations (5.277) impose a much stronger

194   5. Equivariant Localization on Simply Connected Phase Spaces

requirement, this time on the actual coordinate transformation (5.274). If we impose the required single-valuedness property on the metric components above, then the requirement that $g_{r\theta}(r,\theta) = g_{r\theta}(r,\theta+2\pi)$ is equivalent to the condition

$$\frac{\partial}{\partial r} \int_0^{2\pi} \frac{d\theta}{V^\theta} = 0 \qquad (5.278)$$

or equivalently that

$$\int_0^{2\pi} d\theta \, \frac{dq}{dy} = \text{constant} \qquad (5.279)$$

However, the only solution to (5.279) is when the function $\frac{dq}{dy}$ is independent of the radial coordinate r, which from (5.274) is possible only when $y = -q$, so that $U(q) = \frac{1}{2}q^2$ and H_{QM} is the harmonic oscillator Hamiltonian. Thus, with the exception of the harmonic oscillator, equivariant localization fails for all 1-dimensional quantum mechanical Hamiltonians with static potentials which are bounded below, due to the non-existence of a single-valued metric satisfying the Lie derivative constraint in this case.

It is instructive to examine the localization formulas for the harmonic oscillator, which is considered trivial from the point of view of localization theory, to see what role is played by the degree of freedom remaining in the metric tensor which is not determined by the equivariant localization constraints. The Hamiltonian vector field (5.275) in this case is $V^\theta = 1$ which generates a global S^1-action on $\mathcal{M} = \mathbb{R}^2$ given by translations of the angle coordinate θ. Thus the localization formulas should be exact for the harmonic oscillator using any radially symmetric geometry (5.49) to make manifest the localization principle. This is certainly true of the WKB formula (4.115) which does not involve the metric tensor at all, but the more general localization formulas, such as the Niemi-Tirkkonen formula (4.130), are explicitly metric dependent through, e.g. the $\hat{A}$-genus terms, although not manifestly so. Explicitly, the non-vanishing components of the metric tensor (5.49) under the coordinate transformation (5.274) in the case at hand are

$$g_{rr} = e^{\varphi(r)} \quad , \quad g_{\theta\theta} = r^2 \, e^{\varphi(r)} \qquad (5.280)$$

and it is straightforward to work out the Riemann moment map and curvature tensor which with $V^\theta = 1$ lead to the non-vanishing components

$$(\Omega_V)_{\theta r} = -(\Omega_V)_{r\theta} = \frac{r}{2} \, e^{\varphi(r)} \left(2 + r \frac{d\varphi(r)}{dr}\right)$$
$$R_{\theta r \theta r} = -\frac{1}{2}(\Omega_V)_{\theta r} \frac{d}{dr} \log \lambda(r) \qquad (5.281)$$

where we have introduced the function

5.9 Quantization on Non-Homogeneous Phase Spaces

$$\lambda(r) = e^{-\varphi(r)} (\Omega_V)_{\theta r} / 2r \tag{5.282}$$

Substituting the above quantities into the Niemi-Tirkkonen formula (4.130) with $\omega_{r\theta} = r$ and working out the Grassmann and θ integrals there, after some algebra we find the following expression for the harmonic oscillator partition function,

$$Z_{\text{harm}}(T) \sim \frac{1}{i} \int_0^\infty dr \, \frac{d}{dr} \left(\frac{\lambda(r)}{\sin T\lambda(r)} e^{-iTr^2/2} \right) = \frac{1}{i} \lim_{r \to 0} \frac{\lambda(r)}{\sin T\lambda(r)} \tag{5.283}$$

Comparing with (5.75), we see that this result coincides with the exact result for the harmonic oscillator partition function only if the function (5.282) behaves at the origin $r = 0$ as

$$\lim_{r \to 0} \lambda(r) = \frac{1}{2} \tag{5.284}$$

which using (5.281) and (5.282) means that the phase space metric must satisfy, in addition to the radial symmetry constraint, the additional constraint

$$\lim_{r \to 0} r \frac{d}{dr} \varphi(r) = 0 \tag{5.285}$$

The requirement (5.285) means that the conformal factor $\varphi(r)$ of the Riemannian geometry must be an analytic function of r about $r = 0$, and this restriction on the general form of the metric (5.49) (i.e. on the functional properties of the conformal factor φ) ensures that the partition function is independent of this phase space metric, as it should be.

This analyticity requirement, however, simply means that the metric should be chosen so as to eliminate the singularity at the origin of the coordinate transformation to polar coordinates (r, θ) on the plane. That this transformation is singular at $p = q = 0$ is easily seen by computing the Jacobian for the change of variables (5.274) with the harmonic oscillator potential (or by noting that ω and g are degenerate at $r = 0$ in these coordinates). Since the equivariant Atiyah-Singer index which appears as the Niemi-Tirkkonen formula for the quantum mechanical path integral is an integral over characteristic classes, it is manifestly invariant under C^∞ deformations of the metric on $\mathcal{M}$. The transformation to polar coordinates is a diffeomorphism only on the punctured plane $\mathbb{R}^2 - \{0\}$, which destroys the manifest topological invariance of the partition function (at $r = 0$ anyway). For $\lambda(0) \neq 0$, the metric tensor describes a conical geometry [102] for which the parameter $\lambda(0)$ represents the tip angle of the cone. This example shows that for the localization to work the choice of metric tensor is not completely arbitrary, since it has to respect the topology of the phase space on which the problem is defined. As discussed in [40] and [159], this appears to be a general feature of the generalized localization formalisms, in that they detect explicitly the topology of the phase space and this can be used to eliminate some of the arbitrariness of the metric (5.49). Indeed, in the set of prefered coordinates for V it has no zeroes and so the critical points are "absorbed" into the symplectic 2-form ω and in general also the metric g. Thus the prefered coordinate transformation for V is a diffeomorphism only on $\mathcal{M} - \mathcal{M}_V$ in general. Nonetheless, this simple example illustrates that quite general, non-homogeneous geometries can still be used to carry out the equivariant localization framework for path integrals and describe the equivariant Hamiltonian systems which lead to topological quantum theories in terms of the generic phase space geometry.

Although the above arguments appear to have eliminated a large number of interesting physical problems, owing to the fact that their Hamiltonian vector fields do not generate well-defined orbits on the θ-circle, it is still possible that quantum mechanical Hamiltonians with *unbounded* static potentials could fit the localization framework. Such dynamical systems indeed do represent a rather large class of physically interesting quantum systems. The first such attempt was carried out by Dykstra, Lykken and Raiten [40] who showed that the Niemi-Tirkkonen localization formula for such models can be reduced to a relatively simple contour integral. For example, consider the equivariant localization formalism applied to the 1-dimensional hydrogen atom Hamiltonian [95]

$$H_h(p,q) = \frac{1}{2}p^2 - \frac{1}{|q|} \tag{5.286}$$

The eigenvalues of the associated quantum Hamiltonian form a discrete spectrum with energies

$$E_n = -1/2n^2 \quad , \quad n = 1, 2, \ldots \tag{5.287}$$

which resembles the bound state spectrum of the more familiar 3-dimensional hydrogen atom [101]. What is even more interesting about this dynamical system is that the classical bound state orbits all coalesce at the phase space points $q = 0$, $p = \pm\infty$ on $\mathbb{R}^2$, so that a localization onto classical trajectories (like the WKB formula) is highly unsuitable for this quantum mechanical problem. This problem could therefore provide an example wherein although the standard WKB approximation cannot be employed, the more general localization formulas, like the Niemi-Tirkkonen formula, which seem to have no constraints on them other than the usual isometry restrictions on the phase space $\mathcal{M}$, could prove of use in describing the exact quantum theory of the dynamical system.

The key to evaluating the localization formulas for the Darboux Hamiltonian (5.286) is the transformation to the hyperbolic coordinates (r, τ) with $-\infty \leq r, \tau \leq \infty$,

$$p = |r| \sinh \tau \quad , \quad q = 2/r|r| \cosh^2 \tau \tag{5.288}$$

so that the Hamiltonian is again $H_h = -\frac{1}{2}r^2$ and the Hamiltonian vector field has the single non-vanishing component

5.9 Quantization on Non-Homogeneous Phase Spaces

$$V^\tau = -\frac{1}{4}r^3 \cosh^3 \tau \qquad (5.289)$$

Now the Killing equations have precisely the same form as in (5.276), with (r,θ) replaced by (r,τ) there, and thus the general solutions for the metric tensor have precisely the same form as in (5.277). However, because of the non-compact range of the hyperbolic coordinate τ in the case at hand, we do not encounter a single-valuedness problem in defining the components $g_{r\tau}$ as C^∞ functions on $\mathbb{R}^2$ and from (5.277) and (5.289) we find that it is given explicitly by the perfectly well-defined function

$$g_{r\tau} = \frac{12f(r)}{r^4 V^\tau}\left(\frac{\sinh \tau}{2\cosh^2 \tau} + \frac{1}{2}\arctan(\sinh \tau)\right) + \frac{h(r)}{V^\tau} \qquad (5.290)$$

In the context of our isometry analysis above, we again choose the coordinate transformation function χ^2 so that $\mathcal{F}(H) = \sqrt{-2H}$ in (5.272). The other coordinate function $\chi^1 \equiv x''^1$ is determined by noting that the above (r,τ) coordinates are the x'-coordinates in (5.43) from which we wish to define the prefered set of x''-coordinates for the Hamiltonian vector field V. There we identify $(x'^1, x'^2) = (\tau, r)$ according to that prescription. Carrying out the explicit integration over $x'^1 = \tau$ using (5.289), and then substituting in the transformation (5.288) back to the original Darboux coordinates, after some algebra we find

$$\chi^1(p,q) = -\left|\frac{2}{|q|} - p^2\right|^{-3/2}\left[p|q|\left|\frac{2}{|q|} - p^2\right|^{1/2} + 2\arctan\left(\frac{p}{\left|\frac{2}{|q|} - p^2\right|^{1/2}}\right)\right] \qquad (5.291)$$

Thus the Hamiltonian (5.286) is associated with the phase space metric tensor (5.49) which is invariant under the translations $\chi^1 \to \chi^1 + a_0$ of the coordinate (5.291). The analysis above shows explicitly that the phase space indeed does admit a globally well-defined metric which is translation invariant in the variable (5.291). It is also possible to evaluate the Niemi-Tirkkonen localization formula for this quantum problem in a similar fashion as the harmonic oscillator example above. We shall not go into this computation here, but refer to [40] for the technical details. The only other point we wish to make here is that one needs to impose again certain regularity requirements on the conformal factor of the metric (5.49). These conditions are far more complicated than above because of the more complicated form of the translation function (5.291), but they are again associated with the cancelling of the coordinate singularities in (5.288) which make the equivariant Atiyah-Singer index in (4.130) an explicitly metric dependent quantity. With these appropriate geometric restrictions it is enough to argue that the quantum partition function for the Darboux Hamiltonian (5.286) has the form [40]

$$Z_h(T) \sim \sum_{n=1}^{\infty} e^{iT/2n^2} \qquad (5.292)$$

which from (5.287) we see is indeed the exact spectral propagator for the 1-dimensional hydrogen atom [95].

This example shows that more complicated quantum systems can be studied within the equivariant localization framework on a simply connected phase space, but only for those phase spaces which admit Riemannian geometries which have complicated and unusual symmetries, such as translations in the coordinate (5.291) above. Thus besides having to find a metric tensor appropriate to the geometry and topology of a phase space, there is the further general problem as to whether or not a geometry can in fact possess the required symmetry (e.g. for Hamiltonians associated with bounded potentials, there is no such geometry). It is not expected, of course, that any Hamiltonian will have an exactly solvable path integral, and from the point of view of this Chapter the cases where the Feynman path integral fails to be effectively computable within the framework of equivariant localization will be those cases where a required symmetry of the phase space geometry does not lead to a globally well-defined metric tensor appropriate to the given topology. Nonetheless, the analysis in [40] for the 1-dimensional hydrogen atom is a highly non-trivial success of the equivariant localization formulas for path integrals which goes beyond the range of the standard WKB method.

We conclude this Chapter by showing that it is possible to relate the path integrals for generic dynamical systems on non-homogenous phase spaces which fall into the framework of loop space equivariant localization to character formulas for the associated 1-parameter isometry groups $G^{(1)}$ [159]. For this, we need to introduce a formalism for constructing coherent states associated with non-transitive group actions on manifolds [89, 159]. We consider the isothermal metric (5.49) in the prefered x'-coordinates for a Hamiltonian vector field V on $\mathcal{M}$. Using these coordinates, we define the complex coordinates $z = x'^2 \, e^{ix'^1}$, in analogy with the case where V defines a rotationally symmetric geometry (as for the harmonic oscillator). Let $f(z\bar{z})$ be a $G^{(1)}$-invariant analytic solution of the ordinary differential equation

$$\frac{d}{d(z\bar{z})} z\bar{z} \frac{d}{d(z\bar{z})} \log f(z\bar{z}) = \frac{1}{2} \, e^{\varphi(z\bar{z})} \tag{5.293}$$

For the symplectic 2-form of the phase space, we take the $G^{(1)}$-invariant volume form associated with $(\mathcal{M}, g)$,

$$\omega^{(\varphi)} = i \frac{d}{d(z\bar{z})} z\bar{z} \frac{d}{d(z\bar{z})} \log f(z\bar{z}) \, dz \wedge d\bar{z} \tag{5.294}$$

whose associated symplectic potential is

$$\theta^{(\varphi)} = \frac{i}{2} \frac{d}{d(z\bar{z})} \log f(z\bar{z}) \, (\bar{z} dz - z d\bar{z}) \tag{5.295}$$

This definition turns the phase space into a non-homogeneous Kähler manifold with Kähler potential

5.9 Quantization on Non-Homogeneous Phase Spaces

$$F^{(\varphi)}(z, \bar{z}) = \log f(z\bar{z}) \tag{5.296}$$

such that $\omega^{(\varphi)}$ determines the first Chern class of the usual symplectic line bundle $L^{(\varphi)} \to \mathcal{M}$.

Let N_φ, $0 < N_\varphi \le \infty$, be the integer such that the function $f(z\bar{z})$ admits the Taylor series expansion

$$f(z\bar{z}) = \sum_{n=0}^{N_\varphi} (z\bar{z})^n f_n \tag{5.297}$$

and let $\rho(z\bar{z})$ be a $G^{(1)}$-invariant integrable function whose moments are

$$\int_0^P d(z\bar{z})\,(z\bar{z})^n \rho(z\bar{z}) = \frac{1}{f_n}, \qquad 0 \le n \le N_\varphi \tag{5.298}$$

where P is a real number with $0 < P \le \infty$. Let $\hat{a}^\dagger$ and $\hat{a}$ be bosonic creation and annihilation operators on some representation space of the isometry group (as in Section 5.3 above), and let $|n\rangle$, $n \in \mathbb{Z}^+$, be the complete system of orthonormal eigenstates of the corresponding number operator, $\hat{a}^\dagger \hat{a} |n\rangle = n |n\rangle$. The desired coherent states are then defined as

$$|z\rangle = \sum_{n=0}^{N_\varphi} \sqrt{f_n}\, z^n |n\rangle \tag{5.299}$$

The states (5.299) have the normalization

$$\langle z|z\rangle = f(z\bar{z}) = e^{F^{(\varphi)}(z,\bar{z})} \tag{5.300}$$

and they obey a completeness relation analogous to (5.142) in the isometry invariant measure

$$d\mu^{(\varphi)}(z, \bar{z}) = \frac{i}{2\pi} f(z\bar{z}) \rho(z\bar{z}) \Theta(P - z\bar{z})\, dz \wedge d\bar{z} \tag{5.301}$$

where $\Theta(x)$ denotes the step function for $x \in \mathbb{R}$. The completeness of the coherent states (5.299) follows from a calculation analogous to that in (5.85) using the definitions (5.297)–(5.299) above.

Notice that for the functional values $f(z\bar{z}) = e^{z\bar{z}}$, $(1 + z\bar{z})^{2j}$ and $(1 - z\bar{z})^{-2k}$, (5.299) reduces to, respectively, the Heisenberg-Weyl group, spin-j $SU(2)$ and level-k $SU(1,1)$ coherent states that we described earlier. Moreover, in that case we consistently find, respectively, the weight functions $\rho(z\bar{z}) = e^{-z\bar{z}}$ with $P = \infty$, $\rho(z\bar{z}) = (2j+1)(1 + z\bar{z})^{-2(j+1)}$ with $P = \infty$, and $\rho(z\bar{z}) = (2k-1)(1 - z\bar{z})^{2(k-1)}$ with $P = 1$. This is anticipated from (5.293), as then the isothermal metrics in (5.49) correspond to the standard maximally symmetric Kähler geometries. Here the isometry group acts on the states (5.299) as $h^{(\tau)}|z\rangle = |\,\mathrm{e}^{i\tau}z\rangle$, $h^{(\tau)} \in \mathcal{I}(\mathcal{M},g) \equiv G^{(1)}$, $\tau \in \mathbb{R}^1$, which ensures that a Hamiltonian exists (as we shall see explicitly below) such that a time-evolved coherent state remains coherent in this sense, regardless of the choice of ρ [89]. The (holomorphic) dependence of the non-normalized coherent state vectors $|z\rangle$ on only the single complex variable z is, as usual, what makes them amenable to the study of the isometry situation at hand. Notice also that the metric tensor (5.49) and canonical 1-form (5.295) can as usual be represented in the standard coherent state forms (5.92) and (5.91), respectively.

Considering as usual the coherent state matrix elements (5.95) with respect to (5.299), using (5.295) and (5.301) we can construct the usual coherent state path integral

$$Z^{(\varphi)}(T|\mathcal{F}(H)) = \int_{L\mathcal{M}} \prod_{t\in[0,T]} d\mu^{(\varphi)}(z(t),\bar{z}(t))$$

$$\times \exp\left\{i\int_0^T dt\, \left[\frac{1}{2}\frac{d}{d(z\bar{z})}\log f(z\bar{z})\,(z\dot{\bar{z}} - \bar{z}\dot{z}) - \mathcal{F}(H)\right]\right\}$$
(5.302)

where we have again allowed for a possible functional $\mathcal{F}(H)$ of the isometry generator H. The observable $H(z,\bar{z})$ in (5.302) can be found by substituting (5.294), written back in the x'-coordinates using the standard radial form for $z = x'^2\,\mathrm{e}^{ix'^1}$ given in (5.280), and $V'^1 = a_0$, $V'^2 = 0$ into the Hamiltonian equations. Thus the equivariant localization constraints in these cases determine H in terms of the phase space metric as

$$H^{(\varphi)}(z,\bar{z}) = a_0 \cdot z\bar{z}\frac{d}{d(z\bar{z})}\log f(z\bar{z}) + C_0 = a_0 \cdot (\!(z|\hat{a}^\dagger\hat{a}|z)\!) + C_0 = i_V\theta^{(\varphi)}$$
(5.303)

where the function $f(z\bar{z})$ is related to the metric (5.49) by (5.293). Notice that (5.303) reduces to the usual harmonic oscillator height functions in the maximally symmetric cases of Sections 5.3, 5.5 and 5.6 above. Thus (5.303) can be considered as the general localizable Hamiltonian valid for *any* phase space Riemannian geometry, be it maximally symmetric or otherwise (the same is true, of course, for the coherent state path integral (5.302)). This is to be expected, because the localizable Hamiltonian functions in the case of a homogeneous symmetry are simply displaced harmonic oscillators, and these oscillator Hamiltonians correspond to the rotation generators of the isometry groups, i.e. translations in $\arg(z) = x'^1$ (this also agrees with the usual integrability arguments). In fact, (5.303) shows explicitly that the function H is essentially just a harmonic oscillator Hamiltonian written in terms of some generalized phase space geometry.

The main difference in the present context between the homogeneous and non-homogeneous cases lies in the path integral (5.302) itself. In the

5.9 Quantization on Non-Homogeneous Phase Spaces

former case the coherent state measure $d\mu^{(\varphi)}(z,\bar{z})$ which must be used in the Feynman measure in (5.302) coincides with the volume form (5.294), because as mentioned earlier if the isometry group acts transitively on the Riemannian manifold $(\mathcal{M},g)$ then there is a unique left-invariant measure (i.e. a unique solution to (5.70)) and so $d\mu^{(\varphi)} = \omega^{(\varphi)}$ yields the standard Liouville measure on the loop space $L\mathcal{M}$. In the latter case $d\mu^{(\varphi)} \neq \omega^{(\varphi)}$, and (5.302) is not in the canonical form (4.145) for the quantum partition function associated with the loop space symplectic geometry. Nonetheless, by a suitable modification of the loop space supersymmetry associated with the dynamical system by noting that the coherent state measure in (5.301) is invariant under the action of the isometry group on $\mathcal{M}$, it is still possible to derive appropriate versions of the standard localization formulas with the obvious replacements corresponding to this change of integration measure. Of course, we can alternatively follow the analysis of the former part of this Section and use the standard Liouville path integral measure, but then we lose the formal analogies with the Duistermaat-Heckman theorem and its generalizations. It is essentially this non-uniqueness of an invariant symplectic 2-form in the case of non-transitive isometry group actions which leads to numerous possibilities for the localizable Hamiltonian systems defined on such spaces, in marked contrast to the homogeneous cases where everything was uniquely fixed. If one consistently makes the "natural" choice for ω as the Kähler 2-form (5.294), then indeed the only admissible Hamiltonian functions H are generalized harmonic oscillators.

6. Equivariant Localization on Multiply Connected Phase Spaces: Applications to Homology and Modular Representations

In the last Chapter we deduced the general features of the localization formalism on a simply-connected symplectic manifold. We found general forms for the Hamiltonian functions in terms of the underlying phase space Riemannian geometry which is required for their Feynman path integrals to manifestly localize. This feature is quite interesting from the point of view that, as the quantum theory is always *ab initio* metric-independent, this analysis probes the role that the geometry and topology plays towards the understanding of quantum integrability. For instance, we saw that the classical trajectories of a harmonic oscillator must be embedded into a rotationally-invariant geometry and that as such its orbits were always circular trajectories. For more complicated systems these quantum geometries are less familiar and endow the phase space with unusual Riemannian structures (i.e. complicated forms of the localization supersymmetries). In any case, all the localizable Hamiltonians were essentially harmonic oscillators (e.g. the height function for a spherical phase space geometry) in some form or another, and their quantum partition functions could be represented naturally using coherent state formalisms associated with the Poisson-Lie group actions of the isometry groups of the phase space. In the non-homogeneous cases we saw, in particular, that to investigate equivariant localization in general one needs to determine if a Riemannian geometry can possess certain symmetries imposed by some rather ad-hoc restrictions from the dynamical system. In practice, the introduction of such a definite geometry into the problem is highly non-trivial, although we saw that it was possible in some non-trivial examples. These results also impose restrictions on the classes of topological quantum field theories and supersymmetric models which fall into the framework of these geometric localization principles, as we shall discuss at greater length in Chapter 8.

In this Chapter we shall extend the analysis of Chapter 5 to the case when the phase space $\mathcal{M}$ is multiply-connected [153]. We shall primarily focus on the case where $\mathcal{M}$ is a compact Riemann surface of genus $h \geq 1$, again because of the wealth of mathematical characterizations that are available for such spaces. We shall explore how the localization formalism differs from that on a simply-connected manifold. Recall that much of the formalism developed in Chapter 4, in particular that of Section 4.10, relied quite heavily on this topological restriction. We shall see that now the topological quantum field theories that appear also describe the non-trivial first homology group of the Riemann surface, and that it is completely independent of the geometrical structures that are used to carry out the equivariant localization on $\mathcal{M}$, such as the conformal factors and the modular parameters. This is typically what a topological field theory should do (i.e. have only global features), and therefore the equivariant Hamiltonian systems that one obtains in these cases are nice examples of how the localization formalism is especially suited to describe the characteristics of topological quantum field theories on spaces with much larger topological degrees of freedom. Again the common feature will be the description of the quantum dynamics using a coherent state formalism, this time associated with a non-symmetric spin system and some ideas from geometric quantization [22, 172]. We shall in addition see that the coherent states span a multi- but finite-dimensional Hilbert space in which the wavefunctions carry a non-trivial representation of the discrete first homology group of the phase space. We shall verify the localization formulas of Chapter 4 in a slightly modified setting, pointing out the important subtleties that arise in trying to apply them directly on a multiply-connected phase space.

Although we shall attempt to give a quite general description of the localizable dynamics on such spaces, most of our analysis will only be carried out explicitly for genus 1, i.e. on the 2-torus $T^2 = S^1 \times S^1$. In particular, we shall view the torus in a way best suited to describe its complex algebraic geometry, i.e. in the parallelogram representation of Section 3.5, so that we can examine the topological properties of the quantum theory we find and get a good idea of the features of the localization formalism on multiply-connected spaces in general. Another more explicit way to view the torus is by embedding it in $\mathbb{R}^3$ by revolving the circle $(y-a)^2 + x^2 = b^2$ on the xy-plane around the x-axis, where $0 < b < a$, i.e. embedding T^2 in 3-space by $x = b\sin\phi_1$, $y = (a + b\cos\phi_1)\sin\phi_2$ and $z = (a + b\cos\phi_1)\cos\phi_2$. The induced metric on the surface from the flat Euclidean metric of $\mathbb{R}^3$ is then $b^2 d\phi_1 \otimes d\phi_1 + (a + b\cos\phi_1)^2 d\phi_2 \otimes d\phi_2$, and the modular parameter $\tau \in \mathbb{C}^+$ of the parallelogram representation of T^2 is (c.f. Section 3.5)

$$\tau = ib/\sqrt{a^2 - b^2} \tag{6.1}$$

If we now introduce the coordinate

$$\theta = \theta(\phi_1) = \int_0^{\phi_1} d\phi_1' \, \frac{b}{a + b\cos\phi_1'} \tag{6.2}$$

then it is straightforward to verify that $w = \phi_2 + i\theta$ is an isothermal coordinate for the induced metric on T^2 for which its isothermal form is $\rho(\theta)(d\phi_2 \otimes d\phi_2 + d\theta \otimes d\theta)$. This defines a complex structure on T^2. Since this metric is invariant under translations in ϕ_2, we could heuristically follow the analysis of Section

5.9 to deduce that one class of localizable Hamiltonians are those which are functions only of ϕ_1. In order that these Hamiltonians be well-defined globally on $T^2 = S^1 \times S^1$, we require in addition that these be periodic functions of ϕ_1. As we shall soon see, this is consistent with the general localizable dynamical systems we shall find. Topological invariance of the associated quantum theory in this context would be something like the invariance of it under certain rescalings of the modular parameter (6.1), i.e. under rescalings of the radius parameters a or b corresponding to a uniform 'shift' in the local geometry of T^2. A topological quantum theory shouldn't detect such shifts which aren't considered as ones modifying the topological properties of the torus. In other words, the topological quantum theory should be independent of the phase space complex structure. We shall see this in a more algebraic form later on in this Chapter.

6.1 Isometry Groups of Multiply Connected Spaces

To describe the isometries of a generic path connected, multiply-connected Riemannian manifold $(\mathcal{M}, g)$, we lift these isometries up into what is known as the universal covering space of the manifold. The multiple-connectivity of $\mathcal{M}$ means that it has loops in it which cannot be contracted to a point (i.e. $\mathcal{M}$ has 'holes' in it). This is measured algebraically by what is called the fundamental homotopy group $\pi_1(\mathcal{M})$ of $\mathcal{M}$, a similar but rather different mathematical entity as the first homology group $H_1(\mathcal{M}; \mathbb{Z})$. Roughly speaking, this group is defined as follows. We fix a basepoint $x_0 \in \mathcal{M}$ and consider the loop space of periodic maps $\sigma : [0, 1] \to \mathcal{M}$ with $\sigma(0) = \sigma(1) = x_0$. For any 2 loops σ and τ based at x_0 in this way, the product loop $\sigma \cdot \tau$ is defined to be the loop obtained by first going around σ, and then going around τ. The set $\pi_1(\mathcal{M})$ is the space of all equivalence classes $[\sigma]$ of loops, where 2 loops are equivalent if and only if they are homotopic to each other, i.e. there exists a continuous deformation between the loops. It can be shown that the above multiplication of loops then gives a well defined multiplication in $\pi_1(\mathcal{M})$ and turns it into a group with identity the homotopy class of the trivial loop $[0, 1] \to x_0$ and with inverse defined by reversing the orientation of a loop. In general, this group is non-abelian and discrete, and it is related to the first homology group $H_1(\mathcal{M}; \mathbb{Z})$ as follows. Let $[G, G]$ denote the commutator subgroup of any group G, i.e. $[G, G]$ is the normal subgroup of G generated by the products $ghg^{-1}h^{-1}$, $g, h \in G$. The homology group $H_1(\mathcal{M}; \mathbb{Z})$ is then the abelianization of the fundamental group,

$$H_1(\mathcal{M}; \mathbb{Z}) = \pi_1(\mathcal{M})_{\text{ab}} \equiv \pi_1(\mathcal{M}) / [\pi_1(\mathcal{M}), \pi_1(\mathcal{M})] \tag{6.3}$$

If $\pi_1(\mathcal{M})$ is itself abelian, then the homology and homotopy of $\mathcal{M}$ coincide. We refer to [98] for a more complete exposition of homotopy theory and how homology, in the sense of (6.3), is the natural approximation of homotopy.

206   6. Equivariant Localization on Multiply Connected Phase Spaces

The universal covering space of $\mathcal{M}$ is now defined as the smallest simply connected manifold $\tilde{\mathcal{M}}$ covering $\mathcal{M}$. By a covering space we mean that there is a surjective continuous projection map $\pi : \tilde{\mathcal{M}} \to \mathcal{M}$ such that its restriction to any neighbourhood of $\mathcal{M}$ defines a local diffeomorphism. This means that locally on $\mathcal{M}$ we can lift any quantity defined on it to its universal cover and study it on the simply connected space $\tilde{\mathcal{M}}$. The manifold $\mathcal{M}$ and its universal covering space $\tilde{\mathcal{M}}$ are related by the homeomorphism

$$\mathcal{M} \simeq \tilde{\mathcal{M}}/\pi_1(\mathcal{M}) \tag{6.4}$$

where the fundamental group acts freely on $\tilde{\mathcal{M}}$ through what are known as deck or covering transformations [98], i.e. the diffeomorphisms $\sigma : \tilde{\mathcal{M}} \to \tilde{\mathcal{M}}$ such that $\pi(\sigma(x)) = \pi(x)$, $\forall x \in \tilde{\mathcal{M}}$. Thus in this setting, the universal covering space is a principal fiber bundle where the total space $\tilde{\mathcal{M}}$ is locally regarded as the space of all pairs $(x, [C_x])$, where C_x is a curve in $\mathcal{M}$ from x_0 to x and $[C_x]$ is its homotopy class[1]. The structure group of the bundle is $\pi_1(\mathcal{M})$ and the bundle projection $\tilde{\mathcal{M}} \xrightarrow{\pi} \mathcal{M}$ takes a homotopy class of curves to their endpoint, $\pi : [C_x] \to x$. Clearly, $\mathcal{M}$ is its own universal cover if it is simply connected, i.e. $\pi_1(\mathcal{M}) = 0$. We shall see some examples in due course.

Consider now a Riemannian metric g defined on $\mathcal{M}$, and let $\pi^* g$ be its inverse image under the canonical bundle projection of $\tilde{\mathcal{M}}$ onto $\mathcal{M}$. Then $(\tilde{\mathcal{M}}, \pi^* g)$ is a simply-connected Riemannian manifold, and from the analysis of the last Chapter we are well acquainted with the structure of its isometry groups. It is possible to show [94], from the principal fiber bundle interpretation (6.4) above, that to every isometry $h \in \mathcal{I}(\mathcal{M}, g)$ one can associate an isometry $\tilde{h} \in \mathcal{I}(\tilde{\mathcal{M}}, \pi^* g)$ which is compatible with the universal covering projection in the sense that

$$\pi \circ \tilde{h} = h \circ \pi \tag{6.5}$$

To prove this one needs to show that the lifting $\tilde{h} \equiv \pi^* h$ gives a diffeomorphism of $\tilde{\mathcal{M}}$ which is a well-defined function on the homotopy classes of curves used for the definition of $\tilde{\mathcal{M}}$ [94]. Thus the isometries of the Riemannian manifold $(\mathcal{M}, g)$ lift to isometries of the simply connected space $(\tilde{\mathcal{M}}, \pi^* g)$ of which we have a complete description from the last Chapter. It should be kept in mind though that there can be global obstructions from the homotopy of $\mathcal{M}$ to extending an isometry of $\tilde{\mathcal{M}}$ projected locally down onto $\mathcal{M}$ by the bundle projection π. We shall see how this works in the next Section.

---
[1] Here a homotopy class of curves $[C_x]$ can be identified with an element of $\pi_1(\mathcal{M})$ by choosing another basepoint x'_0 and a grid of standard paths from x'_0 to any other point in $\mathcal{M}$. Then the associated homotopy class is represented by the loop $[x'_0, x_0] \cup C_x \cup [x, x'_0]$.

6.2 Equivariant Hamiltonian Systems in Genus One

Our prototypical model for a multiply-connected symplectic manifold will be the 2-torus $T^2 = S^1 \times S^1$ which we first studied in Section 3.5. Notice that the circle is multiply-connected with $\pi_1(S^1) = \mathbb{Z}$ with the integers labelling the number of times that a map $\sigma : S^1 \to S^1$ 'winds' around the circle, i.e. to each homotopy class $[\sigma] \in \pi_1(S^1)$ we can associated an integer which we call the winding number of the loop σ (where a change of sign signifies a change in the direction of traversing the loop). We can describe the homotopy of the torus by introducing 2 loops a and b, both fixed at the same basepoint on $S^1 \times S^1$, with a encircling once the inner circle of the torus (i.e. $a : S^1 \to (\phi_1, 0) \in S^1 \times S^1$) and b encircling once the outer circle (i.e. $b : S^1 \to (0, \phi_2) \in S^1 \times S^1$). Since clearly any other loop in T^2 is homotopic to some combination of the loops a and b, it follows that they generate the fundamental group $\pi_1(T^2)$ of the torus, and furthermore they obey the relation

$$aba^{-1}b^{-1} = 1 \qquad (6.6)$$

which is easily seen by simply tracing the loop product in (6.6) around $S^1 \times S^1$. (6.6) means that $\pi_1(T^2)$ is abelian and therefore coincides with the first homology group (3.81). Thus the loops a and b defined above are also generators of the first homology group $H_1(\Sigma^1; \mathbb{Z})$, and they will henceforth be refered to as the canonical homology cycles of the torus. Note that any homology cycle in Σ^1 which defines the homology class of a (respectively b) can be labelled by the ϕ_1 angle coordinates (respectively ϕ_2). Thus any homology class of a genus 1 compact Riemann surface is labelled by a pair of integers (n, m) which represents the winding numbers around the canonical homology cycles a and b.

Recall from Section 3.5 the description of the torus as a parallelogram with its opposite edges identified in the plane, and with modular parameter $\tau \in \mathbb{C}^+$ which labels the inequivalent complex analytic structures on the torus (or equivalently the conformal equivalence classes of metrics on T^2) [111, 145]. This means that it can be represented as the quotient space

$$\Sigma^1 = \mathbb{C}/(\mathbb{Z} \oplus \tau \mathbb{Z}) \qquad (6.7)$$

where the quotient is by the free bi-holomorphic action of the lattice group $\mathbb{Z} \oplus \tau \mathbb{Z}$ on the simply-connected complex plane $\mathbb{C}$. In other words, the lattice group is the discrete automorphism group of the complex plane and it acts on $\mathbb{C}$ by the translations[2]

$$z \to z + 2\pi(n + \tau m) \quad , \quad \bar{z} \to \bar{z} + 2\pi(n + \bar{\tau} m) \quad ; \quad n, m \in \mathbb{Z} \qquad (6.8)$$

under which the canonical bundle projection $\mathbb{C} \xrightarrow{\pi} \Sigma^1$ is invariant. That the plane is the universal cover of the torus is easily seen by observing that the

---

[2] For an exposition of the various equivalent ways, such as above, of describing compact Riemann surfaces in different geometric forms, see [110].

real line $\mathbb{R}^1$ is the universal cover of the circle S^1 with the bundle projection $\pi(x) = e^{2\pi i x}$ for $x \in \mathbb{R}^1$.

With the identification (6.7), we can now consider the most general Euclidean signature metric on Σ^1. From our discussion in Section 5.2, we know that the most general metric on $\mathbb{C}$ can be written in the global isothermal form (5.49). The covering projection in (6.7) in this way induces the most general metric on the torus, which can therefore be written in terms of a flat Kähler metric as

$$g_\tau = \frac{e^{\varphi(z,\bar{z})}}{\operatorname{Im}\tau} dz \otimes d\bar{z} \tag{6.9}$$

or in terms of the angle coordinates $(\phi_1, \phi_2) \in S^1 \times S^1$

$$[g_{\phi_\mu \phi_\nu}] = \frac{e^{\varphi(\phi_1,\phi_2)}}{\operatorname{Im}\tau} \begin{pmatrix} 1 & \operatorname{Re}\tau \\ \operatorname{Re}\tau & |\tau|^2 \end{pmatrix} \tag{6.10}$$

The complex structure on Σ^1 is now defined by the complex coordinates $z = \phi_1 + \tau \phi_2$, $\bar{z} = \phi_1 + \bar{\tau}\phi_2$ which are therefore considered invariant under the transformations (6.8). The conformal factor $\varphi(z,\bar{z})$ is now a globally defined real-valued function on Σ^1 (i.e. invariant under the translations (6.8)), and the normalization in (6.9) is chosen for simplicity so that the associated metric volume of the torus

$$\operatorname{vol}_{g_\tau}(\Sigma^1) = \int_{\Sigma^1} d^2\phi \sqrt{\det g_\tau} = \int_{\Sigma^1} d^2\phi \; e^{\varphi(\phi_1,\phi_2)} \equiv (2\pi)^2 v \tag{6.11}$$

is finite and independent of the complex structure of Σ^1 with $v \in \mathbb{R}$ a fixed volume parameter of the torus. The metric (6.9) is further constrained by its Gaussian curvature scalar

$$K(g_\tau) = -\frac{1}{2} \operatorname{Im}(\tau) \, e^{-\varphi} \nabla_\tau^2 \varphi \tag{6.12}$$

which by the Gauss-Bonnet-Chern theorem (5.54) for genus $h = 1$ must obey

$$\int_{\Sigma^1} d^2\phi \; \nabla_\tau^2 \varphi(\phi_1, \phi_2) = 0 \tag{6.13}$$

where $\nabla_\tau^2 = \partial\bar{\partial}$ is the scalar Laplacian

$$\nabla_\tau^2 = \partial_{\phi_1}^2 + |\tau|^{-2}\partial_{\phi_2}^2 + 2\operatorname{Re}(\tau)|\tau|^{-2}\partial_{\phi_1}\partial_{\phi_2} \tag{6.14}$$

associated with the Kähler structure in (6.9).

Given this general geometric structure of the 2-torus, following the analysis of the last Chapter we would like to find the most general Hamiltonian system on it which obeys the localization criteria. First of all, the condition that the Hamiltonian H generates a globally integrable isometry of the metric (6.9) implies that the associated Hamiltonian vector fields $V^\mu(x)$ must be

6.2 Equivariant Hamiltonian Systems in Genus One

single-valued functions under the windings (6.8) around the non-trivial homology cycles of Σ^1. This means that these functions must admit convergent 2-dimensional harmonic mode expansions

$$V^\mu(\phi_1, \phi_2) = \sum_{n,m=-\infty}^{\infty} V^\mu_{n,m} e^{i(n\phi_1 + m\phi_2)} \tag{6.15}$$

In other words, the components of V must be C^∞-functions which admit a 2-dimensional Fourier series plane wave expansion (6.15) appropriate to globally-defined periodic functions on $S^1 \times S^1$. As we shall now demonstrate, these topological restrictions from the underlying phase space severely limit the possible Hamiltonian systems to which the equivariant localization constraints apply.

From (2.112) it follows that the Killing equations for the metric (6.10) are

$$2\partial_{\phi_1} V^1 + 2 \operatorname{Re}(\tau)\partial_{\phi_1} V^2 + V^\mu \partial_{\phi_\mu} \varphi = 0$$
$$2 \operatorname{Re}(\tau)\partial_{\phi_2} V^1 + 2|\tau|^2 \partial_{\phi_2} V^2 + |\tau|^2 V^\mu \partial_{\phi_\mu} \varphi = 0 \tag{6.16}$$
$$\partial_{\phi_2} V^1 + \operatorname{Re}(\tau)(\partial_{\phi_2} V^2 + \partial_{\phi_1} V^1) + |\tau|^2 \partial_{\phi_1} V^2 + \operatorname{Re}(\tau) V^\mu \partial_{\phi_\mu} \varphi = 0$$

Substituting in the harmonic expansions (6.15) and using the completeness of the plane waves there to equate the various components of the expansions in (6.16), we find after some algebra that (6.16) generates 2 coupled equations for the Fourier components of the Hamiltonian vector field,

$$(|\tau|^2 n - \operatorname{Re}(\tau)m)V^1_{n,m} = |\tau|^2 (m - \operatorname{Re}(\tau)n)V^2_{n,m}$$
$$(m - \operatorname{Re}(\tau)n)V^1_{n,m} = \left[(\operatorname{Re}(\tau)^2 - \operatorname{Im}(\tau)^2)n - \operatorname{Re}(\tau)m\right] V^2_{n,m} \tag{6.17}$$

which hold for all integers n and m. It is straightforward to show from the coupled equations (6.17) that for $\tau \in \mathbb{C}^+$, $V^1_{n,m} = V^2_{n,m} = 0$ unless $n = m = 0$. Thus the only non-vanishing components of the harmonic expansions (6.15) are the constant modes,

$$V^\mu_{\Sigma^1}(x) = V^\mu_0 \tag{6.18}$$

and the only Killing vectors of the metric (6.9) are the generators of translations (by $V^\mu_0 \in \mathbb{R}$) along the 2 independent homology cycles of Σ^1. Notice that this result is completely independent of the structure of the conformal factor φ in (6.9), and it simply means that although the torus inherits *locally* 3 isometries from the maximally symmetric plane, i.e. local rotations and translations, only the 2 associated translations on Σ^1 are *global* isometries. The independence of this result on the conformal factor is not too surprising, since this just reflects the fact that given any metric on a compact phase space we can make it invariant under a compact group action by averaging it over the group in its Haar measure. The above derivation gives an explicit geometric view of how the non-trivial topology of Σ^1 restricts the allowed global circle actions on the phase space, and we see therefore that the isometry group of any globally-defined Riemannian geometry on the torus is $U(1) \times U(1)$.

The invariance condition (5.70) for the symplectic structure can be solved by imposing the requirement of invariance of ω_{Σ^1} independently under the 2 Killing vectors (6.18). This implies that the components $\omega_{\phi_\mu \phi_\nu}$ must be constant functions, i.e. that ω must be proportional to the Darboux 2-form ω_D, and thus we take

$$\omega_{\Sigma^1} = v \, d\phi_1 \wedge d\phi_2 \qquad (6.19)$$

to be an associated metric-volume form on Σ^1 for the present Riemannian geometry (c.f. (6.11)). The symplectic structure here is thus the symplectic reduction of that from the universal bundle projection $\mathbb{C} \xrightarrow{\pi} \Sigma^1$. It is straightforward to now integrate up the Hamiltonian equations with (6.18) and (6.19), and we find that the Hamiltonian H_{Σ^1} is given by displacements along the homology cycles of Σ^1,

$$H_{\Sigma^1}(\phi_1, \phi_2) = h^1 \phi_1 + h^2 \phi_2 \qquad (6.20)$$

where

$$h^1 = v V_0^2 \quad , \quad h^2 = -v V_0^1 \qquad (6.21)$$

are real-valued constants. Note that, as anticipated from (6.7), the invariant symplectic structure here is uniquely determined just as for 2-dimensional maximally symmetric phase spaces which have 3 (as opposed to just 2 as above) linearly independent Killing vectors. Thus we see here that the localizable Hamiltonian systems in genus 1 are even more severely restricted by the equivariant localization constraints as compared to the simply-connected cases. Note that the Hamiltonian (6.20) does not determine a globally-defined single-valued function on Σ^1, a point which we shall return to shortly.

6.3 Homology Representations and Topological Quantum Field Theory

The Hamiltonian (6.20) defines a rather odd dynamical system on the torus, but besides this feature we see that the allotted Hamiltonians as determined from the geometric localization constraints are in effect completely *independent* of the explicit form of the phase space geometry and depend only on the topological properties of the manifold Σ^1, i.e. (6.20) is explicitly independent of both the complex structure τ and the conformal factor φ appearing in (6.9). From the analysis of the last Chapter, we see that this is in marked contrast to what occurs in the case of a simply connected phase space, where the conformal factor of the metric entered into the final expression for the observable H and the equivariant Hamiltonian systems so obtained depended

6.3 Homology Representations and Topological Quantum Field Theory

on the phase space geometry explicitly. In the present case the partition function with Hamiltonian (6.20) and symplectic 2-form (6.19) obtained as the unique solutions of the equivariant localization constraints can be thought of in this way as defining a topological quantum theory on the torus which is completely independent of any Riemannian geometry on Σ^1. Furthermore, the symplectic potential associated with (6.19) is

$$\theta_{\Sigma^1} = \frac{v}{2}(\phi_1 d\phi_2 - \phi_2 d\phi_1) \tag{6.22}$$

which we note is only locally defined because it involves multi-valued functions in this local form, so that ω_{Σ^1} is a non-trivial element of $H^2(\Sigma^1; \mathbb{Z}) = \mathbb{Z}$. The Hamiltonian (6.20) thus admits the local topological form $H_{\Sigma^1} = i_{V_{\Sigma^1}} \theta_{\Sigma^1}$, so that the corresponding partition function defines a cohomological field theory and it will be a topological invariant of the manifold Σ^1.

To explore some of the features of this topological quantum field theory, we note first that (6.20) is not defined as a global C^∞-function on Σ^1. However, this is not a problem from the point of view of localization theory. Although for the classical dynamics the Hamiltonian can be a multi-valued function on Σ^1, to obtain a well-defined quantum theory we require single-valuedness, under the windings (6.8) around the homology cycles of Σ^1, of the time evolution operator $e^{-iT\hat{H}_{\Sigma^1}}$ which defines the quantum propagator (and also of the Boltzmann weight $e^{iTH_{\Sigma^1}}$ if we wish to have a well-defined classical statistical mechanics). This implies that the constants h^μ in (6.20) must be quantized, i.e. $h^\mu \in h\mathbb{Z}$ for some $h \in \mathbb{R}$, and then time propagation in this quantum system can only be defined in discretized intervals of the base time h^{-1}, i.e. $T = N_T h^{-1}$ where $N_T \in \mathbb{Z}^+$. Such quantizations of coupling parameters in topological gauge theories is a rather common occurence to ensure the invariance of a quantum theory under 'large gauge transformations' when the underlying space has non-trivial topology [22].

In the quantum theory, the Hamiltonian (6.20) therefore represents the winding numbers around the homology cycles of the torus, and therefore to each homology class of Σ^1 we can associate a corresponding Hamiltonian system obeying the equivariant localization constraints. The partition function is now denoted as

$$Z^v_{\Sigma^1}(k,\ell;N_T) \sim \int_{L\Sigma^1} [d^2\phi] \exp\left\{ i \int_0^{N_T h^{-1}} dt \, \left(v\phi_2 \dot{\phi}_1 + h(k\phi_1 + \ell\phi_2) \right) \right\} \tag{6.23}$$

where k and ℓ are integers and we have integrated the kinetic term in (6.23) by parts. This path integral can be evaluated directly by first integrating over the loops $\phi_2(t)$, which gives

$$Z_{\Sigma^1}^v(k,\ell;N_T) \sim \int_{LS^1} [d\phi_1]\, \delta(v\dot\phi_1+h\ell)\exp\left\{i\int_0^{N_T h^{-1}} dt\, hk\phi_1(t)\right\} \sim e^{-ik\ell N_T^2/2v} \quad (6.24)$$

Thus the partition function of this quantum system represents the non-trivial homology classes of the torus, through the winding numbers k and ℓ and the time evolution integer N_T. In fact, (6.24) defines a family of 1-dimensional unitary irreducible representations of the first homology group of Σ^1 through the family of homomorphisms

$$Z_{\Sigma^1}^v(\cdot,\cdot;N_T) : H_1(\Sigma^1;\mathbb{Z}) \to U(1) \otimes U(1) \quad (6.25)$$

from the additive first homology group (3.81) into a multiplicative circle group. Notice that the associated homologically-invariant quantum theory is trivial, in that the sum over all winding numbers of the partition function (6.24) vanishes,

$$\sum_{k,\ell=-\infty}^{\infty} Z_{\Sigma^1}^v(k,\ell;N_T) = \sum_{k=-\infty}^{\infty}\left(\frac{1}{1-e^{-ikN_T^2/2v}} + \frac{1}{1-e^{ikN_T^2/2v}} - 1\right) = 0 \quad (6.26)$$

This sum over all winding numbers is analogous to what one would do in 4-dimensional Yang-Mills theory to include all instanton sectors into the quantum theory [151].

However, it is possible to modify slightly the definition of the quantum propagator on a multiply-connected phase space so that we obtain a partition function which is independent of the homology class defined by the Hamiltonian using a modification of the definition of the path integral over a multiply connected space [147]. In general, if the phase space $\mathcal{M}$ is multiply connected, i.e. $\pi_1(\mathcal{M}) \neq 0$, then the Feynman path integral representation of the quantum propagator can contain parameters $\chi([\sigma])$ which are not present in the classical theory and which weight the homotopy classes $[\sigma]$ of topologically inequivalent time evolutions of the system[3],

$$Z_{\text{hom}}(T) = \sum_{[\sigma]\in\pi_1(\mathcal{M})} \chi([\sigma]) \int_{x(t)\in[\sigma]} [d^{2n}x]\, \sqrt{\det\|\Omega\|}\, e^{iS[x]} \quad (6.27)$$

Unitarity and completeness of the quantum theory (i.e. of the propagator (4.13)) yield, respectively, the constraints that the parameters $\chi([\sigma])$ are phases,

$$\chi([\sigma])^*\chi([\sigma]) = 1 \quad (6.28)$$

---

[3] This definition could also be applied to the full quantum propagator $\mathcal{K}(x',x;T)$ between 2 phase space points. Then the sum in (6.27) is over all homotopy classes of curves $[C_{xx'}]$ from x to x' which are identified with elements of $\pi_1(\mathcal{M})$ using a standard mesh of paths.

and that they form a 1-dimensional unitary representation of $\pi_1(\mathcal{M})$,

$$\chi([\sigma])\chi([\sigma']) = \chi([\sigma \cdot \sigma']) \tag{6.29}$$

Note that the restriction of the path integration to homotopy classes as in (6.27) makes well-defined the representation of the partition function action S with a local symplectic potential following the Wess-Zumino-Witten prescription of Section 4.10. In particular, we can invoke the argument there to conclude that over each homotopy class $[\sigma] \in \pi_1(\mathcal{M})$, the path integral depends only on the second cohomology class defined by ω.

In the case at hand, the partition function (6.24) is regarded as that obtained by restricting the path integration in (6.23) to loops in the homology class labelled by $(k, \ell) \in \mathbb{Z}^2$. In particular, we can add to the sum in (6.26) the phases $\chi(k,\ell) = e^{i\alpha(k,\ell)}$ for each $(k,\ell) \in \mathbb{Z}^2$, which from (6.29) would then have to satisfy

$$\alpha(k + k', \ell + \ell') = \alpha(k, \ell) + \alpha(k', \ell') \tag{6.30}$$

The condition (6.30) means that the phase $\alpha(k,\ell)$ defines a **u(1)**-valued 1-cocycle of the fundamental (or homology) group $\mathbb{Z} \oplus \mathbb{Z}$ of Σ^1 (see Appendix A) as required for them to form a representation of it in the circle group S^1. When they are combined with the character representation (6.24) and the resulting quantity is summed as in (6.26), we can obtain a propagator which is a non-trivial homological invariant of Σ^1 and which yields a character formula for the non-trivial topological groups of the phase space. We shall see how to interpret these character formulas in a group-theoretic setting, as we did in the last Chapter, in the Section 6.5. Notice that, strictly speaking, the volume parameter v in (6.24) should be quantized in terms of h, k and ℓ so that the partition function yields a non-zero result when integrated over the moduli space of T-periodic trajectories. In this way, (6.24) also represents the cohomology class defined by the symplectic 2-form (6.19) through the parameter v. We recall from Section 4.10 that for a simply-connected phase space, the localizable partition functions depend only on the second cohomology class defined by ω. Here we find that the multiple-connectivity of the phase space makes it depend in addition on the first homology group of the manifold. Thus the partition function of the localizable quantum systems on the torus yield topological invariants of the phase space representing its (co-)homology groups.

6.4 Integrability Properties and Localization Formulas

We now turn to a discussion of the structure of the localization formulas for these localizable Hamiltonian systems. Of course, since the canonical $U(1) \times U(1) \sim T^2$ action on the torus generated by (6.20) has no fixed points, this means that the classical partition function

214   6. Equivariant Localization on Multiply Connected Phase Spaces

$$\int_{T^2} v\, d\phi_1\, d\phi_2\ e^{iTH_{\Sigma^1}(\phi_1,\phi_2)} = \frac{1}{T^2 v V_0^1 V_0^2}\left(e^{-2\pi i T v V_0^2} - 1\right)\left(1 - e^{2\pi i T v V_0^1}\right)$$
(6.31)

isn't given by the Duistermaat-Heckman formula. The reason can be traced back to the Poisson algebra $\{\phi_1,\phi_2\}_{\omega_{\Sigma^1}} = -1/v$ which shows that the full Hamiltonian (6.20) is not a functional of action variables in involution. Even if we choose a Hamiltonian which is a Morse function on T^2 given by a functional of the components of the localizable isometry generators on the torus (like the height function), Kirwan's theorem forbids the exactness of the stationary phase approximation for the associated classical partition functions.

We can also view this failure at the level of a Hamiltonian reduction as discussed in Sections 5.7 and 5.8 above. For instance, consider the following partition function that does not fulfill the Duistermaat-Heckman theorem,

$$Z_M^{T^2}(T) \equiv \int_{T^2} v\, d\phi_1\, d\phi_2\ e^{iT(a\cos\phi_1 + b\cos\phi_2)} = (2\pi)^2 v J_0(-iTa) J_0(-iTb)$$
(6.32)

We can write the left-hand side of (6.32) as a reduction from a larger integral on $\mathbb{C}^2$ by introducing 2 complex coordinates $z_1 = |z_1|\,e^{i\phi_1}$ and $z_2 = |z_2|\,e^{i\phi_2}$ with the constraints

$$P(z,\bar z) \equiv z\bar z - 1 = 0 \qquad (6.33)$$

for $z = z_1, z_2$. Introducing Lagrange multipliers λ_1 and λ_2 whose integration over $\mathbb{R}$ produces delta-functions enforcing these constraints, we can write the integration in (6.32) as

$$Z_M^{T^2}(T) = v \int_{-\infty}^{\infty} d\lambda_1\, d\lambda_2 \int_{\mathbb{R}^4} \frac{d^2 z_1\, d^2 z_2}{\pi^2}\ \exp\left\{iT\left(\frac{a}{2}(z_1 + \bar z_1) + \frac{b}{2}(z_2 + \bar z_2)\right)\right.$$
$$\left. + i\lambda_1(z_1\bar z_1 - 1) + i\lambda_2(z_2\bar z_2 - 1)\right\}$$
(6.34)

This shows that, unlike the cases mentioned in Sections 5.7 and 5.8, the Hamiltonian in (6.34) from which the dynamical system in (6.32) is obtained by Hamiltonian reduction is not a bilinear function and it does not commute with the constraint functions defined by (6.33), i.e. the constraints of the reduction are not first class constraints. These alternative integrability arguments therefore also serve as an indication of the breakdown of the localization formalism when applied to multiply connected phase spaces, in addition to the usual topological criteria provided by Kirwan's theorem. We shall see again in Chapter 7 this sort of interplay between integrability and Kirwan's theorem.

The situation is better for the quantum localizations, even as far as the possibilities of using functionals $\mathcal{F}(H_{\Sigma^1})$ of the isometry generator (6.20) for localization as in Section 4.9. Here the arbitrariness of these functionals is

6.4 Integrability Properties and Localization Formulas

not as great as it was in the simply connected cases of Chapter 5. There we required generally only that $\mathcal{F}$ be bounded from below, while in the case at hand the discussion of Section 6.3 above shows that we need in addition the requirement that $\mathcal{F}$ be formally a periodic functional of the observable (6.20). In general, this will not impose any quantization condition on the time translation T, as it did before. For such functionals, however, it is in general rather difficult to determine explicitly the Nicolai transform in (4.148) required for the localization (4.149). Alternatively, one can try to localize the system using (4.142) and the above description of the quantum theory as a topological one, but then we lose the interpretation of the independent Hamiltonians in (6.20) as conserved charges of some integrable dynamical system with phase space the torus. These remarks imply, for example, that one cannot equivariantly quantize a free particle or harmonic oscillator (with compactified momentum and position ranges) on the torus, so that the localizable dynamical systems do not represent generalized harmonic oscillators as they did in the simply connected cases. The same is true of the torus height function (3.78), as anticipated. However, in these cases the periodicity of the Hamiltonian function leads to a much better defined propagator in the sense of it being a tempered distribution represented by a functional integral. Notice that this also shows explicitly, in a rather transparent way, how the Hamiltonian functions on T^2 are restricted by Kirwan's theorem, which essentially means in the above context that the localization formalism loses its interpretation in terms of integrability arguments on a multiply connected phase space. The topological field theory interpretations do, however, carry through from the simply-connected cases but with a much richer structure now.

The expression (6.24) for the quantum partition function also follows directly from substituting into the Boltzmann weight $e^{iS[x]}$ the value of the action in (6.23) evaluated on the classical trajectories $\dot{x}^\mu(t) = V_{\Sigma^1}^\mu$ for the above quantum system, which here are defined by

$$\dot{\phi}_1(t) = V_0^1 \quad , \quad \dot{\phi}_2(t) = V_0^2 \tag{6.35}$$

Thus the path integral (6.23) (trivially) localizes onto the classical loops as in the WKB localization formula (4.115), except that now even the 1-loop fluctuation term vanishes and the path integral is given exactly by its tree-level value. This also independently establishes the quantizations of the propagation time T and the volume parameter v, in that T-periodic solutions to the classical equations of motion with the degenerate structure of the Hamiltonian (6.20) only exist with the discretizations of the parameters h^μ and T above. This is consistent with the discussion at the beginning of Section 4.6 concerning the structure of the moduli space of classical solutions, and again for these discretizations the path integral can be evaluated using the degenerate localization formula (4.122) while for the non-discretized values the critical trajectory set (trivially) coincides with the critical point set $\mathcal{M}_V$

6. Equivariant Localization on Multiply Connected Phase Spaces

of the Hamiltonian. Furthermore, the fact that the conformal factor φ is not involved at all in the solutions of the localization constraints just reflects the fact that the torus is locally flat (as is immediate from its parallelogram representation) and any global 'curving' of its geometry represented by φ in (6.9) can only be done in a uniform periodic fashion around the canonical homology cycles of Σ^1 (c.f. eq. (6.13)). However, the Niemi-Tirkkonen formula (4.130) does depend explicitly on φ. It is here that the geometry of the phase space enters explicitly into the quantum theory, as it did in Chapter 5, if we demand that the metric (6.9) obey the appropriate regularity conditions and therefore make the equivariant localization manifest. This ensures that the localization formula (4.130) coincides with the exact result (6.24), as it should.

In the case at hand (4.130) becomes

$$Z^v_{\Sigma^1}(k,\ell;N_T) \sim \int_{\Sigma^1} \mathrm{ch}_{V_{\Sigma^1}}(-iN_T\omega_{\Sigma^1}/h) \wedge \hat{A}_{V_{\Sigma^1}}(N_T R_\tau/h)$$

$$= \int_{\Sigma^1} d^2\phi \int d^2\eta \; \exp\left[-\frac{iN_T}{h}\left(H_{\Sigma^1}(k,\ell) - \frac{1}{2}(\omega_{\Sigma^1})_{\mu\nu}\eta^\mu\eta^\nu\right)\right]$$

$$\times \sqrt{\det\left[\frac{N_T(2\nabla_\tau)_\mu V^\nu_{\Sigma^1} + (R_\tau)^\nu_{\mu\lambda\rho}\eta^\lambda\eta^\rho)/4h}{\sinh\left(N_T(2\nabla_\tau)_\mu V^\nu_{\Sigma^1} + (R_\tau)^\nu_{\mu\lambda\rho}\eta^\lambda\eta^\rho)/4h\right)}\right]}$$

(6.36)

Again, because of the Kähler structure of (6.9), the Riemann moment map and curvature 2-form have the non-vanishing components

$$(\mu_{V_{\Sigma^1}})^z_{\bar{z}} = -(\mu_{V_{\Sigma^1}})^{\bar{z}}_z = V^z_{\Sigma^1}\partial\varphi + V^{\bar{z}}_{\Sigma^1}\bar{\partial}\varphi \;,\;\; R^z_z = -R^{\bar{z}}_{\bar{z}} = \mathrm{Im}(\tau)\, e^{-\varphi}\nabla^2_z\varphi\, \eta\bar{\eta}$$

(6.37)

We substitute (6.18)–(6.21) and (6.37) into (6.36) and carry out the Berezin integrations there. Comparing the resulting expression with the exact one (6.24) for the partition function, we arrive after some algebra at a condition on the conformal factor of the metric (6.9),

$$\int_{\Sigma^1} d^2\phi \; e^{-iN_T(k\phi_1+\ell\phi_2)} \sqrt{1 - \frac{N_T^2(\ell\partial_{\phi_1}\varphi - k\partial_{\phi_2}\varphi)^2}{4v^2\sinh^2\left(\frac{N_T}{2v}(\ell\partial_{\phi_1}\varphi - k\partial_{\phi_2}\varphi)\right)}}$$

$$= -\frac{2i}{N_T v}\, e^{-ik\ell N_T^2/2v}$$

(6.38)

The Fourier series constraint (6.38) on the metric is rather complicated and it represents a similar sort of metric regularity condition that we encountered in Section 5.9 before. It fixes the harmonic modes of the square-root integrand in (6.38) which should have an expansion such as (6.15). Notice, however, that (6.38) is independent of the phase space complex structure τ, and thus it only depends on the representative of the conformal equivalence class of the metric (6.9). This is typical of a topological field theory path integral [22].

The condition (6.38) can be used to check if a given phase space metric really does result in the correct quantum theory (6.24), and this procedure then tells us what (representatives of the conformal equivalence classes of) quantum geometries in this sense are applicable to the equivariant localization of path integrals on the torus. For example, suppose we tried to quantize a flat torus using equivariant localization. Then from (6.12) the conformal factor would have to solve the Laplace equation $\nabla_\tau^2 \varphi = 0$ globally on Σ^1. Since φ is assumed to be a globally-defined function on Σ^1, it must admit a harmonic mode expansion over Σ^1 as in (6.15). From (6.14) and this Fourier series for φ we see that the Laplace equation implies that all Fourier modes of φ except the constant modes vanish, and so the left-hand side of (6.38) is zero. Thus a flat torus cannot be used to localize the quantum mechanical path integral (6.23) onto the equivariant Atiyah-Singer index in (4.130). This means that a flat Kähler metric (6.9) on Σ^1 does not lead to a homotopically trivial localization 1-form $\psi = i_{V_{\Sigma^1}} g_\tau$ on the loop space $L\Sigma^1$ within any homotopy class (c.f. Section 4.4). This simple example shows that the condition (6.38), along with the Riemannian restrictions (6.11) and (6.13), give a very strong probe of the quantum geometry of the torus. Moreover, when (6.38) does hold, we can represent the equivariant characteristic classes in (4.130) in terms of the homomorphism (6.24) of the first homology group of Σ^1.

6.5 Holomorphic Quantization and Non-Symmetric Coadjoint Orbits

In this Section we shall show that it is possible to interpret the topological path integral (6.23) as a character formula associated with the quantization of a coadjoint orbit corresponding to some novel sort of spin system described by Σ^1, as was the situation in all of the simply connected cases of the last Chapter. For this, we examine the canonical quantum theory defined by the symplectic structure (6.19) in the Schrödinger picture representation. We first rewrite the symplectic 2-form (6.19) in complex coordinates to get the Kähler structure

$$\omega_{\Sigma^1} = \frac{v}{2i \, \mathrm{Im}\, \tau} dz \wedge d\bar{z} = -i\partial\bar{\partial} F_{\Sigma^1} \tag{6.39}$$

with corresponding local Kähler potential

$$F_{\Sigma^1}(z, \bar{z}) = v z \bar{z} / 2 \, \mathrm{Im}\, \tau \tag{6.40}$$

We then map the corresponding Poisson algebra onto the associated Heisenberg algebra by the standard commutator prescription (c.f. beginning of Section 5.1). With this we obtain the quantum commutator

$$[\hat{z}, \hat{\bar{z}}] = 2\,\mathrm{Im}(\tau)/v \qquad (6.41)$$

We can represent the algebra (6.41) on the space $\mathrm{Hol}(\Sigma^1;\tau)$ of holomorphic functions $\Psi(z)$ on Σ^1 by letting $\hat{z}$ act as multipication by the complex coordinate $z = \phi_1 + \tau\phi_2$ and $\hat{\bar{z}}$ as the derivative operator

$$\hat{\bar{z}} = -\frac{2\,\mathrm{Im}\,\tau}{v}\frac{\partial}{\partial z} \qquad (6.42)$$

With this holomorphic Schrödinger polarization, the operators $\hat{z}$ and $\hat{\bar{z}}$ with the commutator algebra (6.41) resemble the creation and annihilation operators (5.79) of the Heisenberg-Weyl algebra with the commutation relation (5.80). In analogy with that situation, we can construct the corresponding coherent states

$$|z\rangle = \mathrm{e}^{(-v/2\,\mathrm{Im}\,\tau)z\hat{\bar{z}}}|0\rangle \quad ; \quad z \in \Sigma^1 \qquad (6.43)$$

which are normalized as

$$(z|z) = \mathrm{e}^{(v/2\,\mathrm{Im}\,\tau)z\bar{z}} = \mathrm{e}^{F_{\Sigma^1}(z,\bar{z})} \qquad (6.44)$$

and obey the completeness relation

$$\int_{\Sigma^1} \frac{d^2z}{(2\pi)^2}\,|z))((z| = \mathbf{1} \qquad (6.45)$$

These coherent states are associated with the quantization of the coadjoint orbit $U(1) \times U(1) = S^1 \times S^1$. However, since Σ^1 is a non-symmetric space, it cannot be considered as a Kähler manifold associated with the coadjoint orbit of a semi-simple Lie group, as was the case in the last Chapter. The orbits above are, however, associated with the action of the isometry group $U(1) \times U(1)$ on Σ^1, which has an interesting Lie algebraic structure that we shall discuss below.

In the Schrödinger representation (6.42), we consistently find the action of the operator $\hat{\bar{z}}$ on the states (6.43) as $\bar{z}|z\rangle$. The holomorphic representation space $\mathrm{Hol}(\Sigma^1;\tau)$ in this context is then regarded as the space of entire functions $\Psi(z) = (z|\Psi)$ for each state $|\Psi)$ in the span of the coherent states (6.43). An inner product on $\mathrm{Hol}(\Sigma^1;\tau)$ is then determined from the completeness relation (6.45) and the normalization (6.44) as

$$(\Psi_1|\Psi_2) = \int_{\Sigma^1} \frac{d^2z}{(2\pi)^2}\,\frac{(\Psi_1|z)(z|\Psi_2)}{(z|z)} = \int_{\Sigma^1}\frac{d^2z}{(2\pi)^2}\,\mathrm{e}^{-(v/2\,\mathrm{Im}\,\tau)z\bar{z}}\Psi_1^\dagger(\bar{z})\Psi_2(z) \qquad (6.46)$$

With the inner product (6.46), we find that the operator $\hat{\bar{z}} = \hat{z}^\dagger$ is the adjoint of $\hat{z}$, as it consistently should be. An operator $\hat{\mathcal{H}}$ acting on the space of coherent states (6.43) can now be represented on $\mathrm{Hol}(\Sigma^1;\tau)$ as usual by

6.5 Holomorphic Quantization and Non-Symmetric Coadjoint Orbits 219

an integral kernel as in (5.95) with the identification of $\bar{z}$ as the derivative operator (6.42). Furthermore, the quantum propagator associated with such an operator

$$\mathcal{K}_{\Sigma^1}(\bar{z}', z; T) = (\!(z'| \, e^{-iT\hat{\mathcal{H}}} |z)\!) \tag{6.47}$$

determines the corresponding time-evolution of the coherent state wavefunctions as

$$\Psi(z; T) = \int_{\Sigma^1} \frac{d^2 z'}{(2\pi)^2} \, e^{-F_{\Sigma^1}(z', \bar{z}')} \mathcal{K}_{\Sigma^1}(\bar{z}', z; T) \Psi(z') \tag{6.48}$$

with $\Psi(z;t) \equiv (z,t|\Psi)$ and $\Psi(z) \equiv \Psi(z;0)$. In the following we shall build up the initial states $\Psi(z)$, and then the associated time evolution determined by the localizable Hamiltonians on Σ^1 (and hence the solutions of the Schrödinger wave equation) are determined by the path integral above for the propagator (6.47).

The advantage of working with the holomorphic representation space $\text{Hol}(\Sigma^1; \tau)$ is that we shall want to discuss the explicit structure of the Hilbert space associated with the localizable quantum systems we found above. With the Kähler structure defined by the symplectic 2-form ω_{Σ^1} above, the Hilbert space of the quantum theory is then the space of holomorphic sections of the usual symplectic complex line bundle $L \to \Sigma^1$, which in this context is usually called the prequantum line bundle over Σ^1. As such, ω_{Σ^1} represents the first Chern characteristic class of L, and so such a bundle exists only if ω_{Σ^1} is an integral 2-form on Σ^1. This method of quantizing the Hamiltonian dynamics in terms of the geometry of fiber bundles is called geometric quantization [172] and it is equivalent to the Borel-Weil-Bott method of constructing coherent states that we encountered in the last Chapter. In light of the requirement of single-valuedness of the quantum propagator that we discussed in the Section 6.3, we require, from the point of view of equivariant localization, that the wavefunctions $\Psi(z)$ change only by a unitary transformation under the winding transformations (6.8) on Σ^1, so that all physical quantities, such as the probability density $\Psi^\dagger \Psi$, are well-defined C^∞-functions on the phase space Σ^1 and respect the symmetries of the quantum theory as defined by the quantum Hamiltonian, i.e. by the supersymmetry making the dynamical system a localizable one. In this setting, the multivalued wavefunctions, regarded as sections of the associated line bundle $L \to \Sigma^1$ where the structure group $\pi_1(\Sigma^1) = \mathbb{Z} \oplus \mathbb{Z}$ acts through a unitary representation, are single-valued functions on the universal cover $\mathbb{C}$ of the torus and so they can be thought of as single-valued functions of homotopy classes $[\sigma]$ of loops on Σ^1. This also ensures that the coherent states (6.43) remain coherent under the time evolution determined by the localizable Hamiltonians of the last Section (i.e. under the action of $\mathcal{I}(\Sigma^1; g_\tau)$) which will lead to a consistent coherent state path integral representation of (6.23).

To explore this in more detail, we need a representation for the discretized equivariant Hamiltonian generators above of the isometry group $\mathcal{I}(\Sigma^1; g_\tau)$

220   6. Equivariant Localization on Multiply Connected Phase Spaces

on the space $\mathrm{Hol}(\Sigma^1;\tau)$ [18, 153]. This group action then coincides with the automorphisms of the symplectic line bundle above in the usual way. Note that translations by $a \in \mathbb{C}$ on z are generated on functions of z by the action of the operator $\mathrm{e}^{a\frac{\partial}{\partial z}}$, and likewise on functions of $\bar z$ by $\mathrm{e}^{\bar a\frac{\partial}{\partial \bar z}}$. On the holomorphic representation space $\mathrm{Hol}(\Sigma^1;\tau)$, we represent the latter operator using the commutation relation (6.41) as $\mathrm{e}^{(v/2\,\mathrm{Im}\,\tau)\bar a z}$, in accordance with the coherent state representation above. Thus the generators of large $U(1)$ transformations around the homology cycles of Σ^1 in the holomorphic Schrödinger polarization above are the unitary quantum operators

$$U(n,m) = \exp\left(2\pi(n+m\tau)\frac{\partial}{\partial z} + \frac{\pi v}{\mathrm{Im}\,\tau}(n+m\bar\tau)z\right) \quad ; \quad n,m \in \mathbb{Z} \quad (6.49)$$

which generate simultaneously both of the winding transformations in (6.8). By the above arguments, the quantum states should be invariant (up to unitary equivalence) under their action on the Hilbert space. Solving this invariance condition will then give a representation of the equivariant localization symmetry constraints (i.e. of the pertinent cohomological supersymmetry) and of the coadjoint orbit system directly in the Hilbert space of the canonical quantum theory.

In contrast with their classical counterparts, the quantum operators (6.49) do not commute among themselves in general and products of them differ from their reverse-ordered products by a $\mathbf{u}(1)$-valued 2-cocycle (see Appendix A). The Baker-Campbell-Hausdorff formula,

$$\mathrm{e}^{X+Y} = \mathrm{e}^{-[X,Y]/2}\,\mathrm{e}^X\,\mathrm{e}^Y \quad \text{when} \quad [X,[X,Y]] = [Y,[X,Y]] = 0 \quad (6.50)$$

implies

$$\mathrm{e}^X\,\mathrm{e}^Y = \mathrm{e}^{[X,Y]}\,\mathrm{e}^Y\,\mathrm{e}^X \quad (6.51)$$

Applying (6.51) to products of the operators (6.49) and using the commutation relation (6.41) with (6.42), we find that they obey what is called a clock algebra,

$$U(n_1,m_1)U(n_2,m_2) = \mathrm{e}^{2\pi i v(n_2 m_1 - n_1 m_2)} U(n_2,m_2)U(n_1,m_1) \quad (6.52)$$

To determine the action of the operators (6.49) explicitly on the wavefunctions $\Psi(z)$, we apply the Baker-Campbell-Hausdorff formula (6.50) to get

$$U(n,m) = \exp\left[\frac{\pi v}{\mathrm{Im}\,\tau}\left(\pi|n+m\tau|^2 + (n+m\bar\tau)z\right)\right] \mathrm{e}^{2\pi(n+m\tau)\frac{\partial}{\partial z}} \quad (6.53)$$

so that the action of (6.53) on the quantum states of the theory is

$$U(n,m)\Psi(z) = \exp\left[\frac{\pi v}{\mathrm{Im}\,\tau}\left(\pi|n+m\tau|^2 + (n+m\bar\tau)z\right)\right] \Psi(z + 2\pi(n+m\tau)) \quad (6.54)$$

If the volume parameter $v = \mathrm{vol}_{g_\tau}(\Sigma^1)/(2\pi)^2$ is an irrational number, then it follows from the clock algebra (6.52) that the $U(1)$ generators above

6.5 Holomorphic Quantization and Non-Symmetric Coadjoint Orbits

act as infinite-dimensional raising operators in (6.54) and so the Hilbert space of quantum states in this case is infinite-dimensional. However, we recall the necessary quantization requirements for the parameters of the Hamiltonian system required for a consistent quantum theory. With this in mind, we instead consider the case where the volume of the torus is quantized so that

$$v = v_1/v_2 \quad ; \quad v_1, v_2 \in \mathbb{Z}^+ \tag{6.55}$$

is rational-valued. Alternatively, such a discretization of v is required in order that the symplectic 2-form ω_{Σ^1} define an integer cohomology class, as in (4.161). In this case, the cocycle relation (6.52) shows that the operator $U(v_2 n, v_2 m)$ commutes with all of the other $U(1)$ generators and the time evolution operator, and so they can be simultaneously diagonalized over the same basis of states. This means that their action (6.54) on the wavefunctions must produce a state that lies on the same ray in the Hilbert space as that defined by $\Psi(z)$, i.e.

$$U(v_2 n, v_2 m)\Psi(z) = e^{i\eta(n,m)}\Psi(z) \tag{6.56}$$

for some phases $\eta(n, m) \in S^1$. The invariance condition (6.56), expressing the symmetry of the wavefunctions under the action of the (non-simple) Lie group $U(1) \times U(1)$, is called a projective representation of the symmetry group. It must obey a particular consistency condition. The composition law for the group operations induces a composition law for the phases in (6.56),

$$\begin{aligned}
U(v_2(n_1 + n_2), v_2(m_1 + m_2))\Psi(z) \\
= U(v_2 n_1, v_2 m_1) U(v_2 n_2, v_2 m_2)\Psi(z) \\
= e^{i\eta(n_1+n_2, m_1+m_2)}\Psi(z) \\
\times \exp\left[i\{\eta(n_1, m_1) + \eta(n_2, m_2) - \eta(n_1 + n_2, m_1 + m_2)\}\right]
\end{aligned} \tag{6.57}$$

If the last phase in (6.57) vanishes, as in (6.30), then the projective phase $\eta(n, m)$ is a 1-cocycle of the symmetry group $U(1) \times U(1)$ and the wavefunctions carry a unitary representation of the group, as required [77]. The determination of these 1-cocycles explicitly below will then yield an explicit representation of the homologically-invariant partition function (6.27).

Comparing (6.56) and (6.54), we see that the invariance of the quantum states under the $U(1)$ action on the phase space can be expressed as

$$\Psi(z + 2\pi v_2(n + m\tau))$$
$$= \exp\left[i\eta(n, m) - \frac{\pi v_1}{\operatorname{Im}\tau}\left(\pi v_2 |n + m\tau|^2 + (n + m\bar{\tau})z\right)\right]\Psi(z) \tag{6.58}$$

The only functions which obey quasi-periodic conditions like (6.58) are combinations of the Jacobi theta functions [61, 109, 145]

$$\Theta^{(D)}\begin{pmatrix}c\\d\end{pmatrix}[z|\Pi]$$
$$= \sum_{\{n^\ell\}\in\mathbb{Z}^D} \exp\left[i\pi(n^\ell + c^\ell)\Pi_{\ell p}(n^p + c^p) + 2\pi i(n^\ell + c^\ell)(z_\ell + d_\ell)\right] \quad (6.59)$$

where $c^\ell, d_\ell \in [0,1]$. The functions (6.59) are well-defined holomorphic functions of $\{z_\ell\} \in \mathbb{C}^D$ for $D \times D$ complex-valued matrices $\Pi = [\Pi_{\ell p}]$ in the Siegal upper half-plane (i.e. Im $\Pi > 0$). They obey the doubly semi-periodic conditions

$$\Theta^{(D)}\begin{pmatrix}c\\d\end{pmatrix}[z + s + \Pi \cdot t|\Pi]$$
$$= \exp\left[2\pi i c^\ell s_\ell - i\pi t^\ell \Pi_{\ell p} t^p - 2\pi i t^\ell (z_\ell + d_\ell)\right] \Theta^{(D)}\begin{pmatrix}c\\d\end{pmatrix}[z|\Pi] \quad (6.60)$$

where $s = \{s_\ell\}$ and $t = \{t^\ell\}$ are integer-valued vectors, and

$$\Theta^{(D)}\begin{pmatrix}c\\d\end{pmatrix}[z + \alpha\Pi \cdot t|\Pi]$$
$$= \exp\left[-i\pi\alpha^2 t^\ell \Pi_{\ell p} t^p - 2\pi i\alpha t^\ell (z_\ell + d_\ell)\right] \Theta^{(D)}\begin{pmatrix}c - \alpha t\\d\end{pmatrix}[z|\Pi] \quad (6.61)$$

for any non-integer constant $\alpha \in \mathbb{R}$. We remark here that the transformations in (6.60) can be applied in many different steps with the same final result, but successive applications of (6.60) and (6.61) do not commute [18]. In the context of the action of the unitary operators $U(n,m)$ above, when these transformations are applied in different orders in (6.58), the final results differ by a phase which forms a representation of the clock algebra (6.52). To avoid this minor ambiguity, we simply *define* the operators $U(n,m)$ by their action on the states $\Psi(z)$ with the convention that the transformation (6.60) is applied before (6.61).

After some algebra, we find that the algebraic constraints (6.58) are uniquely solved by the $v_1 v_2$ independent holomorphic wavefunctions

$$\Psi_{p,r}\begin{pmatrix}c\\d\end{pmatrix}(z) = e^{-(v/4\,\mathrm{Im}\,\tau)z^2} \Theta^{(1)}\begin{pmatrix}\frac{c + 2\pi v_1 p + v_2 r}{2\pi v_1 v_2}\\d\end{pmatrix}[v_1 z | 2\pi v_1 v_2 \tau] \quad (6.62)$$

where $p = 1, 2, \ldots, v_2$ and $r = 1, 2, \ldots, v_1$. The phases in (6.56) are then found to be the non-trivial 1-cocycles

$$\eta(n,m)/2\pi = cn - dm + \pi v_1 v_2 nm \quad (6.63)$$

of the global $U(1) \times U(1)$ group acting on Σ^1 here. Furthermore, the winding transformations (6.54) can be written as

$$U(n,m)\Psi_{p,r}\begin{pmatrix}c\\d\end{pmatrix}(z) = \sum_{p'=1}^{v_2}[U(n,m)]_{pp'}\Psi_{p',r}\begin{pmatrix}c\\d\end{pmatrix}(z) \quad (6.64)$$

6.5 Holomorphic Quantization and Non-Symmetric Coadjoint Orbits

where the finite-dimensional unitary matrices

$$[U(n,m)]_{pp'} = \exp\left[\frac{2\pi i}{v_2}(cn - dm + \pi v_1 n(m+2p))\right]\delta_{p+m,p'} \tag{6.65}$$

form a v_2-dimensional projective representation, which is cyclic of period v_2, of the clock algebra (6.52). The projective phase here is the non-trivial $U(1) \times U(1)$ 1-cocycle

$$\eta^{(p)}(n,m)/2\pi = (cn - dm + \pi v_1 n(m+2p))/v_2 \tag{6.66}$$

which could also therefore be used to construct an unambiguous partition function as in the Section 6.3 by

$$Z_{\text{hom}}(T) = \sum_{k,\ell=-\infty}^{\infty} e^{i\eta(k,\ell)} Z_{\Sigma^1}^v(k,\ell;N_T) \tag{6.67}$$

Thus the Hilbert space is $v_1 v_2$-dimensional and the quantum states carry a v_2-dimensional projective representation of the equivariant localization symmetries via the clock algebra (6.52) which involves the $\mathbf{u}(1)$-valued 2-cocycle

$$\xi(n_1,m_1;n_2,m_2)/2\pi = v_1(n_2 m_1 - n_1 m_2)/v_2 \tag{6.68}$$

of the $U(1) \times U(1)$ isometry group of Σ^1. This shows explicitly how the $U(1)$ equivariant localization constraints and the topological toroidal restrictions are realized in the canonical quantum theory, as then these conditions imply that the only invariant operators on the Hilbert space here are essentially combinations of the generators (6.49). In particular, this implies, by construction, that the coherent state wavefunctions (6.62) are complete. This is much different than the situation for the coherent states associated with simply-connected phase spaces where there are no such topological symmetries to be respected for the supersymmetric localization of the path integral and the Hilbert space is 1-dimensional. Intuitively, the finite-dimensionality of the Hilbert space of physical states is expected from the compactness of the phase space Σ^1.

Notice though that the wavefunctions (6.62) contain the 2 free parameters c and d. We can eliminate one of them by requiring that the Hamiltonian (6.20) in this basis of states does indeed lead to the correct propagator (6.24), i.e. that (6.24) be equal to the trace of the time evolution operator on the finite dimensional vector space spanned by the coherent states (6.62),

$$\text{tr } e^{-iN_T \hat{H}_{\Sigma^1}(k,\ell)/\hbar} = \sum_{p=1}^{v_2}\sum_{r=1}^{v_1} \langle \Psi_{p,r}| e^{-iN_T \hat{H}_{\Sigma^1}(k,\ell)/\hbar} |\Psi_{p,r}\rangle \tag{6.69}$$

where the coherent state inner product is given by (6.46). With this inner product, it is straightforward to show that the states (6.62) define an orthonormal basis of the Hilbert space,

224   6. Equivariant Localization on Multiply Connected Phase Spaces

$$(\Psi_{p_1,r_1}|\Psi_{p_2,r_2}) = \delta_{p_1,p_2}\delta_{r_1,r_2} \tag{6.70}$$

Substituting the identity $e^{-iN_T \hat{H}_{\Sigma^1}(k,\ell)/\hbar} = [U(\ell,-k)]^{N_T v_2/2\pi v_1}$ into (6.69) and using (6.64), (6.65) and (6.70) we find

$$\operatorname{tr} e^{-iN_T \hat{H}_{\Sigma^1}(k,\ell)/\hbar} = (-1)^{k\ell N_T} e^{iN_T(ck+d\ell)/v_1} \tag{6.71}$$

Comparing the result (6.71) with the exact one (6.24), we find that the parameter d appearing in the wavefunctions (6.62) can be determined as

$$d_{k\ell} = (k\ell N_T - 2ck)/2\ell \tag{6.72}$$

Another way to eliminate the parameters c and d appearing in (6.62) is to regard the quantum theory as a topological field theory. The above construction produces a Hilbert space $\mathcal{H}^\tau$ of holomorphic sections of a complex line bundle $L^\tau \to \Sigma^1$ for each modular parameter τ. If we smoothly vary the complex structure τ, then this gives a family of finite-dimensional Hilbert spaces which can be regarded as forming in this way a holomorphic vector bundle (i.e. one with a holomorphic projection map) over the Teichmüller space $\mathbb{C}^+$ of the torus for which the projective representations above define a canonical projectively-flat connection. This is a typical feature of the Hilbert space for a Schwarz-type topological gauge theory [22]. Equivalent complex structures (i.e. those which generate the same conformal equivalence classes as (6.9)) in the sense of the topological field theory of this Chapter should be regarded as leading to the same quantum theory, and this should be inherent in both the homological partition functions of the Section 6.3 and in the canonical quantum theory above. It can be shown [111] that 2 toroidal complex structures $\tau, \tau' \in \mathbb{C}^+$ define conformally equivalent metrics (i.e. $g_\tau = \rho g_{\tau'}$ for some smooth function $\rho > 0$) if and only if they are related by the projective transformation[4]

$$\tau' = \frac{\alpha\tau + \beta}{\gamma\tau + \delta} \quad \text{with} \quad \alpha,\beta,\gamma,\delta \in \mathbb{Z} \;, \quad \alpha\delta - \beta\gamma = 1 \tag{6.73}$$

on $\mathbb{C}^+ \to \mathbb{C}^+$. The transformations (6.73) generate the action of the group $SL(2,\mathbb{Z})/\mathbb{Z}_2$ on $\mathbb{C}^+$, which is a discrete subgroup of the Möbius group $SL(2,\mathbb{C})/\mathbb{Z}_2$ of linear fractional transformations of $\mathbb{C}$ wherein we take $\alpha, \beta, \gamma, \delta \in \mathbb{C}$ in (6.73). We call this discrete group the modular or mapping class group Γ_{Σ^1} of the Riemann surface Σ^1 and it consists of the discrete automorphisms of Σ^1 (i.e. the conformal diffeomorphisms of Σ^1 which aren't connected to the identity and so cannot be represented as global flows of vector fields). The Teichmüller space $\mathbb{C}^+$ modulo this group action, i.e. the space of inequivalent complex structures on Σ^1, is called the moduli space $\mathcal{M}_{\Sigma^1} \equiv \mathbb{C}^+/\Gamma_{\Sigma^1}$ of Σ^1.

---
[4] That the 2 associated tori are conformally isomorphic can be seen intuitively by representing each as a parallelogram in the complex plane and tracing out this transformation.

6.5 Holomorphic Quantization and Non-Symmetric Coadjoint Orbits

The topological quantum theory above therefore should also reflect this sort of full topological invariance on the torus, because it is independent of the conformal factor φ in (6.9).

Under the modular transformation (6.73), it is possible to show that, up to an overall phase, the 1-dimensional Jacobi theta functions in (6.59) transform as [109]

$$\Theta^{(1)}\begin{pmatrix}c\\d\end{pmatrix}[z|\tau] \to \Theta^{(1)}\begin{pmatrix}c'\\d'\end{pmatrix}[z'|\tau'] = \sqrt{\gamma\tau+\delta}\; e^{i\pi z^2/(\gamma\tau+\delta)}\Theta^{(1)}\begin{pmatrix}c\\d\end{pmatrix}[z|\tau] \tag{6.74}$$

where

$$z' = \phi_1 + \tau'\phi_2 = z/(\gamma\tau+\delta) \tag{6.75}$$

is the new (but equivalent) complex structure defined by (6.73) and the new parameters c' and d' are given by

$$c' = \delta c - \gamma d - \gamma\delta/2 \quad , \quad d' = \alpha d - \beta c - \alpha\beta/2 \tag{6.76}$$

Using (6.74) we find after some algebra that the wavefunctions (6.62) transform under the modular transformation of isomorphic complex structures as

$$\Psi_{p,r}\begin{pmatrix}c\\d\end{pmatrix}(z) \to \frac{1}{\sqrt{\gamma\tau+\delta}}\Psi_{p,r}\begin{pmatrix}\tilde c'\\\tilde d'\end{pmatrix}(z') \tag{6.77}$$

with

$$\tilde c' = \delta c - \gamma d - \pi v_1 v_2 \gamma\delta \quad , \quad \tilde d' = \alpha d - \beta c - \pi v_1 v_2 \alpha\beta \tag{6.78}$$

It follows that a set of modular invariant wavefunctions can exist only when the combination $v_1 v_2$ is an even integer, in which case the invariance condition requires $c = d = 0$. For $v_1 v_2$ an odd integer, we can take $c, d \in \{0, \frac{1}{2}\}$, and then the holomorphic wavefunctions carry a non-trivial spinor representation of the modular group as defined by (6.77). These choices of c and d correspond to the 4 possible choices of spin structure on the torus [61] (i.e. representations of the 2-dimensional spinor group $U(1)$ in the tangent bundle of Σ^1) which are determined by the mod 2 cohomology $H^1(\Sigma^1; \mathbb{Z}) \otimes \mathbb{Z}_2 = \mathbb{Z}_2 \oplus \mathbb{Z}_2$ [41, 61]. This increases the number of basis wavefunctions (6.62) by a factor of 4.

It is in this way that one may adjust the parameters c and d so that the wavefunctions (6.62) are modular invariants, as they should be since the topological quantum theory defined by (6.23) is independent of the phase space complex structure. We note also that these specific choices of the parameters in turn then fix the propagation time integers N_T by (6.72), so that these topological requirements completely determine the topological quantum field theory in this case. Thus one can remove all apparent ambiguities here and obtain a situation that parallels the topological quantum theories in the simply connected cases, although now the emerging topological and group theoretical structures are far more complicated. With these appropriate choices of parameter values, the propagator (6.69) then coincides with the coherent state path integral

$$Z_{\Sigma^1}^v(k,\ell;N_T) \sim \int_{L\Sigma^1} \prod_{t\in[0,T]} \frac{dz(t)\,d\bar{z}(t)}{(2\pi)^2}$$

$$\times \exp\left\{\frac{1}{2i\,\text{Im}\,\tau}\int_0^{N_T h^{-1}} dt\left[\frac{v_1}{2v_2}(\bar{z}\dot{z} - z\dot{\bar{z}}) + ih\left((\ell - \tau k)\bar{z} - (\ell - \bar{\tau}k)z\right)\right]\right\}$$
(6.79)

The coherent state path integral (6.79) models the quantization of some novel, unusual spin system defined by the Hamiltonians (6.20) which are associated with the quantized, non-symmetric coadjoint Lie group orbit $U(1) \times U(1) = S^1 \times S^1$. This abelian orbit is an unreduced one as it already is its own maximal torus, and we can therefore think of this spin system as 2 independent planar spins each tracing out a circle. The points on this orbit are in one-to-one correspondence with the coherent state representations above of the projective clock algebra (6.52) of the discrete first homology group of the torus. The associated character formula represented by (6.79) gives path integral representations of the homology classes of Σ^1, in accordance with the fact that it defines a topological quantum field theory, and these localizable quantum systems are exactly solvable via both the functional integral and canonical quantization formalisms, as above. In this latter formalism, the Hilbert space of physical states is finite-dimensional and the basis states carry a non-trivial projective representation of the first homology group of the phase space, in addition to the usual representation of $H^2(\mathcal{M};\mathbb{Z})$.

6.6 Generalization to Hyperbolic Riemann Surfaces

We conclude this Chapter by indicating how the above features of equivariant localization could generalize to the case where the phase space is a hyperbolic Riemann surface [153], although our conclusions are somewhat heuristic and more care needs to be exercised in order to study these examples in detail. Since for $h > 1$, Σ^h can be regarded as h tori stuck together, its homotopy can be described by the $2h$ loops a_i, b_i, $i = 1,\ldots,h$, where each pair a_i, b_i encircle the 2 holes of the i-th torus in the connected sum representation of Σ^h. The constraint (6.6) on the fundamental homotopy generators now generalizes to

$$\prod_{i=1}^h a_i b_i a_i^{-1} b_i^{-1} = \mathbf{1} \tag{6.80}$$

and so the commutator subgroup of $\pi_1(\Sigma^h)$ for $h > 1$ is non-trivial and the fundamental group of a hyperbolic Riemann surface is non-abelian. Its first homology group is given by (3.85), and, using an abusive notation, we shall denote its generators as well by a_i, b_i, $i = 1, \ldots, h$ and call them a canonical basis of homology cycles for Σ^h.

According to the Riemann uniformization theorem [111], there are only 3 (compact or non-compact) simply-connected Riemann surfaces – the 2-sphere S^2, the plane $\mathbb{C}$ and the Poincaré upper half-plane $\mathcal{H}^2$, each equipped with their standard metrics as discussed in the last Chapter. The sphere is its own universal cover (being simply-connected and having a unique complex structure), while $\mathbb{C}$ is the universal cover of the torus. The hyperbolic plane $\mathcal{H}^2$ is always the universal cover of a Riemann surface of genus $h \geq 2$, which is represented as

$$\Sigma^h = \mathcal{H}^2 / F_h \tag{6.81}$$

where $F_h = \pi_1(\Sigma^h)$ is in this context refered to as a discrete Fuchsian group. The quotient in (6.81) is by the fixed-point free bi-holomorphic action of F_h on $\mathcal{H}^2$. The group of analytic automorphisms of the upper half-plane $\mathcal{H}^2$ is $PSL(2,\mathbb{R}) = SL(2,\mathbb{R})/\mathbb{Z}_2$, the group of projective linear fractional transformations as in (6.73) except that now the coefficients $\alpha, \beta, \gamma, \delta$ are taken to be real-valued. Then $\pi_1(\Sigma^h)$ is taken as a discrete subgroup of this $PSL(2,\mathbb{R})$-action on $\mathcal{H}^2$ and the different isomorphism classes of complex analytic structures of Σ^h are essentially the different possible classes of discrete subgroups. Note that this generalizes the genus 1 situation above, where the automorphism group of $\mathbb{C}$ was the group $PSL(2,\mathbb{C})$ of global conformal transformations in 2-dimensions and $\pi_1(\Sigma^1)$ was taken to be the lattice subgroup. Indeed, it is possible to regard Σ^h as a $4h$-gon in the plane with edges identified appropriately to generate the h 'holes' in Σ^h.

It is difficult to generalize the explicit constructions of the last few Sections because of the complicated, abstract fashion in (6.81) that the complex coordinatization of Σ^h occurs. For the various ways of describing the Teichmüller space and Fuchsian groups of hyperbolic Riemann surfaces without the explicit introduction of local coordinates, see [74]. The Teichmüller space of Σ^h can be naturally given the geometric structure of a non-compact complex manifold which is homeomorphic to $\mathbb{C}^{3h-3}$, so that the coordinatization of Σ^h is far more intricate for $h \geq 2$ because it now involves $3h - 3$ complex parameters, as opposed to just 1 as before. Nonetheless, it is still possible to deduce the unique localizable Hamiltonian system on a hyperbolic Riemann surface and deduce some general features of the ensuing topological quantum field theory just as we did above.

We choose a complex structure on Σ^h for which the universal bundle projection in (6.81) is holomorphic (as for the torus), and then the metric g_{Σ^h} induced on Σ^h by this projection involves a globally-defined conformal factor φ as in (6.9) and a constant negative curvature Kähler metric (the hyperbolic Poincaré metric – see Section 5.6). The condition now that the Killing vectors of this metric be globally-defined on Σ^h means that they must be single-valued under windings around the canonical homology cycles

6. Equivariant Localization on Multiply Connected Phase Spaces

$\{a_\ell, b_\ell\}_{\ell=1}^h \in H_1(\Sigma^h; \mathbb{Z})$, or equivalently that

$$\oint_{a_\ell} dV^\mu = \oint_{b_\ell} dV^\mu = 0 \quad , \quad \ell = 1, \ldots, h \tag{6.82}$$

Using this single-valued condition and the Killing equations

$$g_{\Sigma^h}(dV, \cdot) = -i_V dg_{\Sigma^h} \tag{6.83}$$

we can now deduce the general form of the Killing vectors of Σ^h. For this, we apply the Hodge decomposition theorem [22, 32] to the metric-dual 1-form $g(V, \cdot) \in \Lambda^1 \Sigma^h$,

$$g(V, \cdot) = d\chi + \star d\xi + \lambda_h \tag{6.84}$$

where χ and ξ are C^∞-functions on Σ^h and λ_h is a harmonic 1-form, i.e. a solution of the zero-mode Laplace equation for 1-forms,

$$\Delta_1 \lambda_h \equiv (\star d \star d + d \star d \star) \lambda_h = 0 \tag{6.85}$$

In the above, $\star$ denotes the Hodge duality operator and, on a general d-dimensional Riemannian manifold $(\mathcal{M}, g)$, it encodes the Riemannian geometry directly into the DeRham cohomology. It is defined as the map

$$\star : \Lambda^k \mathcal{M} \to \Lambda^{d-k} \mathcal{M} \tag{6.86}$$

which is given locally by

$$\star \alpha = \frac{1}{(d-k)!} \sqrt{\det g(x)} \, g_{j_1 \lambda_1}(x) \cdots g_{j_{d-k} \lambda_{d-k}}(x)$$
$$\times \epsilon^{\lambda_1 \ldots \lambda_{d-k} i_1 \ldots i_k} \alpha_{i_1 \ldots i_k}(x) \, dx^{j_1} \wedge \cdots \wedge dx^{j_{d-k}} \tag{6.87}$$

and satisfies

$$(\star)^2 = (-1)^{(d-1)k} \tag{6.88}$$

on $\Lambda^k \mathcal{M}$. It defines an inner product $\int_\mathcal{M} \alpha \wedge \star \beta$ on each vector space $\Lambda^k \mathcal{M}$. Using this inner product it is possible to show that a differential form λ_h as above is harmonic if and only if

$$d\lambda_h = d \star \lambda_h = 0 \tag{6.89}$$

and the Hodge decomposition theorem (6.84) (which can be generalized to arbitrary degree differential forms in the general case) implies that the DeRham cohomology groups of $\mathcal{M}$ are spanned by a basis of harmonic forms.

The Hodge decomposition (6.84) is unique and the components involved there are explicitly given by

$$\chi = \frac{1}{\nabla^2_{\Sigma^h}} \star d \star g(V, \cdot) \quad , \quad \xi = \frac{1}{\nabla^2_{\Sigma^h}} \star dg(V, \cdot) \tag{6.90}$$

6.6 Generalization to Hyperbolic Riemann Surfaces

where the scalar Laplacians $\nabla^2_{\Sigma^h} \equiv \star d \star d$ in (6.90) are assumed to have their zero modes removed. The 1-form λ_h in (6.84) can be written as a linear combination of basis elements of the DeRham cohomology group $H^1(\Sigma^h;\mathbb{R})$. According to the Poincaré-Hodge duality theorem [32], we can in particular choose an orthonormal basis of harmonic 1-forms $\{\alpha_\ell, \beta_\ell\}_{\ell=1}^h \in H^1(\Sigma^h;\mathbb{R})$ which are Poincaré-dual to the chosen canonical homology basis $\{a_\ell, b_\ell\}_{\ell=1}^h \in H_1(\mathcal{M};\mathbb{Z})$ above, i.e.

$$\oint_{a_\ell} \alpha_{\ell'} = \oint_{b_\ell} \beta_{\ell'} = \delta_{\ell\ell'} \quad , \quad \oint_{b_\ell} \alpha_{\ell'} = \oint_{a_\ell} \beta_{\ell'} = 0 \qquad (6.91)$$

We remark here that the local parts of the decomposition (6.84) simply form the decomposition of the vector $g_{\Sigma^h}(V, \cdot)$ into its curl-free, longitudinal and divergence-free, transverse pieces as $\nabla_{\Sigma^h}\chi + \nabla_{\Sigma^h} \times \xi$. The harmonic part λ_h accounts for the fact that this 1-form may sit in a non-trivial DeRham cohomology class of $H^1(\Sigma^h;\mathbb{R})$.

We can now write the general form of the isometries of Σ^h. As before, Σ^h inherits 3 local isometries via the bundle projection in (6.81) from the maximally symmetric Poincaré upper half-plane. However, only the 2 quasi-translations on $\mathcal{H}^2$ become global isometries of Σ^h, and they can be expressed in terms of the canonical homology basis using the above relations. This global isometry condition along with (6.82) and (6.83) imply that Hodge decomposition (6.84) of the metric-dual 1-form to the Hamiltonian vector field V_{Σ^h} is simply given by its harmonic part which can be written as

$$g_{\Sigma^h}(V_{\Sigma^h}, \cdot) = \sum_{\ell=1}^h (V_1^\ell \alpha_\ell + V_2^\ell \beta_\ell) \qquad (6.92)$$

The harmonic decomposition (6.92) is the generalization of (6.18). Indeed, on the torus we can identify the canonical harmonic forms above as $\alpha = d\phi_1/2\pi$ and $\beta = d\phi_2/2\pi$. The Killing vectors dual to (6.92) generate translations along the homology cycles of Σ^h, and the isometry group of Σ^h is $\prod_{i=1}^{2h} U(1)$. The usual equivariance condition $\mathcal{L}_{V_{\Sigma^h}}\omega = 0$ on the symplectic 2-form of Σ^h now becomes

$$div_{V_{\Sigma^h}}\omega = \sum_{\ell=1}^h d\left(\bar{\omega} \star \{V_1^\ell \alpha_\ell + V_2^\ell \beta_\ell\}\right) = 0 \qquad (6.93)$$

where $\bar{\omega}(x)$ is the C^∞-function on Σ^h defined by $\omega_{\mu\nu}(x) = \bar{\omega}(x)\epsilon_{\mu\nu}$, and (6.93) implies that it is constant on Σ^h, just as in (6.19).

Integrating up the Hamiltonian equations we see therefore that the unique equivariant (Darboux) Hamiltonians have the form

$$H_{\Sigma^h}(x) = \sum_{\ell=1}^h \int_{C_x} (h_1^\ell \alpha_\ell + h_2^\ell \beta_\ell) \qquad (6.94)$$

where h_μ^ℓ are real-valued constants and $C_x \subset \Sigma^h$ is a simple curve from some fixed basepoint to x. The Hamiltonian (6.94) is multi-valued because it depends explicitly on the particular representatives α_ℓ, β_ℓ of the DeRham cohomology classes in $H^1(\Sigma^h; \mathbb{R}) = \mathbb{R}^{2h}$. As before, single-valuedness of the time-evolution operator requires that $h_\mu^\ell = n_\mu^\ell h$, for some $n_\mu^\ell \in \mathbb{Z}$ and $h \in \mathbb{R}$, and the propagation times are again the discrete intervals $T = N_T h^{-1}$. Thus the Hamiltonian (6.94) represents the windings around the non-trivial homology cycles of Σ^h and the partition function defines a topological quantum field theory which again represents the homology classes of Σ^h through a family of homomorphisms from $\bigoplus_{i=1}^{2h} \mathbb{Z}$ into $U(1)^{\otimes 2h}$. Again, the partition function path integral should be properly defined in the homologically-invariant form (6.27) to make the usual quantities appearing in the associated action S well-defined by restricting the functional integrations to homotopically equivalent loops. We note that again the general conformal factor involved in the metric g_{Σ^h} obeys Riemannian restrictions from the Gauss-Bonnet-Chern theorem and a volume constraint similar to those in Section 6.2 above. When the volume parameter is quantized as in (6.55), we expect that the Hilbert space of physical states will be $(v_1 v_2)^{3h-3}$ dimensional (one copy of the genus 1 Hilbert spaces for each of the $3h - 3$ modular degrees of freedom in this case) and the coherent state wavefunctions, which can be expressed in terms of $D = 3h - 3$ dimensional Jacobi theta functions (6.59), will in addition carry a $(v_2)^{3h-3}$ dimensional projective representation of the discrete first homology group of Σ^h (i.e. of the equivariant localization constraint algebra). The explicit proofs of all of the above facts appear to be difficult, because of the lack of complex coordinatization for these manifolds which is required for the definition of coherent states associated with the isometry group action $\prod_{i=1}^{2h} U(1) = \prod_{i=1}^{2h} S^1$ on the non-symmetric space $\Sigma^h = (S^1 \times S^1)^{\#h}$.

Thus the general feature of abelian equivariant localization of path integrals on multiply connected compact Riemann surfaces is that it leads to a topological quantum theory whose associated topologically invariant partition function represents the non-trivial homology classes of the phase space. The coherent states in the finite-dimensional Hilbert space also carry a multi-dimensional representation of the discrete first homology group, and the localizable Hamiltonians on these phases spaces are rather unusual and even more restricted than in the simply-connected cases. The invariant symplectic 2-forms in these cases are non-trivial elements of $H^2(\Sigma^h; \mathbb{Z}) = \mathbb{Z}$, as in the maximally-symmetric cases, and it is essentially the global topological features of these multiply-connected phase spaces which leads to these rather severe restrictions. The coherent state quantization of these systems shows that the path integral describes the coadjoint orbit quantization of an unusual spin system described by the Riemann surface. These spin systems are exactly solvable both from the point of view of path integral quantization on the loop space and of canonical holomorphic quantization in the Schrödinger polarization. The localizable systems that one obtains in these cases are rather

6.6 Generalization to Hyperbolic Riemann Surfaces

trivial in appearence and are associated with *abelian* isometry groups acting on the phase spaces. However, these quantum theories probe deep geometric and topological features of the phase spaces, such as their complex algebraic geometry and their homology. This is in contrast with the topological quantum field theories that we found in the simply-connected cases, where at best the topological path integral could only represent the possible non-trivial cohomology classes in $H^2(\mathcal{M};\mathbb{Z})$. It is not completely clear though how these path integral representations correspond to analogs of the standard character formulas on homogeneous symplectic manifolds which are associated with semi-simple Lie groups, since, for instance, the usual Kähler structure between the Riemannian and symplectic geometries is absent in these non-symmetric cases.

7. Beyond the Semi-Classical Approximation

In this Chapter we shall examine a different approach to the problem of localization [157]. We return to the general finite-dimensional analysis of Chapter 3 and consider a Hamiltonian system whose Hamiltonian function is a Morse function[1]. From this we will construct the full $\frac{1}{T}$-expansion for the classical partition function, as we described briefly in Section 3.3. A proper covariantization of this expansion will then allow us to determine somewhat general geometrical characteristics of dynamical systems whose partition functions localize, which in this context will be the vanishing of all terms in the perturbative loop expansion beyond 1-loop order. The possible advantages of this analysis are numerous. For instance, we can analyse the fundamental isometry condition required for equivariant localization and see more precisely what mechanism or symmetry makes the higher-order terms disappear. This could then expand the set of localizable systems beyond the ones we have already encountered that are predicted from localization theory, and at the same time probe deeper into the geometrical structures of the phase space or the whole dynamical system thus providing richer examples of topological field theories. Indeed we shall find some noteworthy geometrical significances of when a partition function is given exactly by its semi-classical approximation as well as new geometric criteria for localization which expand the previous isometry conditions.

This approach to the Duistermaat-Heckman integration formula using the perturbative loop-expansion has been discussed in a different context recently in [174] where the classical partition function was evaluated for the dynamical system describing the kinematics of thin vortex tubes in a 3-dimensional fluid whose Hamiltonian is similar to that considered at the end of Section 5.8 for geodesic motion on group manifolds. For such fluid mechanics problems, the phase space $\mathcal{M}$ is neither finite-dimensional nor compact and the Hamiltonian flows need not be periodic, but the dynamical system admits an infinite sequence of constants of motion which are in involution so that it should be localizable [174]. The standard localization analyses therefore do not apply and one needs to resort to an analysis of the sort which will follow here. We

---
[1] The extension to degenerate Hamiltonians is fairly straightforward. In what follows all statements made concerning the structure of the discrete critical point set $\mathcal{M}_V$ of H will then apply to the full critical submanifold.

shall indeed find extensions of the localization formalisms which cover certain such cases.

Recalling that the isometry condition can always be satisfied at least *locally* on $\mathcal{M}$, we then present some ideas towards developing a novel geometric method for systematically constructing corrections to the Duistermaat-Heckman formula. Given that a particular system does not localize, the idea is that we can "localize" in local neighbourhoods on $\mathcal{M}$ where the Killing equation can be satisfied. The correction terms are then picked up when these open sets are patched back together on the manifold, as then there are non-trivial singular contributions to the usual 1-loop term owing to the fact that the Lie derived metric tensor cannot be defined globally in a smooth way over the entire manifold $\mathcal{M}$. Recalling from Section 3.6 that the properties of such a metric tensor are intimately related to the integrability properties of the dynamical system, we can explore the integrability problem again in a (different) geometric setting now by closely examining these correction terms. This will provide a highly non-trivial geometric classification of the localizability of a dynamical system which is related to the homology of $\mathcal{M}$, the integrability of the dynamical system, and is moreover completely consistent with Kirwan's theorem. Although these ideas are not yet fully developed, they do provide a first step to a full analysis of corrections to path integral localization formulas (e.g. corrections to the WKB approximation), and to uncovering systematically the reasons why these approximations aren't exact for certain dynamical systems [157]. The generalizations of these ideas to path integrals are not yet known, but we discuss the situation somewhat heuristically at the end of this Chapter.

7.1 Geometrical Characterizations of the Loop Expansion

Throughout this Chapter we return to the situation of Section 3.3 where the Hamiltonian H is a Morse function on a (usually compact) symplectic manifold $\mathcal{M}$. For now we assume that $\partial \mathcal{M} = \emptyset$, but later we shall also consider manifolds with boundary. We now explicitly work out the full stationary phase series whose construction we briefly outlined in Section 3.3 [72]. We first expand the C^∞-function H in a neighbourhood U_p of a given critical point $p \in \mathcal{M}_V$ in a Taylor series

$$H(x) = H(p) + \mathcal{H}(p)_{\mu\nu} x_p^\mu x_p^\nu / 2 + g(x;p) \quad , \quad x \in U_p \quad (7.1)$$

where $x_p = x - p \in U_p$ are the fluctuation modes about the extrema of H and $g(x;p)$ is the Gaussian deviation of $H(x)$ in the neighbourhood U_p (i.e. all terms in the Taylor series beyond quadratic order). The determinant of the symplectic 2-form which appears in (3.52) is similarly expanded in U_p as

7.1 Geometrical Characterizations of the Loop Expansion

$$\sqrt{\det \omega(x)} = \sqrt{\det \omega(p)} + \sum_{k=1}^{\infty} \frac{1}{k!} x_p^{\mu_1} \cdots x_p^{\mu_k} \partial_{\mu_1} \cdots \partial_{\mu_k} \sqrt{\det \omega(x)} \Big|_{x=p} \quad (7.2)$$

where $x \in U_p$.

We substitute (7.1) and (7.2) into (3.52), expand the exponential function there in powers of the Gaussian deviation function, and then integrate over each of the neighbourhoods $U_p \simeq \mathbb{R}^{2n}$. In this way we arrive at a series expansion of (3.52) for large-T in terms of Gaussian moment integrals over the fluctuations x_p, with Gaussian weight $e^{iT\mathcal{H}(p)_{\mu\nu} x_p^\mu x_p^\nu /2}$, associated with each open neighbourhood U_p for $p \in \mathcal{M}_V$. The Gaussian moments $\langle x^{\mu_1} \cdots x^{\mu_k} \rangle$ can be found from the Gaussian integration formula (1.2) in the usual way by applying the operator $\frac{\partial}{\partial \lambda_{\mu_1}} \cdots \frac{\partial}{\partial \lambda_{\mu_k}}$ to both sides of (1.2) and then setting all the λ's equal to 0. The odd-order moments vanish, since these integrands are odd functions, and the $2k$-th order moment contributes a term of order $\mathcal{O}(1/T^{n+k})$. Rearranging terms carefully, taking into account the signature of the Hessian at each critical point, and noting that for large-T the integral will localize around each of the disjoint neighbourhoods U_p, we arrive at the standard stationary-phase expansion[2]

$$Z(T) = \left(\frac{2\pi i}{T}\right)^n \sum_{p \in \mathcal{M}_V} (-i)^{\lambda(p)} e^{iTH(p)} \sum_{\ell=0}^{\infty} \frac{A_\ell(p)}{(-2T)^\ell} \quad (7.3)$$

where

$$A_\ell(p) = \frac{1}{\sqrt{\det \mathcal{H}(p)}} \sum_{j=0}^{2\ell} \frac{(-1)^j}{2^j j!(\ell+j)!} \left(\mathcal{H}(p)^{\mu\nu} \partial_\mu \partial_\nu\right)^{\ell+j} \\ \times \left(g(x;p)^j \sqrt{\det \omega(x)}\right)\Big|_{x=p} \quad (7.4)$$

and $\mathcal{H}(x)^{\mu\nu}$ is the matrix inverse of $\mathcal{H}(x)_{\mu\nu}$.

If the stationary-phase series diverges (e.g. applying Kirwan's theorem in appropriate instances), then (7.3) is to be understood formally as an asymptotic expansion order by order in $\frac{1}{T}$. Borrowing terminology from quantum field theory, we shall refer to the series (7.3) as the loop-expansion of this zero-dimensional quantum field theory, because each of the $2\ell + 1$ terms in (7.4) can be understood from pairing fluctuation modes $x_p^\mu x_p^\nu$ (i.e. a loop) associated with each derivative operator there. Indeed, the expansion (7.3),(7.4) is just the finite-dimensional version of the perturbation expansion (for large-T) in quantum field theory. We shall call the $\mathcal{O}(1/T^{n+\ell})$ contribution to the series (7.3) the $(\ell+1)$-loop term.

In this Section we shall be interested in extracting information about the dynamical system under consideration from the loop-expansion with the hope

---

[2] See [72] for the generalization of this formula to the case where H is a degenerate function.

of being able to understand the vanishing or non-vanishing of the k-loop contributions for $k > 1$ in terms of geometrical and topological features of the phase space. Given our experience now with the Duistermaat-Heckman theorem, we will remove any requirements on the flows of the Hamiltonian vector field and leave these as quite arbitrary for now. When these orbits describe tori, we already have a thorough understanding of the localization in terms of equivariant cohomology, and we shall therefore look at dynamical systems which do not necessarily obey this requirement. Thus any classification that we obtain below that is described solely by the vanishing of higher-loop contributions will for the most part be of a different geometrical nature than the situation that prevails in Duistermaat-Heckman localization. This then has the possibility of expanding the cohomological symmetries usually resposible for localization.

The perturbative series (7.3), however, must be appropriately modified before we can put it to use. The partition function (3.52) is invariant under arbitrary smooth changes of local coordinates on $\mathcal{M}$ (i.e. $Z(T)$ is manifestly a topological invariant), and this should be reflected order by order in the $1/T$-expansion (7.3). This is explicitly observed in the lowest-order Duistermaat-Heckman term $A_0(p)$ above, but the higher-order terms (7.4) in the loop expansion are not manifestly scalar quantities under local diffeomorphisms of the coordinates. This is a result of having to pick local coordinates on $\mathcal{M}$ to carry out explicitly the Gaussian integrations in $\mathbb{R}^{2n}$. At each order of the $\frac{1}{T}$-expansion we should have a manifestly coordinate independent quantity, i.e. a scalar. To write the contributions (7.4) in such a fashion so as to be manifestly invariant under local diffeomorphisms of $\mathcal{M}$, we have to introduce a Christoffel connection $\Gamma^\lambda_{\mu\nu}$ of the tangent bundle of $\mathcal{M}$ which makes the derivative operators appearing in (7.4) manifestly covariant objects, i.e. we write them in terms of covariant derivatives $\nabla = d + \Gamma$. Because $dH(p) = 0$ at a critical point $p \in \mathcal{M}_V$, the Hessian evaluated at a critical point is automatically covariant, i.e. $\nabla\nabla H(p) = \mathcal{H}(p)$. This process, which we shall call 'covariantization', will then ensure that each term (7.4) is manifestly a scalar. We note that the Morse index of any critical point is a topological invariant in this sense.

First, we cycle out the symplectic factors in (7.4) to get

$$A_\ell(p) = A_0(p) \sum_{j=0}^{2\ell} \frac{(-1)^j}{2^j j!(\ell+j)!} \left(\mathcal{H}(p)^{\mu\nu}\mathcal{D}_\mu\mathcal{D}_\nu\right)^{\ell+j} g(x;p)^j \bigg|_{x=p} \qquad (7.5)$$

where

$$A_0(p) = \sqrt{\frac{\det \omega(p)}{\det \mathcal{H}(p)}} \qquad (7.6)$$

is the Duistermaat-Heckman (1-loop) contribution to (7.3), and

$$\mathcal{D} = d + \gamma \qquad (7.7)$$

7.1 Geometrical Characterizations of the Loop Expansion

where we have introduced the one-component connection

$$\gamma = h_L^{-1} dh_L \tag{7.8}$$

and

$$h_L(x) = \sqrt{\det \omega(x)} \tag{7.9}$$

is the Liouville volume density. The derivative operator $\mathcal{D}$ transforms like an abelian gauge connection under local diffeomorphisms $x \to x'(x)$ of $\mathcal{M}$,

$$\mathcal{D}_\mu(x) \xrightarrow{\Lambda} \mathcal{D}'_\mu(x') = \Lambda^\nu_\mu(x') \left[\mathcal{D}_\nu(x') + \text{tr } \Lambda^{-1}(x') \partial'_\nu \Lambda(x') \right] \tag{7.10}$$

where

$$\Lambda^{-1}(x) = \left[\frac{\partial x'^\mu(x)}{\partial x^\nu} \right] \in GL(2n, \mathbb{R}) \tag{7.11}$$

is the induced change of basis transformation on the tangent bundle.

For example, consider the expression (7.5) for $\ell = 1$. We expand out the 3 terms there in higher-order derivatives of H and the connection γ, noting that only third- and higher-order derivatives of $g(x;p)$ when evaluated at $x = p$ are non-vanishing. After some algebra, we arrive at

$$A_1(p) = \frac{A_0(p)}{2} \mathcal{H}(p)^{\mu\nu} \left\{ \mathcal{D}_\mu(x)\gamma_\nu(x) - \frac{\mathcal{H}(p)^{\lambda\rho}}{4} \left(\partial_\mu \partial_\lambda \partial_\nu \partial_\rho H(x) \right. \right.$$
$$+ 4\gamma_\mu(x)\partial_\lambda \partial_\nu \partial_\rho H(x) + \frac{\mathcal{H}(p)^{\alpha\beta}}{12} \left[3\partial_\mu \partial_\lambda \partial_\nu H(x) \partial_\rho \partial_\alpha \partial_\beta H(x) \right. \tag{7.12}$$
$$\left. \left. + 2\partial_\mu \partial_\lambda \partial_\alpha H(x) \partial_\nu \partial_\rho \partial_\beta H(x) \right] \right) \right\} \bigg|_{x=p}$$

It is readily checked, after some algebra, that this expression is indeed invariant under local diffeomorphisms of $\mathcal{M}$. To manifestly covariantize it, we introduce an arbitrary torsion-free connection $\Gamma^\lambda_{\mu\nu}$ of the tangent bundle $T\mathcal{M}$. For now, we need not assume that Γ is the Levi-Civita connection associated with a Riemannian metric on $\mathcal{M}$. Indeed, as the original dynamical problem is defined only in terms of a symplectic geometry, not a Riemannian geometry, the expression (7.12) should be manifestly covariant in its own right without reference to any geometry that is external to the problem. All that is required is some connection that specifies parallel transport along the fibers of the tangent bundle and allows us to extend derivatives of quantities to an entire neighbourhood, rather than just at a point, in a covariant way.

The Hessian of H can be written in terms of this connection and the associated covariant derivative as

$$\mathcal{H}(x)_{\mu\nu} = \nabla_\mu \nabla_\nu H(x) + \Gamma^\lambda_{\mu\nu}(x) \partial_\lambda H(x) \tag{7.13}$$

and, using $d \equiv \nabla - \Gamma$, we can write the third and fourth order derivatives appearing in (7.12) in terms of ∇ and Γ by taking derivatives of (7.13).

238   7. Beyond the Semi-Classical Approximation

Substituting these complicated derivative expressions into (7.12) and using the fact that $dH = 0$ on $\mathcal{M}_V$, after a long and quite tedious calculation we arrive at a manifestly covariant and coordinate-independent expression for the 2-loop correction,

$$A_1(p) = \frac{A_0(p)}{8} \mathcal{H}(p)^{\mu\nu} \left\{ \frac{\mathcal{H}(p)^{\lambda\rho}\mathcal{H}(p)^{\alpha\beta}}{3} \left[3\nabla_\mu \nabla_\nu \nabla_\lambda H(x) \nabla_\alpha \nabla_\beta \nabla_\rho H(x) \right. \right.$$
$$\left. + 2\nabla_\mu \nabla_\lambda \nabla_\alpha H(x) \nabla_\nu \nabla_\rho \nabla_\beta H(x) \right] - \mathcal{H}(p)^{\lambda\rho} \nabla_\rho \nabla_\mu \nabla_\nu \nabla_\lambda H(x)$$
$$\left. + 4 \left(\nabla_\mu + \Delta_\mu - \mathcal{H}(p)^{\lambda\rho} \nabla_\rho \nabla_\mu \nabla_\lambda H(x) \right) \Delta_\nu(x) + R_{\mu\nu}(\Gamma) \right\} \bigg|_{x=p}$$
(7.14)

where

$$R_{\mu\nu}(\Gamma) \equiv R^\lambda_{\mu\lambda\nu} = \partial_\nu \Gamma^\lambda_{\mu\lambda} - \partial_\lambda \Gamma^\lambda_{\mu\nu} + \Gamma^\alpha_{\mu\lambda} \Gamma^\lambda_{\alpha\nu} - \Gamma^\alpha_{\mu\nu} \Gamma^\lambda_{\alpha\lambda} \qquad (7.15)$$

is the (symmetric) Ricci curvature tensor of Γ and we have introduced the 1-form $\Delta = \Delta_\mu(x)dx^\mu$ with the local components

$$\Delta_\mu(x) = \gamma_\mu(x) - \Gamma^\lambda_{\mu\lambda}(x) = \nabla_\mu \log h_L(x) \qquad (7.16)$$

It is intriguing that the covariantization of the 2-loop expression simply involves replacing ordinary derivatives d with covariant ones ∇, non-covariant connection terms γ with the 1-form Δ, and then the remainder terms from this process are simply determined by the curvature of the Christoffel connection Γ which realizes the covariantization. Note that if Γ is in addition chosen as the Levi-Civita connection compatible with a metric g, i.e. $\nabla g = 0$, then $\Gamma^\lambda_{\mu\lambda} = \partial_\mu \log \sqrt{\det g}$ and the 1-form components (7.16) become $\Delta_\mu = \partial_\mu \log \sqrt{\det(g^{-1} \cdot \omega)}$.

The covariantization of the higher-loop terms is hopelessly complicated. We note that if, however, the 2-loop correction $A_1(x)$ vanishes in an entire neighbourhood $U_p \subset \mathcal{M}$ of each critical point p, then this is enough to imply the vanishing of the corrections to the Duistermaat-Heckman formula to all orders in the loop-expansion. To see this, we exploit the topological invariance of (7.14) and apply the Morse lemma [111] to the correction terms (7.5). This says that there exists a sufficiently small neighbourhood U_p about each critical point p in which the Morse function H looks like a "harmonic oscillator",

$$H(x) = H(p) - (x^1)^2 - (x^2)^2 - \ldots - (x^{\lambda(p)})^2 + (x^{\lambda(p)+1})^2 + \ldots + (x^{2n})^2 \quad , \quad x \in U_p$$
(7.17)

so that the critical point p is at $x = 0$ in this open set in $\mathcal{M}$. We shall call these 'harmonic coordinates', and this result simply means that the symmetric matrix $\mathcal{H}(x)$ can be diagonalized constantly in an entire neighbourhood of the critical point. Given that the quantity (7.5) must be independent of coordinates (although not manifestly), we can evaluate it in a harmonic coordinate

7.1 Geometrical Characterizations of the Loop Expansion 239

system. Then the Gaussian deviation function $g(x;p)$ vanishes identically in the neighbourhood U_p and only the $j=0$ term contributes to (7.5). Thus the series (7.3) is simply

$$Z(T) = \left(\frac{2\pi i}{T}\right)^n \sum_{p \in \mathcal{M}_V} (-i)^{\lambda(p)}\, e^{iTH(p)} A_0(p) \left(e^{-\frac{\mathcal{H}(p)^{\mu\nu}}{2T}\mathcal{D}_\mu(x)\mathcal{D}_\nu(x)} \right) \cdot 1 \Big|_{x=p} \tag{7.18}$$

It follows that if the 2-loop term vanishes in the entire *neighbourhood* of the critical point p (and not just *at* $x=p$), i.e.

$$\mathcal{H}(p)^{\mu\nu}\mathcal{D}_\mu(x)\mathcal{D}_\nu(x) \equiv 0 \quad \text{for} \quad x \in U_p \tag{7.19}$$

then, as all higher-loop terms in these coordinates can be written as derivative operators acting on the 2-loop contribution $A_1(x)$ as prescribed by (7.18), all corrections to the semi-classical approximation vanish. In the case of the Duistermaat-Heckman theorem, it is for this reason that the Lie derivative condition $\mathcal{L}_V g = 0$ is generally required to hold globally on $\mathcal{M}$. In general, though, whether or not the vanishing of $A_1(p)$ implies the vanishing of all loop orders is not that clear, because there is a large ambiguity in the structure of a function in a neighbourhood of $\mathcal{M}_V$ which vanishes at each critical point p (any functional of ∇H will do). The above vanishing property in an entire neighbourhood therefore need not be true. It is hard to imagine though that the vanishing of the 2-loop correction term would not imply the vanishing of all higher-orders, because then there would be an infinite set of conditions that a dynamical system would have to obey in order for its partition function to be WKB-exact. This would then greatly limit the possibilities for localization.

In any case, from the point of view of localization, we can consider the vanishing of the 2-loop contribution in (7.14) as an infinitesimal Duistermaat-Heckman localization of the partition function. The expression (7.14) in general is extremely complicated. However, besides being manifestly independent of the choice of coordinates, (7.14) is independent of the chosen connection Γ, because by construction it simply reduces to the original connection-independent term (7.12). We can exploit this degree of freedom by choosing a connection that simplifies the correction (7.14) to a form that is amenable to explicit analysis. We shall now describe 2 geometrical characteristics of the loop-expansion above which can be used to classify the localizable or non-localizable dynamical systems [157].

The first such general geometric localization symmetry is a symmetry implied by (7.14) between Hessian and metric tensors, i.e. that the Hessian essentially defines a metric on $\mathcal{M}$. This is evident in the correction term (7.14), where the inverse Hessians contract with the other tensorial terms to form scalars, i.e. the Hessians in that expression act just like metrics. This suggests that the non-degenerate Hessian of H could be used to define a metric which is compatible with the connection Γ used in (7.14). This in general cannot be done globally on the manifold $\mathcal{M}$, because the signature of $\mathcal{H}(x)$ varies over $\mathcal{M}$ in general, but for a C^∞ Hamiltonian H it can at least be done locally in a sufficiently small neighbourhood surrounding each critical point. For now, we concentrate on the case of a 2-dimensional phase space. We *define* a Riemannian metric tensor g that is proportional to the covariant Hessian in the neighbourhood U_p of each critical point p,

$$\nabla \nabla H = \mathcal{G} g \tag{7.20}$$

where $\mathcal{G}(x)$ is some globally-defined C^∞-function on $\mathcal{M}$, and for which the connection Γ used in the covariant derivatives ∇ is the Levi-Civita connection for g. This means that, given a Hamiltonian H on $\mathcal{M}$, we try to locally solve the coupled system of non-linear partial differential equations

$$\mathcal{G}(x) g_{\mu\nu}(x) = \partial_\mu \partial_\nu H(x) - \Gamma^\lambda_{\mu\nu}(x) \partial_\lambda H(x)$$
$$\Gamma^\lambda_{\mu\nu} = \frac{1}{2} g^{\lambda\rho} (\partial_\mu g_{\rho\nu} + \partial_\nu g_{\rho\mu} - \partial_\rho g_{\mu\nu}) \tag{7.21}$$

consistently for g (or Γ).

This sort of "metric ansatz" may seem somewhat peculiar, and indeed impossible to solve in the general case. The covariant constancy condition on g in (7.21) implies that

$$\partial_\mu \mathcal{G} = R^\lambda_\mu \partial_\lambda H = R \partial_\mu H \tag{7.22}$$

where $R = g^{\mu\nu} R_{\mu\nu}$ is the scalar curvature of g. (7.22) follows from the defining identity for the Riemann curvature tensor $R = d\Gamma + \Gamma \wedge \Gamma$,

$$\nabla_\mu \nabla_\nu \nabla_\lambda H = \nabla_\nu \nabla_\mu \nabla_\lambda H + R^\rho_{\lambda\mu\nu} \nabla_\rho H \tag{7.23}$$

Given H, (7.22) determines $\mathcal{G}$ locally in terms of g. This means that the above ansatz can be written as an equation for the associated connection coefficients $\Gamma^\lambda_{\mu\nu}$,

$$\nabla_\lambda \nabla_\mu \nabla_\nu H = R_{\mu\nu} \nabla_\lambda H \tag{7.24}$$

The existence of a local solution to (7.24) 'almost' all of the time can now be argued by analysing this set of differential equations in local isothermal coordinates (5.49) for the connection Γ above [157]. Notice, in particular, that if the metric (7.20) actually has a constant curvature R, then the equation (7.22) can be integrated in each neighbourhood U_p to give

$$\mathcal{G}(x) = C_0 + R \cdot H(x) \tag{7.25}$$

We shall see examples and other evidences for this sort of geometric structure throughout this Chapter.

The main advantage of using the inductively-defined metric in (7.21) is that it allows a relatively straightforward analysis of the 2-loop correction

7.1 Geometrical Characterizations of the Loop Expansion   241

terms to the Duistermaat-Heckman formula. With this definition, (7.22) implies that the first order derivatives of $\mathcal{G}$ vanish at each critical point $p \in \mathcal{M}_V$, and hence so do all third order covariant derivatives of H in (7.14). The fourth order covariant derivatives contribute curvature terms according to (7.24), which are then cancelled by the curvature tensor already present in (7.14). The final result is an expression involving only the Liouville and Levi-Civita connections expressed in terms of the 1-form Δ, which after some algebra we find can be written in the simple form

$$A_1(p) = \frac{A_0(p)}{2\mathcal{G}} g^{\mu\nu} \left[\nabla_\mu + \Delta_\mu(x)\right] \Delta_\nu(x) \bigg|_{x=p} \qquad (7.26)$$

Requiring this correction term to vanish in an entire neighbourhood of each critical point implies, from the definition (7.16), that the connection γ associated with the symplectic structure coincides with the connection Γ associated with the Riemannian structure which solves the relation (7.21). Thus the components of the symplectic 2-form ω and the metric tensor g are proportional to each other in local complex isothermal coordinates for g. The proportionality factor must be a constant so that this be consistent with the existence of local Darboux coordinates for ω [157]. In other words, ω and g together locally define a Kähler structure on the phase space $\mathcal{M}$.

Conversely, suppose that Γ is the Levi-Civita connection associated to some generic, globally-defined metric tensor g on $\mathcal{M}$, and consider the rank (1,1) tensor field

$$J_\mu^\nu = \sqrt{\frac{\det \omega}{\det g}} g_{\mu\lambda} \omega^{\lambda\nu} \qquad (7.27)$$

In 2-dimensions, it is easily seen that (7.27) defines a linear isomorphism $J : T\mathcal{M} \to T\mathcal{M}$ satisfying $J^2 = -\mathbf{1}$. In general, if such a linear transformation J exists then it is called an almost complex structure of the manifold $\mathcal{M}$ [41, 61]. This means that there is a local basis of tangent vectors in which the only non-vanishing components of J are given by (5.9), so that there is "almost" a separation of the tangent bundle into holomorphic and anti-holomorphic components. However, an almost complex structure does not necessarily lead to a complex structure – there are certain sufficiency requirements to be met before J can be used to define local complex coordinates in which the overlap transition functions can be taken to be holomorphic [22]. One such case is when J is covariantly constant, $\nabla J = 0$ – actually this condition only ensures that a sub-collection of subsets of the differentiable structure determine a local complex structure (but recall that any Riemann surface is a complex manifold). Again in 2-dimensions this means that then $\nabla \omega = 0$ and the pair (g, ω) define a Kähler structure on $\mathcal{M}$ (again note that any 2-dimensional symplectic manifold is automatically a Kähler manifold for *some* metric defined by ω).

Given these facts, suppose now that g and ω define a Kähler structure on the $2n$-dimensional manifold $\mathcal{M}$ with respect to an almost complex structure

J, i.e. $\det \omega = \det g$, g is Hermitian with respect to J,

$$g_{\mu\nu} = J_\mu^\lambda g_{\lambda\rho} J_\nu^\rho , \qquad (7.28)$$

and ω is determined from g by (7.27). In the local coordinates (5.9), this means the usual Kähler conditions that we encountered before, i.e. $g_{\mu\nu} = g_{\bar{\mu}\bar{\nu}} = 0$, $g_{\mu\bar{\nu}} = g^*_{\bar{\nu}\mu}$ and $\omega_{\mu\bar{\nu}} = -ig_{\mu\bar{\nu}}$. In this case, the flows of g under the action of the Hamiltonian vector field V,

$$(\mathcal{L}_V g)_{\mu\nu} = g_{\mu\lambda}\nabla_\nu \omega^{\lambda\rho}\partial_\rho H + g_{\nu\lambda}\nabla_\mu \omega^{\lambda\rho}\partial_\rho H + \omega^{\lambda\rho}(g_{\mu\lambda}\nabla_\nu\nabla_\rho H + g_{\nu\lambda}\nabla_\mu\nabla_\rho H) \qquad (7.29)$$

can be written using the almost complex structure as the anti-commutator

$$\mathcal{L}_V g = [\nabla\nabla H, J]_+ \qquad (7.30)$$

Thus if V is a global Killing vector of a Kähler metric on $\mathcal{M}$, then the covariant Hessian of H is also Hermitian with respect to J, as in (7.28). Since the Kähler metric of a Kähler manifold is essentially the unique Hermitian rank (2,0) tensor, it follows that the covariant Hessian is related to the Kähler metric by a transformation of the form

$$\nabla_\mu\nabla_\nu H = K_\mu^\lambda g_{\lambda\rho} K_\nu^\rho \qquad (7.31)$$

where K is some non-singular (1,1) tensor which commutes with J. In 2-dimensions, the Hermiticity conditions imply that both the Hessian and g have only 1 degree of freedom and (7.31) gets replaced by the much simpler condition (7.21). From the fundamental equivariant localization principle we know that this implies the vanishing of the 2-loop correction term, i.e. the Duistermaat-Heckman theorem. Indeed, from the analysis of the last 2 Chapters we have seen that most of the localizable examples fall into these Kähler-type scenarios. Notice that the covariant Hessian determined from the Hamiltonian equations is

$$\nabla_\mu\nabla_\nu H = V^\lambda \nabla_\mu \omega_{\nu\lambda} + \omega_{\nu\lambda}\nabla_\mu V^\lambda \qquad (7.32)$$

so that in the Kähler case, when $\nabla\omega = 0$, the proportionality function $\mathcal{G}$ is determined in terms of the Riemann moment map $\mu_V = \nabla V$ as

$$\mathcal{G}(x) = \sqrt{\det \mu_V(x)} \qquad (7.33)$$

On a homogeneous Kähler manifold, when $\mathcal{G}$ is integrated to be (7.25), this relation generalizes that observed for the height function of the sphere in Section 5.5.

What is particularly interesting here is that conversely the localization of the partition function determines this sort of Kähler structure on $\mathcal{M}$. The Lie derivative (7.29) of the metric (7.21) is easily seen to be zero in a neighbourhood of the critical point. Conversely, if the Lie derivative of the

7.1 Geometrical Characterizations of the Loop Expansion 243

metric in (7.21) vanishes on $\mathcal{M}$, then it induces a Kähler structure (i.e. $\nabla \omega = 0$). Recalling from Section 3.3 the proof of the Duistermaat-Heckman theorem using solely symplectic geometry arguments, we see that the main feature of the localization is the possibility of simultaneously choosing harmonic and Darboux coordinates. This same feature occurs similarly above, when we map onto local Darboux coordinates [157]. The new insight gained here is the geometric manner in which this occurs – the vanishing of the loop expansion beyond leading order gives the dynamical system a local Kähler structure (see (7.19)). Whether or not this extends to a *global* Kähler geometry depends on many things. If the topology of $\mathcal{M}$ allows this to be globally extended away from $\mathcal{M}_V$, then the Riemannian geometry so introduced induces a global Kähler structure with respect to the canonical symplectic structure of $\mathcal{M}$ [3]. Furthermore, the coefficient function $\mathcal{G}(x)$ in (7.20) must be so that the metric defined by that equation has a constant signature on the whole of $\mathcal{M}$. If H has odd Morse indices $\lambda(p)$, then it is impossible to choose the function $\mathcal{G}$ in (7.20) such that, say, g has a uniform Euclidean signature on the whole of $\mathcal{M}$. But if the correction terms above vanish, then Kirwan's theorem implies that H has only even Morse indices and it may be possible to extend this geometry globally. In this way, the examination of the vanishing of the loop expansion beyond leading order gives insights into some novel geometrical structures on the phase space representing symmetries of the localization. Moreover, if such a metric is globally-defined on $\mathcal{M}$, then $\mathcal{L}_V g = 0$, and these classes of localizable systems fall into the same framework as those we studied before.

The second general geometric symmetry of the loop expansion that we wish to point out here is based on the observation that the symplectic connection (7.8) is reminiscent of the connection that appears when one constructs the Fubini-Study metric using the geometry of a holomorphic line bundle $L \to \mathcal{M}$ [41]. If we choose such a line bundle as the standard symplectic line bundle over $\mathcal{M}$ and view the Liouville density (7.9) as a metric in the fibers of this bundle, then from it one can construct a Kähler structure on $\mathcal{M}$ from the curvature 2-forms of the associated connections (7.8), i.e.

$$\Omega = -i(\partial + \bar{\partial})\gamma_{\bar{z}} = -i\partial\bar{\partial} \log h_L \qquad (7.34)$$

where we have restricted to the holomorphic and anti-holomorphic components of the connection (7.8). For instance, the symplectic structure (5.208) is the Fubini-Study metric associated with the natural line bundle $L \to \mathbb{C}P^N$

---

[3] The existence of an almost complex structure J for which the symplectic 2-form ω is Hermitian and for which the associated Kähler metric $g = J \cdot \omega$ is positive-definite is not really an issue for a symplectic manifold [171]. Such a J always exists (and is unique up to homotopy) because the Siegal upper-half plane is contractible. Thus the existence of a Kähler structure for which $\Delta = \nabla \log h_L = 0$ is not a problem. However, for the Killing equation for $g = J \cdot \omega$ to hold, J itself must be invariant under the flows of the Hamiltonian vector field V, i.e. $\mathcal{L}_V J = 0$.

244   7. Beyond the Semi-Classical Approximation

which is a sub-bundle of the trivial bundle $\mathbb{C}P^N \times \mathbb{C}^{N+1}$, i.e. the fiber of L over a point $p \in \mathbb{C}P^N$ is just the set of points $(z^1, \ldots, z^{N+1}) \in \mathbb{C}^{N+1}$ which belong to the line p. In that case, the natural fiber metric induced by the canonical complex Euclidean metric of $\mathbb{C}^{N+1}$ is $h(p; z^1, \ldots, z^{N+1}) = \sum_{\mu=1}^{N+1} |z^\mu|^2$ and (5.208) is determined as the symplectic reduction of h as in (7.34). This construction generalizes to any holomorphic line bundle once a fiber metric has been chosen. In the general case, if $\mathcal{M}$ itself is already a Kähler manifold, then whether or not (7.34) agrees with the original Kähler 2-form will depend on where these 2-forms sit in the DeRham cohomology of $\mathcal{M}$. If we further adjust the Christoffel connection Γ so that it is related to the Liouville connection by the boundary condition $\Gamma^\lambda_{\mu\lambda} = \gamma_\mu$, i.e. $\Delta = 0$, then the correction term (7.14) will involve only derivatives of the Hamiltonian, but now the vanishing of the correction term can be related to the geometry of a line bundle $L \to \mathcal{M}$.

This is as far as we shall go with a general geometric interpretation of localization from the covariant loop expansion. The above discussion shows what deep structures could be uncovered from further development of such analyses. It is in this way that the localization of the partition function probes the geometry and topology of the phase space $\mathcal{M}$ and thus can lead to interesting topological quantum field theories.

7.2 Conformal and Geodetic Localization Symmetries

In this Section we shall examine some alternative geometric symmetries which lead to a localization of the partition function of a dynamical system and which extend the fundamental symmetry requirement of the localization theorems we encountered earlier on [84, 134]. Let g be a globally-defined metric tensor on $\mathcal{M}$, and consider its flows under the Hamiltonian vector field V. Instead of the usual assumption that V be an infinitesimal isometry generator for g, we weaken this requirement and assume that instead V is globally an infinitesimal generator of *conformal* transformations with respect to g, i.e.

$$\mathcal{L}_V g = \Upsilon g \tag{7.35}$$

where $\Upsilon(x)$ is some C^∞-function on $\mathcal{M}$. Intuitively, this means that the diffeomorphisms generated by V preserve angles in the space, but not lengths. The function Υ can be explicitly determined by contracting both sides of (7.35) with g^{-1} to get

$$\Upsilon = \nabla_\mu V^\mu / n = \nabla_\mu \omega^{\mu\nu} \partial_\nu H / n \tag{7.36}$$

which we note vanishes on the critical point set $\mathcal{M}_V$ of the Hamiltonian H. This implies, in particular, that either $\Upsilon \equiv 0$ almost everywhere on $\mathcal{M}$ (in which case V is an isometry of g) or $\Upsilon(x)$ is a non-constant function on $\mathcal{M}$

7.2 Conformal and Geodetic Localization Symmetries

corresponding to non-homothetic or 'special conformal' transformations (i.e. constant rescalings, or dilatations, of g are not possible under the flow of a Hamiltonian vector field). Ordinary Killing vector fields in this context arise as those which are covariantly divergence-free, tr $\mu_V = \nabla_\mu V^\mu = 0$.

We shall show that this conformal symmetry requirement on the dynamical system also leads to the Duistermaat-Heckman integration formula. First, we establish this at the level of the perturbative loop expansion of the last Section by showing that the (infinitesimal) corrections to the 1-loop term in the stationary phase series vanish. For simplicity we restrict our attention here to the case of $n = 1$ degree of freedom. The extension to arbitrary degrees of freedom is immediate, and indeed we shall shortly see how the condition (7.35) explicitly implies the vanishing of the correction terms to the Duistermaat-Heckman formula in arbitrary dimensions and in a much more general framework. We insert everywhere into the covariant connection-independent expression (7.14) the covariant derivative ∇ associated with g in (7.35). Using the covariant Hessian (7.32) and the conformal Killing equation (7.35) we can solve for the covariant derivatives of the Hamiltonian vector field. Notice that, in contrast to the ordinary Killing equations, one of the 3 components of the conformal Killing equation (7.35) will be an identity since one of the Killing equations tells us that V is covariantly divergence-free with respect to g (see (5.51)). In 2-dimensions, after some algebra we find that the (symmetric) Hessian (7.32) can be written as

$$\nabla_\mu \nabla_\nu H = -\Omega \, \Sigma \, g_{\mu\nu} + \left(\nabla_\mu H \, \nabla_\nu \log \Omega + \nabla_\nu H \, \nabla_\mu \log \Omega\right)/2 \qquad (7.37)$$

where we have introduced the C^∞-functions

$$\Sigma(x) = \sqrt{\det \nabla V(x) - (\nabla_\lambda V^\lambda(x)/2)^2} \quad , \quad \Omega(x) = \sqrt{\frac{\det \omega(x)}{\det g(x)}} \qquad (7.38)$$

The covariant derivatives appearing in (7.14) are now easily found from (7.37) to be

$$\nabla_\rho \nabla_\mu \nabla_\nu H = \frac{1}{\Omega \Sigma} \nabla_\rho(\Omega \Sigma) \nabla_\mu \nabla_\nu H + \frac{1}{2} \nabla_\mu \log \Omega \nabla_\rho \nabla_\nu H \\ + \frac{1}{2} \nabla_\nu \log \Omega \nabla_\rho \nabla_\mu H + \mathcal{F}(\nabla H) \qquad (7.39)$$

246   7. Beyond the Semi-Classical Approximation

$$\nabla_\lambda \nabla_\rho \nabla_\mu \nabla_\nu H$$

$$= \frac{1}{\Omega \Sigma} \left(\nabla_\lambda \nabla_\rho (\Omega \Sigma) \nabla_\mu \nabla_\nu H + \frac{1}{2} \nabla_\lambda (\Sigma \nabla_\mu \Omega) \nabla_\rho \nabla_\nu H \right.$$

$$\left. - \frac{1}{2} \nabla_\lambda (\Sigma \nabla_\nu \Omega) \nabla_\rho \nabla_\mu H \right) + \frac{1}{2} \nabla_\lambda \nabla_\rho H \nabla_\mu \log \Omega \nabla_\nu \log \Omega \qquad (7.40)$$

$$+ \frac{1}{2} \nabla_\lambda \nabla_\nu H \left(\frac{1}{2} \nabla_\mu \log \Omega \nabla_\rho \log \Omega + \nabla_\rho \nabla_\mu \log \Omega \right)$$

$$+ \frac{1}{2} \nabla_\lambda \nabla_\mu H \left(\frac{1}{2} \nabla_\nu \log \Omega \nabla_\rho \log \Omega + \nabla_\rho \nabla_\nu \log \Omega \right) + \mathcal{F}(\nabla H)$$

where $\mathcal{F}(\nabla H)$ denotes terms involving single derivatives of H, and which therefore vanish on $I(H)$. Substituting (7.39) into (7.23) and then contracting (7.23) with $\mathcal{H}(x)^{\mu\lambda}$ leads to a relationship between the symplectic structure and the flows generated by the Hamiltonian vector field,

$$\nabla_\mu \log \Omega + 2 \nabla_\mu \log \Sigma = \mathcal{F}(\nabla H) \qquad (7.41)$$

We now use a covariant derivative of (7.23) to rewrite the curvature term in (7.14) in terms of covariant derivatives of the Hamiltonian, substitute (7.39) and (7.40) into (7.14), and use everywhere the identity (7.41). Since $\Delta_\mu = \nabla_\mu \log \Omega$ for the metric connection Γ, it follows after some algebra that the generally-covariant expression (7.14) for the first-order correction to the Duistermaat-Heckman formula is $\mathcal{F}(\nabla H)$. This establishes the claim above.

We shall now establish quite generally that this conformal localization symmetry is in some sense the most general one that can be construed for a classical partition function [134]. For this, we return to the general localization principle of Section 2.5 and re-evaluate the function $\mathcal{Z}(s)$ in (2.128) for a generic (not necessarily equivariant) metric-dual 1-form $\beta = i_V g$ of the Hamiltonian vector field V (i.e. with no assumptions for now about the symmetries of g) and the equivariantly-closed form $\alpha = \mathrm{e}^{iT(H+\omega)}/(iT)^n$ with respect to V, i.e. $D_V \alpha = 0$, but $\mathcal{L}_V \beta \neq 0$ in general. We also assume, for full generality, that the manifold $\mathcal{M}$ can have a boundary $\partial \mathcal{M}$. The derivative (2.129) need not vanish now and the second equality there can be evaluated using Stokes' theorem and the Cartan-Weil identity. Then using the identity

$$\mathcal{Z}(0) = \lim_{s \to \infty} \mathcal{Z}(s) - \int_0^\infty ds\, \frac{d}{ds} \mathcal{Z}(s) \qquad (7.42)$$

it follows that the partition function $Z(T) = \int_\mathcal{M} \alpha$ can be determined in general by

$$Z(T) = \lim_{s \to \infty} \int_\mathcal{M} \alpha\, \mathrm{e}^{-sD_V \beta} + \int_0^\infty ds \left\{ \oint_{\partial \mathcal{M}} \alpha \beta\, \mathrm{e}^{-sD_V \beta} - s \int_\mathcal{M} \alpha \beta (\mathcal{L}_V \beta)\, \mathrm{e}^{-sD_V \beta} \right\}$$

(7.43)

The large-s limit integral in (7.43) can be worked out in the same way as we did in Section 2.6 to arrive at the same expression (2.142), except that now the 2-form $\Omega_V = d\beta$ there is given quite arbitrarily as

$$\Omega_V = 2g \cdot \mu_V - \mathcal{L}_V g \tag{7.44}$$

so that

$$\lim_{s \to \infty} \int_\mathcal{M} \alpha\, e^{-sD_V \beta} = (-2\pi)^{n/2} \sum_{p \in \mathcal{M}_V} \frac{\alpha^{(0)}(p)}{|\det dV(p)|} \tag{7.45}$$
$$\times \operatorname{Pfaff}\left(dV(p) - g(p)^{-1}\mathcal{L}_V g(p)/2\right)$$

Using (3.59) to rewrite derivatives of V in (7.45) in the usual way, substituting in the definitions in Section 2.5 for the geometric quantities in (7.43), and then expanding exponentials in the 2-forms ω and Ω_V to the highest degrees that the manifold integrations pick up, we arrive at an expression for the partition function in terms of geometrical characteristics of the phase space [134]

$$Z(T) = \left(\frac{2\pi i}{T}\right)^n \sum_{p \in \mathcal{M}_V} (-i)^{\lambda(p)} \sqrt{\frac{\det \omega(p)}{\det \mathcal{H}(p)}}\, e^{iTH(p)}$$
$$\times \sqrt{\det\left(1 - \mathcal{H}^{-1}\omega g^{-1}\mathcal{L}_V g/2\right)(p)}$$
$$+ \frac{1}{(iT)^n} \int_0^\infty ds \oint_{\partial \mathcal{M}} \frac{e^{iTH-sK_V}}{(n-1)!} g(V,\cdot) \wedge (iT\omega - s\Omega_V)^{n-1}$$
$$- \frac{1}{(iT)^n} \int_0^\infty ds\, s \int_\mathcal{M} \frac{e^{iTH-sK_V}}{(n-1)!} g(V,\cdot) \wedge (\mathcal{L}_V g)(V,\cdot) \wedge (iT\omega - s\Omega_V)^{n-1}$$
$$\tag{7.46}$$

which holds for an arbitrary Riemannian metric g on $\mathcal{M}$. From the expression (7.46) and the fact that $g(V,\cdot)^{\wedge 2} \equiv 0$, we can see explicitly now how the conformal Lie derivative condition (7.35) collapses this expression down to the Duistermaat-Heckman formula (3.63) when $\partial\mathcal{M} = \emptyset$, as we saw by more explicit means above (This conformal localization property has also been explained within a more general BRST approach recently in [83]). Note that we cannot naively carry out the s-integrations quite yet, because the function $K_V = g(V,V)$ has zeroes on $\mathcal{M}$. The expression (7.46), although quite complicated, shows explicitly how the Lie derivative conditions make the semi-classical approximation to the partition function exact. This is in contrast to the loop-expansion we studied earlier, where the corrections to the Duistermaat-Heckman formula were not just some combinations of Lie derivatives. (7.46) therefore represents a sort of resummation of the loop-expansion that explicitly takes into account the geometric symmetries that

248   7. Beyond the Semi-Classical Approximation

make the 1-loop approximation exact. We shall see soon that it is quite consistent with the results predicted from the loop-expansion, and moreover that it gives many new insights.

There are several points to make at this stage. First of all we note the appearence of the boundary contribution in (7.43). If we assume that $\mathcal{L}_V \beta = 0$, that the group action represented by the flows of the Hamiltonian vector field V preserves the boundary of $\mathcal{M}$ (i.e. $g \cdot \partial \mathcal{M} = \partial \mathcal{M}$) and that the action is free on $\partial \mathcal{M}$, then the s-integral in the boundary term in (7.43) can be carried out explicitly and we find the extra contribution to the Duistermaat-Heckman formula for manifolds with boundary,

$$Z_{\partial \mathcal{M}}(T) = - \oint_{\partial \mathcal{M}} \frac{\beta \wedge \alpha}{D_V \beta} \qquad (7.47)$$

In this context $\beta = i_V g$ is the connection 1-form for the induced group action on the boundary $\partial \mathcal{M}$, because as we have seen earlier $d\beta$ is the moment map for this action. This boundary term can be determined using the Jeffrey-Kirwan-Kalkman residue that was introduced in Section 3.8, i.e. the coefficient of $\frac{1}{\phi}$ in the quantity $(\beta \wedge \alpha)/D_V \beta$, where ϕ is the element of the symmetric algebra $S(\mathbf{g}^*)$ representing the given circle action [82].

Secondly, notice that the conformal localization symmetry gives an explicit realization in (7.37) of the Hessian-metric ansatz which was discussed in the last Section. In particular, (7.46) establishes that with this Hessian-metric substitution and the appropriate extension away from the critical points of H (the terms proportional to ∇H in (7.37)) the corrections to the Duistermaat-Heckman integration formula vanish to all orders of the loop-expansion (and not just to 2-loop order as was established in the last Section). Notice, however, that because of (7.36) any conformal Killing vector on a Kähler manifold is automatically an isometry. In fact, the generic case of a non-vanishing scaling function $\Upsilon(x)$ in (7.35) is similar to the isometry case from the point of view of the equivariant localization priniciple. Note that away from the critical points of H we can rescale $\beta \to \bar{\beta} \equiv g(V, \cdot)/g(V, V)$. With this choice for the localization 1-form β it is easy to show from (7.35) that away from the critical point set of the Hamiltonian it satisfies $\mathcal{L}_V \bar{\beta} = 0$, i.e. $\bar{\beta}$ is then an equivariant differential form on $\mathcal{M} - \mathcal{M}_V$. Thus away from the subset $\mathcal{M}_V \subset \mathcal{M}$ the conformal Killing condition can be cast into the same equivariant cohomological framework as the isometry condition by a rescaling of the metric tensor in (7.35), $g_{\mu\nu} \to G_{\mu\nu} = g_{\mu\nu}/g(V, V)$, for which $\mathcal{L}_V G = 0$. The rescaled metric $G_{\mu\nu}(x)$ is only defined on $\mathcal{M} - \mathcal{M}_V$, but we recall from Section 2.5 that all that is needed to establish the localization of (3.52) onto the zeroes of the vector field V (i.e. the equivariant localization principle) is an invariant metric tensor (or equivalently an equivariant differential form β) which is defined everywhere on $\mathcal{M}$ except possibly in an arbitrarily small neighbourhood of $\mathcal{M}_V$. The fact that the weaker conformal symmetry condition is equivalent to the isometry condition in this respect is essentially a

7.2 Conformal and Geodetic Localization Symmetries

consequence of the fact that the differential form $\beta = i_V g$ above is still a connection 1-form that specifies a splitting of the tangent bundle into a component over $\mathcal{M}_V$ (represented by the discrete sum over $\mathcal{M}_V$ in (7.46)) and a component orthogonal to $\mathcal{M}_V$ (represented by the Lie derivative integral in (7.46)). This is in fact implicit in the proof by Atiyah and Bott in [9] using the Weil algebra.

One may ask as to the possibilities of using other localization forms to carry out the localization onto $\mathcal{M}_V$, but it is readily seen that, up to components orthogonal to V, $\beta = i_V g$ is the most general localization form up to multiplication by some strictly positive function. This follows from the fact that in order to obtain finite results in the limit $s \to \infty$ in (7.45) we need to ensure that the form $D_V \beta$ has a 0-form component to produce an exponential damping factor, since higher-degree forms will contribute only polynomially in the Taylor expansion of the exponential. This is guaranteed only if β has a 1-form component. Thus it is only the 1-form part of β that is relevant to the localization formula, and so without loss of generality the most general localization principle follows from choosing β to be a 1-form. Furthermore, the 0-form part $V^\mu \beta_\mu$ of $D_V \beta$ must attain its global minimum at zero so that the large-s limit in (7.45) yields a non-zero result. This boundedness requirement is equivalent to the condition that the component of β along V has the same orientation as V, i.e. that β be proportional to the metric dual 1-form of V with respect to a globally-defined Euclidean signature Riemannian structure on the phase space $\mathcal{M}$.

In addition, for a compact phase space $\mathcal{M}$ the conformal group is in general non-compact, so that conformal Killing vectors need not automatically generate circle actions in these cases as opposed to isometry generators where this would be immediate. To explore whether this larger conformal group symmetry of the Duistermaat-Heckman localization leads to *globally* different sorts of dynamical systems, one would like to construct examples of systems with non-trivial (i.e. $\Upsilon \neq 0$) conformal symmetry. For this, one has to look at spaces which have a Riemannian metric g for which the Hamiltonian vector field V is a generator of both the conformal group $\mathrm{Conf}(\mathcal{M}, g)$ *and* the symplectomorphism group $\mathrm{Sp}(\mathcal{M}, \omega)$ of canonical transformations on the phase space. From the analysis thus far we have a relatively good idea of what the latter group looks like. The conformal group for certain Riemannian manifolds is also well-understood [54]. For instance, the conformal group of flat Euclidean space of dimension $d \geq 3$ is locally isomorphic to $SO(d+1, 1)$. The (global) conformal group of the Riemann sphere $\mathbb{C} \cup \{\infty\}$ was already encountered in Chapter 6 above (in a different context), namely the group $SL(2, \mathbb{C})/\mathbb{Z}_2 \simeq SO(3, 1)$ of projective conformal transformations. In these cases, the conformal group consists of the usual isometries of the space, along with dilatations or scale transformations (e.g. translations of r in $z = r\, \mathrm{e}^{i\theta}$) and the d-dimensional subgroup of so-called special conformal transformations.

250   7. Beyond the Semi-Classical Approximation

An interesting example is provided by the flat complex plane $\mathbb{C}$. Here the conformal algebra is infinite-dimensional and its Lie algebra is just the classical Virasoro algebra [54]. Indeed, the conformal Killing equations in this case are just the first set of Cauchy-Riemann equations in (5.51) (the other one represents the divergence-free condition $\operatorname{tr} \mu_V = 0$). This means that the conformal Killing vectors in this case are the holomorphic functions $V^z = f(z), V^{\bar z} = \bar f(\bar z)$. The Hamiltonian flows of these vector fields are therefore the *arbitrary* analytic coordinate transformations

$$\dot z(t) = f(z(t)) \quad , \quad \dot{\bar z}(t) = \bar f(\bar z(t)) \tag{7.48}$$

As an explicit example, consider the conformal Killing vector which describes a Hamiltonian system with $n+1$ distinct stationary points,

$$V^z = i\beta z(1 - \alpha_1 z) \cdots (1 - \alpha_n z) \tag{7.49}$$

at $z = 0$ and $z = 1/\alpha_i$, where $\beta, \alpha_i \in \mathbb{C}$. The associated scaling function in (7.35) is then

$$\Upsilon(z, \bar z) = \partial_z V^z + \partial_{\bar z} V^{\bar z} \tag{7.50}$$

The symplecticity condition $\mathcal{L}_V \omega = 0$ leads to the first-order linear partial differential equation

$$V^z \partial_z \omega_{z\bar z} + V^{\bar z} \partial_{\bar z} \omega_{z\bar z} = -\Upsilon(z, \bar z) \omega_{z\bar z} \tag{7.51}$$

which is easily solved by separation of the variables $z, \bar z$. The solution for the symplectic 2-form with arbitrary separation parameter $\lambda \in \mathbb{R}$ is

$$\omega_{z\bar z}^{(\lambda)}(z, \bar z) = w_\lambda(z) \bar w_\lambda(\bar z) / V^z V^{\bar z} \tag{7.52}$$

where

$$w_\lambda(z) = \mathrm{e}^{i\lambda \int dz/V^z} = \left(\frac{z}{(1 - \alpha_1 z)^{A_1} \cdots (1 - \alpha_n z)^{A_n}} \right)^{\lambda/\beta} \tag{7.53}$$

and the constants

$$A_i(\alpha_1, \ldots, \alpha_n) = (\alpha_i)^{n-1} \prod_{j \neq i} \frac{1}{\alpha_i - \alpha_j} \tag{7.54}$$

are the coefficients of the partial fraction decomposition

$$(V^z)^{-1} = \frac{1}{i\beta} \left(\frac{1}{z} + \sum_{i=1}^n \frac{A_i}{1 - \alpha_i z} \right) \tag{7.55}$$

To ensure that (7.52) is a single-valued function on $\mathbb{C}$, we restrict the α_k's to all have the same phase, so that $A_i(\alpha_1, \ldots, \alpha_n) \in \mathbb{R}$, and the parameter β to be real-valued. The Hamiltonian equations (5.68) can now be integrated

7.2 Conformal and Geodetic Localization Symmetries 251

up with the vector field (7.49) and the symplectic 2-form (7.52), from which we find the family of Hamiltonians

$$H^{(\lambda)}_{\beta,\alpha_i}(z,\bar{z}) = \frac{1}{\lambda}\left(\frac{z}{(1-\alpha_1 z)^{A_1}\cdots(1-\alpha_n z)^{A_n}}\right)^{\lambda/\beta} \times \left(\frac{\bar{z}}{(1-\bar{\alpha}_1 \bar{z})^{A_1}\cdots(1-\bar{\alpha}_n \bar{z})^{A_n}}\right)^{\lambda/\beta} \quad (7.56)$$

To ensure that this Hamiltonian has only non-degenerate critical points we set $\lambda = \beta$. This also guarantees that the level (constant energy) curves of this Hamiltonian coincide with the curves which are the solutions of the equations of motion (7.48) [134].

Since the Hamiltonian (7.56) either vanishes or is infinite on its critical point set, it is easy to show that the partition function (3.52) is independent of α_k and coincides with the anticipated result from the Duistermaat-Heckman integration formula, namely $Z(T) = 2\pi i\beta/T$. This partition function coincides with that of the simple harmonic oscillator $z\bar{z}/\beta$, as expected since for $\alpha_i = 0$, $\omega^{(\beta)}_{z\bar{z}}$ becomes the Darboux 2-form and $H^{(\beta)}_{\beta,0}$ the harmonic oscillator Hamiltonian. In fact, we can integrate up the flow equation (7.48) in the general case and we find that the classical trajectories $z(t)$ are determined by the equation

$$e^{i\beta(t-t_0)} = w_\beta(z(t)) = \frac{z(t)}{(1-\alpha_1 z(t))^{A_1}\cdots(1-\alpha_n z(t))^{A_n}} \quad (7.57)$$

The coordinate change $z \to w_\beta(z)$ is just the finite conformal transformation generated by the vector field (7.49) and it maps the dynamical system $(\omega^{(\beta)}_{z\bar{z}}, H^{(\beta)}_{\beta,\alpha_i})$ onto the harmonic oscillator $H \propto w\bar{w}$, $\omega \propto \omega_D$ with the usual circular classical trajectories $w(t) = e^{i\beta(t-t_0)}$ associated with a $U(1)$ generator corresponding to an isometry. This transformation is in general multi-valued and has singularities at the critical points $z = 1/\alpha_i$ of the Hamiltonian $H^{(\beta)}_{\beta,\alpha_i}$. It is therefore not a diffeomorphism of the plane for $\alpha_i \neq 0$ and the Hamiltonian system $(\mathbb{R}^2, \omega^{(\beta)}_{z\bar{z}}, H^{(\beta)}_{\beta,\alpha_i})$ is not globally isomorphic to the simple harmonic oscillator. For more details about the global differences between these systems and those associated with isometry generators, see [134]. The conformal group structures on phase spaces like S^2 yield novel generalizations of the localizable systems which are associated with coadjoint orbits of isometry groups as we discussed in Chapter 5. The appearance of the larger (non-compact) conformal group may lead to interesting new structures in other instances which usually employ the full isometry group of the phase space, such as the Witten localization formalism of Section 3.8.

Other geometric alternatives to the Lie derivative condition $\mathcal{L}_V g = 0$ have also been discussed by Kärki and Niemi in [84]. For instance, consider the alternative condition

252   7. Beyond the Semi-Classical Approximation

$$V^\lambda \nabla_\lambda V^\mu = 0 \tag{7.58}$$

to the Killing equation, which means that the Hamiltonian flows $\dot{x}^\mu = V^\mu$ are geodetic to g (see Section 2.4). From (7.58) it follows that the Lie derivative of the localization 1-form $\beta = i_V g$ can be written as

$$\mathcal{L}_V \beta = V^\rho (g_{\rho\lambda} \nabla_\mu V^\lambda + g_{\mu\lambda} \nabla_\rho V^\lambda) dx^\mu = V^\rho g_{\rho\lambda} \nabla_\mu V^\lambda dx^\mu = \frac{1}{2} div \beta \tag{7.59}$$

Comparing this with the Cartan-Weil identity $\mathcal{L}_V = di_V + i_V d$ for the Lie derivative acting on differential forms we find the relation

$$i_V d\beta = -\frac{1}{2} div \beta \tag{7.60}$$

which leads to the equivariance condition [84]

$$D_V(K_V/2 + \Omega_V) = 0 \tag{7.61}$$

so that the dynamical systems $(\frac{1}{2} K_V, \Omega_V)$ and (H, ω) determine a bi-Hamiltonian structure. Moreover, in this case it is also possible to explicitly solve the equivariant Poincaré lemma [84], just as we did in Section 3.6. Thus given that $\frac{1}{2} K_V + \Omega_V$ is an equivariantly-closed differential form, the condition (7.58) has the potential of leading to possibly new localization formulas.

However, there are 2 things to note about the geometric condition (7.58). The first is its connection with a non-trivial conformal Killing equation $\mathcal{L}_V g = \Upsilon g$, which follows from the identity

$$V^\nu g^{\alpha\mu} (\mathcal{L}_V g)_{\mu\nu} = V^\lambda \nabla_\lambda V^\alpha + g^{\alpha\mu} \nabla_\mu K_V/2 \tag{7.62}$$

Contracting both sides of (7.62) with $g_{\alpha\rho} V^\rho$ leads to

$$\Upsilon K_V = 2 g_{\mu\nu} V^\nu V^\lambda \nabla_\lambda V^\mu \tag{7.63}$$

when (7.35) holds. This implies that if (7.58) is satisfied, then $\Upsilon \equiv 0$ away from the zeroes of V. Thus $\Upsilon \equiv 0$ almost everywhere on $\mathcal{M}$ and so the geometric condition (7.58) can only be compatible with the Killing equation, and not the inhomogeneous conformal Killing equation.

Secondly, the exact 2-form $\Omega_V \equiv div g$ is degenerate on $\mathcal{M}$, because an application of the Leibniz rule and Stokes' theorem gives

$$n! \int_{\mathcal{M}} d^{2n} x \sqrt{\det \Omega_V(x)} = \int_{\mathcal{M}} \Omega_V^n = \int_{\mathcal{M}} d\left(i_V g \wedge \Omega_V^{n-1}\right) = 0 \tag{7.64}$$

when $\partial \mathcal{M} = \emptyset$. Thus $\det \Omega_V(x) = 0$ on some submanifold of $\mathcal{M}$, and so the Hamiltonian system determined by $(\frac{1}{2} K_V, \Omega_V)$ is degenerate. As mentioned in Section 3.6, this isn't so crucial so long as the support of $\det \Omega_V(x)$ is a submanifold of $\mathcal{M}$ of dimension at least 2 (so that there exists at least 1 degree of freedom from the classical equations of motion). It would be interesting to investigate these geometric structures in more detail and see what localization schemes they lead to.

7.3 Corrections to the Duistermaat-Heckman Formula: A Geometric Approach

The integration formula (7.46) suggests a geometric approach to the evaluation of corrections to the Duistermaat-Heckman formula in the cases where it is known to fail. Recall that there is always *locally* a metric tensor on $\mathcal{M} - \mathcal{M}_V$ for which V is a Killing vector (see the discussion at the beginning of Section 3.6). For the systems where the semi-classical approximation is not exact, there are global obstructions to extending these locally invariant metric tensors to globally-defined geometries on the phase space which are invariant under the full group action generated by the Hamiltonian vector field V on $\mathcal{M}$, i.e. there are no *globally* defined single-valued Riemannian geometries on $\mathcal{M}$ for which V is globally a Killing vector. This means that although the Killing equation $\mathcal{L}_V g = 0$ can be solved for g locally on patches covering the manifold, there is no way to glue the patches together to give a single-valued invariant geometry on the whole of $\mathcal{M}$ (c.f. Section 5.9). In this Section we shall describe how the expression (7.46) could be used in this sense to evaluate the corrections to the sum over critical points there [157], and we shall see that not only does this method encompass much more of the loop-expansion than the term-by-term analysis of Section 7.1 above, but it also characterizes the non-exactness of the Duistermaat-Heckman formula in a much more transparent and geometric way than Kirwan's theorem. In this way we can obtain an explicit geometric picture of the failure of the Duistermaat-Heckman theorem and in addition a systematic, geometric method for approximating the integral (3.52). Furthermore, this analysis will show explicitly the reasons why for certain dynamical systems there are no globally defined Riemannian metrics on the given symplectic manifold for which any given vector field with isolated zeroes is a Killing vector, and as well this will give another geometric description of the integrability properties of the given dynamical system. The analysis presented here is by no means complete and deserves a more careful, detailed investigation.

The idea is to define a set of patches covering $\mathcal{M}$ in each of which we can solve the Killing equations for g, but for which the gluing of these patches together to give a globally defined metric tensor is highly singular. The non-triviality that occurs when these subsets are patched back together will then represent the corrections to the Duistermaat-Heckman formula, and from our earlier arguments we know that this will be connected with the integrability of the Hamiltonian system. We introduce a set of preferred coordinates x'' for the vector field V following Section 5.2. In general, this diffeomorphism can only be defined locally on patches over $\mathcal{M}$ and the failure of this coordinate transformation in producing globally-defined C^∞-coordinates on $\mathcal{M}$ gives an analytic picture of why the Hamiltonian vector field fails to generate global isometries. Notice in particular that these coordinates are only defined on $\mathcal{M} - \mathcal{M}_V$. In this way we shall see geometrically how Kirwan's theorem restricts dynamical systems whose phase spaces have non-trivial odd-degree homology and explicitly what type of flow the Hamiltonian vector field generates.

Recall that the coordinate functions x'' map the constant coordinate lines $(x_0^2, \ldots, x_0^{2n}) \in \mathbb{R}^{2n-1}$ onto the integral curves of the isometry defined by the classical Hamilton equations of motion $\dot{x}^\mu(t) = V^\mu(x(t))$, i.e. in the coordinates $x''(x)$, the flows generated by the Hamiltonian vector field look like

$$x''^1(t) = x_0^1 + t \quad ; \quad x''^\mu(t) = x_0^\mu \quad , \quad \mu > 1 \tag{7.65}$$

In general, this coordinate transformation function will have singularities associated with the fact that there is no Riemannian metric tensor on $\mathcal{M}$ for which the Lie derivative condition $\mathcal{L}_V g = 0$ holds. Otherwise, if these transformation functions were globally defined on $\mathcal{M} - \mathcal{M}_V$, then we could take the metric on $\mathcal{M}$ to be any one whose components in the x''-coordinates are independent of x''^1 thereby solving the Killing equations directly, and hence from (7.46) the WKB approximation would be exact. For a non-integrable system, there must therefore be some sort of obstructions to defining the x''-coordinate system globally over $\mathcal{M}$. In light of the above comments, these singularities will partition the manifold up into patches P, each of which is a $2n$-dimensional contractable submanifold of $\mathcal{M}$ with boundaries ∂P which are some other $(2n-1)$-dimensional submanifolds of $\mathcal{M}$ induced by the constant coordinate line transformation from $\mathbb{R}^{2n-1}$ above. By dropping some of these coordinate surfaces if necessary, we can assume that these patches induced from the singularities of the above coordinate transformation form a disjoint cover of the manifold, $\mathcal{M} = \coprod_P P$ [4]. Then we can write the partition function as

$$Z(T) = \sum_P \int_P \alpha|_P \tag{7.66}$$

where as usual α is the equivariantly-closed differential form (3.57).

By the choice of the patches P, in their interior there is a well-defined (bounded) translation action generated by V''^μ. Since the patches P are diffeomorphic to hypercubes in $\mathbb{R}^{2n}$, we can place a Euclidean metric on them,

$$g_P = e^{\varphi_P(x'')} dx''_\mu \otimes dx''^\mu \tag{7.67}$$

where the conformal factor $\varphi_P(x'')$ is a globally-defined real-valued C^∞-function on P. If we choose it so that it is independent of the coordinate x''^1, then the metric (7.67) satisfies the Killing equation on P. Thus on each patch P, by the given choice of coordinates, we can solve the Lie derivative constraint, even though this cannot be extended to the whole of $\mathcal{M}$. Then each integral over P in (7.66) can be written using the formula (7.46), restricted to the patch P, to get

---

[4] Here we assume that $\mathcal{M}$ is compact, but we shall see that this formalism can also be extended to the phase space $\mathbb{R}^{2n}$.

$$\int_P \alpha|_P = \left(\frac{2\pi i}{T}\right)^n \sum_{p \in \mathcal{M}_V \cap P} (-i)^{\lambda(p)} \sqrt{\frac{\det \omega(p)}{\det \mathcal{H}(p)}} e^{iTH(p)}$$

$$+ \frac{1}{(iT)^n} \int_0^\infty ds \oint_{\partial P} \frac{e^{iTH-sK_V}}{(n-1)!} g(V,\cdot) \wedge (iT\omega - s\Omega_V)^{n-1} \bigg|_{\partial P} \tag{7.68}$$

The first term here, when (7.68) is substituted back into (7.66), represents the lowest-order term in the semi-classical expansion of the partition function over $\mathcal{M}$, i.e. the Duistermaat-Heckman term $Z_0(T)$ in (3.63), while the boundary terms give the general corrections to this formula and represent the non-triviality that occurs rendering inexact the stationary-phase approximation. The result is

$$Z(T) = Z_0(T) + \delta Z(T) \tag{7.69}$$

where

$$\delta Z(T) = \frac{1}{(iT)^n} \int_0^\infty ds \sum_P \oint_{\partial P} \frac{e^{iTH-sK_V}}{(n-1)!} g(V,\cdot) \wedge (iT\omega - s\Omega_V)^{n-1} \bigg|_{\partial P} \tag{7.70}$$

The contributions from the patch terms in (7.70) therefore represent an alternative geometric approach to the loop-expansion of Section 7.1 above.

To evaluate the correction term $\delta Z(T)$, we recall from Section 5.2 (eqs. (5.39),(5.43)) that the coordinate functions $\chi^\mu(x)$ for $\mu = 2,\ldots,2n$ are local conserved charges of the Hamiltonian system, i.e.

$$\{\chi^\mu, H\}_\omega = V^\nu \partial_\nu \chi^\mu = 0 \tag{7.71}$$

Thus we can take one of them, say χ^2, to be a functional of the Hamiltonian, which we choose to be $x''^2(x) = \chi^2(x) = \sqrt{H(x)}$, where by adding an irrelevant constant to H we may assume that it is a positive function on the (compact) manifold $\mathcal{M}$. Then, using the metric tensor transformation law, we find that the metric (7.67) when written back into the original (unprimed) coordinates has the form

$$g_P = e^{\varphi_P(x)} \left(\frac{1}{(V^\lambda \partial_\lambda \chi^1)^2} \partial_\mu \chi^1 \partial_\nu \chi^1 + \frac{1}{4H} \partial_\mu H \partial_\nu H + \sum_{\alpha>2} \partial_\mu \chi^\alpha \partial_\nu \chi^\alpha \right) dx^\mu \otimes dx^\nu \tag{7.72}$$

so that the metric-dependent quantities appearing in (7.70) can be written as

$$g_P(V,\cdot) = \frac{e^{\varphi_P(x)}}{V^\lambda(x)\partial_\lambda \chi^1(x)} \partial_\mu \chi^1(x) dx^\mu \quad , \quad K_V(x)|_P = g_P(V,V) = e^{\varphi_P(x)} \tag{7.73}$$

256   7. Beyond the Semi-Classical Approximation

$$\Omega_V|_P = \frac{e^{\varphi_P(x)}}{2(V^\lambda \partial_\lambda \chi^1)^2} \left\{ \partial_\lambda \chi^1 \left(\partial_\mu V^\lambda \partial_\nu \chi^1 - \partial_\nu V^\lambda \partial_\mu \chi^1 \right) \right.$$
$$\left. + V^\lambda \partial_\lambda \chi^1 \left(\partial_\mu \varphi_P \partial_\nu \chi^1 - \partial_\nu \varphi_P \partial_\mu \chi^1 \right) \right\} dx^\mu \wedge dx^\nu \qquad (7.74)$$

When these expressions are substituted back into the correction term (7.70), we find that the integrands of $\delta Z(T)$ depend only on the coordinate function $\chi^1(x)$. This is not surprising, since the only effect of the other coordinate functions, which define local action variables of the dynamical system, is to make the effect of the partitioning of $\mathcal{M}$ into patches above non-trivial, reflecting the fact that the system is locally integrable, but not globally (otherwise, the partition function localizes).

In general, the correction term (7.70) is extremely complicated, but we recall that there is quite some freedom left in the choice of χ^1. All that is required of this function is that it have no critical points in the given coordinate neighbourhood. We can therefore choose it appropriately so as to simplify the correction $\delta Z(T)$ somewhat. Given this choice, in general singularities will appear from the fact that it cannot be defined globally on $\mathcal{M}$, and we can use these identifications to identify the specific regions P above. The form of the function χ^1 is at the very heart of this approach to evaluating corrections to the Duistermaat-Heckman formula. We shall see how this works in some explicit examples in the next Section. Notice that a similar phenomenon to what occured in Section 5.9 has happened here – the function K_V in (7.73) is non-zero, as the zeroes of the vector field V have been absorbed into the metric term $g_P(V, \cdot)$ thereby making it singular. We can therefore now carry out the explicit s-integral in (7.70), as the singularities on $\mathcal{M}_V$ are already present in the integrand there. Although this may seem to make everything hopelessly singular, we shall see that they can be regulated with special choices of the function χ^1 thereby giving workable forms. We shall see in fact that when such divergences do occur, they are related to those predicted by Kirwan's theorem which we recall dictates also when the full stationary-phase series diverges for a given function H.

There does not seem to be any immediate way of simplifying the patch corrections $\delta Z(T)$ above due to the complicated nature of the integrand forms. However, as usual in 2-dimensions things can be simplified rather nicely and the analysis reveals some very interesting properties of this formalism which could be generalized to higher-dimensional symplectic manifolds. To start, we notice that in 2-dimensions, if $\mathcal{M}$ is a compact manifold, then the union above over all of the patch boundaries $\partial P \subset \mathcal{M}$ will in general form a sum over 1-cycles $a_\ell \in H_1(\mathcal{M}; \mathbb{Z})$. Next, we substitute (7.73) and (7.74) into (7.70) with $n = 1$, and after working out the easy s-integration we find that the 2-dimensional correction terms can be written as

$$\delta Z(T) = \frac{1}{iT} \sum_\ell \oint_{a_\ell} \frac{e^{iTH(x)}}{V^\lambda(x) \partial_\lambda \chi^1(x)} \partial_\mu \chi^1(x) dx^\mu \qquad (7.75)$$

7.3 Corrections to the Duistermaat-Heckman Formula

As for the function χ^1, we need to choose one which is independent of the other coordinate transformation function χ^2 to ensure that these 2 functions truly do define a (local) diffeomorphism of $\mathcal{M}$. The simplest choice, as far as the evaluation of (7.75) is concerned, is to choose χ^1 as the solution of the first-order linear partial differential equation

$$V^1(x)\partial_1\chi^1(x) = V^2(x)\partial_2\chi^1(x) \tag{7.76}$$

With this choice of χ^1, the functions χ^1 and χ^2 are independent of each other wherever $\partial_\mu \chi^\nu(x) \neq 0$, $\mu, \nu = 1, 2$, which follows from working out the Jacobian for the coordinate transformation defined by χ^μ and using their defining partial differential equations above.

With this and the Hamiltonian equations $dH = -i_V\omega$, the correction terms (7.75) become

$$\delta Z(T) = -\frac{1}{2iT}\sum_\ell \oint_{a_\ell} F|_{a_\ell} \tag{7.77}$$

where we have introduced the 1-form

$$F = \omega_{12}(x)\, e^{iTH(x)} \left(\frac{1}{\partial_2 H(x)}dx^1 - \frac{1}{\partial_1 H(x)}dx^2\right) \tag{7.78}$$

The expression (7.77) leads to a nice geometric interpretation of the corrections above to the Duistermaat-Heckman formula. To each of the homology cycles $a_\ell \in H_1(\mathcal{M}; \mathbb{Z})$, there corresponds a cohomology class $\eta_\ell \in H^1(\mathcal{M}; \mathbb{R})$, called their Poincaré dual [32], which has the property that it localizes integrals of 1-forms $\alpha \in \Lambda^1 \mathcal{M}$ to a_ℓ, i.e.

$$\oint_{a_\ell} \alpha|_{a_\ell} = \int_\mathcal{M} \alpha \wedge \eta_\ell \tag{7.79}$$

Defining

$$\eta = \sum_\ell \eta_\ell \in H^1(\mathcal{M}; \mathbb{R}) \tag{7.80}$$

we see that the correction term (7.77) can be written as

$$\delta Z(T) = -\frac{1}{2iT}\int_\mathcal{M} F \wedge \eta \tag{7.81}$$

Noting also that the original partition function itself can be written as

$$Z(T) = \frac{1}{2}\int_\mathcal{M} F \wedge dH \tag{7.82}$$

it then follows from $Z(T) = Z_0(T) + \delta Z(T)$ that

258   7. Beyond the Semi-Classical Approximation

$$\int_{\mathcal{M}} F \wedge (iTdH + \eta) = -4\pi \sum_{p \in \mathcal{M}_V} (-i)^{\lambda(p)} \sqrt{\frac{\det \omega(p)}{\det \mathcal{H}(p)}}\, e^{iTH(p)} \qquad (7.83)$$

Thus in this sense, the partition function represents intersection numbers of $\mathcal{M}$ associated to the homology cycles a_ℓ.

This last equation is particularly interesting. It shows that the corrections to the Duistermaat-Heckman formula generate the Poincaré duals to the homology cycles which signify that the Hamiltonian vector field does not generate a globally well-defined group action on $\mathcal{M}$. When the correction 1-form η/iT is added to the 1-form $dH = -\omega(V, \cdot)$ which defines the flow of the Hamiltonian vector field on $\mathcal{M}$, the resulting 1-form is enough to render the Duistermaat-Heckman formula exact for the new "effective" partition function. This means that although the initial Hamiltonian flow dH doesn't 'close enough' to satisfy the conditions required for the Duistermaat-Heckman theorem, adding the cohomological Poincaré dual to the singular homology cycles of the flow is enough to close the flows so that the partition function is now given exactly by the lowest-order term $Z_0(T)$ of its semi-classical expansion. One now can solve for the vector field W satisfying the "renormalized" Hamiltonian equations

$$dH + \eta/iT = -\omega(W, \cdot) \qquad (7.84)$$

We can consider W as a "renormalization" of the Hamiltonian vector field V which renders the stationary-phase series convergent and the Duistermaat-Heckman formula exact. Note that since the symplectic form ω defines a cohomology class in $H^2(\mathcal{M}; \mathbb{R})$, this just corresponds to choosing a different, possibly non-trivial representative in $H^1(\mathcal{M}; \mathbb{R})$ for $\omega(V, \cdot)$ (recall $\eta \in H^1(\mathcal{M}; \mathbb{R})$). Thus in our approach here, the corrections to the Duistermaat-Heckman formula compute (possibly) non-trivial cohomology classes of the manifold $\mathcal{M}$ and express geometrically what symmetry is missing from the original dynamical system that prevents its saddle-point approximation from being exact. The explicit constructions of the Poincaré duals above are well-known [32] – one takes the embedding $\sigma_\ell : S^1 \to \mathcal{M}$ of S^1 in $\mathcal{M}$ which corresponds to the loop a_ℓ, and constructs its DeRham current which is the Dirac delta-function 1-form $\delta^{(1,1)}(x, \sigma_\ell(y)) \in \Lambda^1 \mathcal{M}(x) \otimes \Lambda^1 \mathcal{M}(y)$ with the property (7.79) [18].

There is one crucial point that needs to be addressed before we turn to some explicit examples. In general we shall see that there are essentially 2 types of homology cycles that appear in the above when examining the singularities of the diffeomorphisms χ^μ that prevent them from being global coordinate transformations of $\mathcal{M}$. The first type we shall call 'pure singular cycles'. These arise solely as a manifestation of the choice of equation satisfied by χ^1. The second type shall be refered to as 'critical cycles'. These are the cycles on which at least one of the components of the Hamiltonian vector field vanish, $V^\mu(x) = 0$ for $\mu = 1$ or 2. On these latter cycles the above integrals in $\delta Z(T)$ become highly singular and require regularization. Notice in particular that if, say, $V^1(x) = 0$ but $V^2(x) \neq 0$ on some cycle a_ℓ, then the equations (7.71) and (7.76) which determine the functions χ^μ imply that $\partial_2 \chi^\mu(x) = 0$ while leaving the derivatives $\partial_1 \chi^\mu(x)$ undetermined. Recall that it was precisely at these points where the Jacobian of the coordinate transformation defined by χ^μ vanished.

In this case one must regulate the 1-form F defined above by letting $\partial_1 \chi^1$ and $\partial_2 \chi^1$ both approach zero on this cycle a_ℓ in a correlated manner so as to cancel the resulting divergence in the integrand of (7.75). Note that this regularization procedure now requires that x^1 and x^2 transform identically, particularly under rescalings, so that the tensorial properties of the differential form F are unaffected by this definition. In this case, the 1-form F which appears above gets replaced by the 1-form

$$F|_{a_\ell} = -\frac{1}{V^2(x)} \left(dx^1 + dx^2 \right) e^{iTH(x)} = \frac{\omega_{12}(x)}{\partial_1 H(x)} \left(dx^1 + dx^2 \right) e^{iTH(x)} \quad (7.85)$$

which follows from the general expression (7.75). This procedure for defining F can be thought of as a quantum field theoretic ultraviolet regularization for the higher-loop corrections to the partition function. In general, we shall always obtain such singularities corresponding to the critical points of the Hamiltonian because, as mentioned before, the diffeomorphism equations above become singular at the points where $V^\mu(x) = 0$. Note that (7.85) will also diverge when the cycle a_ℓ crosses a critical point, i.e. on $a_\ell \cap \mathcal{M}_V$. Such singularities, as we shall see, will be just a geometric manifestation of Kirwan's theorem and the fact that in general the stationary-phase expansion does not converge for the given Hamiltonian system. We shall also see that in general the pure singular cycles do not contribute to the corrections, as anticipated, as they are only a manifestation of the particular coordinate system used, of which the covariant corrections should be independent. It is only the critical cycles that contribute to the corrections and mimick in some sense the sum over critical points series for the partition function.

7.4 Examples

In this Section we illustrate some of the formalism of this Chapter with 2 classes of explicit examples. The first class we shall consider is the height function of a Riemann surface, a set of examples with which we have become well-acquainted. In the case of the Riemann sphere we have little to add at this point since the height function (2.1) localizes. The only point we wish to make here is that the covariant Hessian in this case with respect to the standard Kähler geometry of S^2 (see Section 5.5) is related to the Kähler metric g_{S^2} by

$$\nabla \nabla h_{\Sigma^0} = 2\frac{1 - z\bar{z}}{(1 + z\bar{z})^3} dz \otimes d\bar{z} = 2(1 - h_{\Sigma^0}) g_{S^2} \quad (7.86)$$

260   7. Beyond the Semi-Classical Approximation

which is in agreement with the analysis of Section 7.1 above. This shows the precise mechanism (i.e. the Hessian of h_{Σ^0} generates covariantly the Kähler structure of S^2) that makes the loop corrections vanish.

An interesting check of the above formalisms is provided by a modified version of the height function h_{Σ^0} which is the quadratic functional

$$h^{(2)}_{\Sigma^0} = h_{\Sigma^0} - h^2_{\Sigma^0} = (1-\cos\theta) - (1-\cos\theta)^2 = -\frac{2z\bar{z}}{1+z\bar{z}} - \left(\frac{2z\bar{z}}{1+z\bar{z}}\right)^2 \quad (7.87)$$

which has the same critical behaviour as h_{Σ^0}. Now we find that the metric equations (7.21) are solved by taking the isothermal solution [157]

$$\Gamma^z_{zz} \equiv g^{z\bar{z}}\partial_z g_{z\bar{z}} = \partial_z^2 h^{(2)}_{\Sigma^0}/\partial_z h^{(2)}_{\Sigma^0} \quad , \quad g_{z\bar{z}}(z,\bar{z}) = H'(z\bar{z}) \quad (7.88)$$

for $H = h^{(2)}_{\Sigma^0}$, which follows from (7.24) written in local isothermal coordinates for the implicitly defined metric. Thus the solution to (7.21) is

$$\nabla\nabla h^{(2)}_{\Sigma^0} = \frac{2z\bar{z}}{(1+z\bar{z})^3}(z\bar{z}-1)\,dz\otimes d\bar{z} = g_{z\bar{z}}dz\otimes d\bar{z} \quad (7.89)$$

As (7.89) does not coincide with the standard Kähler geometry of S^2, the 1-loop approximation to the partition function in this case is not exact, as expected. However, the partition function still localizes, in the sense that it can be computed via the Gaussian integral transform

$$Z(T) = \int_{\mathcal{M}} d\mu_L\, e^{iT(H-H^2)} = \int_{-\infty}^{\infty}\frac{d\phi}{\sqrt{2\pi i}}\, e^{-i\phi^2/2}\int_{\mathcal{M}} d\mu_L\, e^{i(T-2i\sqrt{T}\phi)H} \quad (7.90)$$

of the usual equivariant characteristic classes. Thus since (7.87) is a functional of an isometry generator (i.e. a conserved charge), it is still localizable, as anticipated from the discussions in Sections 4.8 and 4.9. This is also consistent with the formalism of the previous Section. In this case, the preferred coordinates for the Hamiltonian vector field are θ and $x = \phi/(1-\cos\theta)$. Although these coordinates are singular at the poles of S^2 (i.e. the critical points of (7.87)), the correction terms $\delta Z(T)$ do not localize onto any cycles and just represent the terms in the characteristic class expansion for $Z(T)$ here. This just reflects the fact that S^2 is simply connected, and also that the geometric terms $\delta Z(T)$ detect the integrability features of a dynamical system (as (7.87) is an integrable Hamiltonian).

Next, we consider the height function on the torus, with the Kähler geometry in Section 6.2 adjusted so that $\varphi = 0$ in (6.9) and $v = 1$ in (6.39). The covariant Hessian of the Hamiltonian (3.78) in this case is

$$\mathcal{H}(\phi_1,\phi_2) = \text{Im}\,\tau\cos\phi_1\cos\phi_2 d\phi_1\otimes d\phi_1 - 2\,\text{Im}\,\tau\sin\phi_1\sin\phi_2 d\phi_1\otimes d\phi_2$$
$$+(r_1+\text{Im}\,\tau\cos\phi_1)\cos\phi_2 d\phi_2\otimes d\phi_2$$
$$(7.91)$$

In the complex coordinatization used to define the Kähler structure this Hessian is not of the standard Hermitian form and the analysis used to show the exactness of the stationary phase approximation in the case of the height function on S^2 using the loop-expansion will not work here. Indeed, we do not expect that any metric on T^2 will be defined from the covariant Hessian here as we did in Section 7.1, and we already know that the Duistermaat-Heckman formula is not exact for this example. This is because of the saddle-points at $(\phi_1, \phi_2) = (0, \pi)$ and (π, π). The Hessian at these points will always determine an indefinite metric which is not admissible as a globally-defined geometry on the torus.

This is also apparent from examination of the connection (7.8) and its associated Fubini-Study geometry defined by (7.34). In this case $\gamma \equiv 0$ and the curvature (7.34) is trivial. The 2-form Ω does not determine the same cohomology class as the Kähler 2-form of Σ^1 does, so that there is not enough "mixing" of the Hessian and Liouville terms in the loop expansion to cancel out higher-order corrections. For the sphere, the Fubini-Study metric coincides with the standard Kähler metric and thus the appropriate mixing is there to make the dynamics integrable (recall that $\mathbb{C}P^1 = S^2$). It is the lack of formation of a non-trivial Kähler structure on the torus here that makes almost all dynamical systems on it non-integrable.

Although the failure of the Duistermaat-Heckman theorem in this case can be understood in terms of the non-trivial first homology of T^2 via Kirwan's theorem, we can examine analytically the obstructions in extending the Hamiltonian vector field (3.80) to a global isometry of the standard Kähler metric (6.9) of T^2 which defines the unique Riemannian geometry for equivariant localization on the torus. We shall find that the local translation action defined by the vector field (3.80) cannot be extended globally in a smooth way to the whole of T^2. The set of coordinates (x, y) on the torus in which the components of the Hamiltonian vector field are $V^x = 1$ and $V^y = 0$ as prescribed before are first defined by taking $\chi^2(\phi_1, \phi_2)$ to be the square root of the height function (3.78) and $\chi^1(\phi_1, \phi_2)$ to be the C^∞-function with non-vanishing first order derivatives which is the solution of the partial differential equation (7.76). In the case at hand (7.76) can be written as

$$-(r_1 + \text{Im } \tau \cos \phi_1)\frac{\partial \chi^1}{\partial \phi_1} = \text{Im } \tau \sin \phi_1 \cot \phi_2 \frac{\partial \chi^1}{\partial \phi_2} \qquad (7.92)$$

which is solved by

$$\chi^1(\phi_1, \phi_2) = \log(r_1 + \text{Im } \tau \cos \phi_1) - \log(\cos \phi_2) \qquad (7.93)$$

and integrating (7.93) as in (5.43) yields the desired set of coordinates (x, y). This gives

$$x(\phi_1, \phi_2) = \frac{1}{2\,\mathrm{Im}\,\tau} \left[\frac{2\,\mathrm{Im}\,\tau}{\sqrt{r_2 |\mathrm{Re}\,\tau|} \sin \phi_2} \arctan\left(\sqrt{\frac{|\mathrm{Re}\,\tau|}{r_2}} \tan \frac{\phi_1}{2} \right) \right.$$
$$\left. - \frac{\log\left(\tan \frac{\phi_2}{2}\right)}{\cos \phi_1} \right] \tag{7.94}$$

$$y(\phi_1, \phi_2) = \sqrt{r_2 - (r_1 + \mathrm{Im}\,\tau \cos \phi_1) \cos \phi_2}$$

which hold provided that $\mathrm{Re}\,\tau \neq 0$.

In the coordinates defined by the diffeomorphism (7.94) the Hamiltonian vector field generates the local action of the group $\mathbb{R}^1$ of translations in x. However, this diffeomorphism cannot be extended globally to the whole of T^2 because it has singularities along the coordinate circles

$$a_1 = \{(\pi/2, \phi) \in T^2\} \quad , \quad a_2 = \{(3\pi/2, \phi) \in T^2\} \tag{7.95}$$

$$b_1 = \{(\phi, 0) \in T^2\} \quad , \quad b_2 = \{(\phi, \pi) \in T^2\} \tag{7.96}$$

This means that V_{Σ^1} cannot globally generate isometries of any Riemannian geometry on T^2. Although translations in the coordinate x represent some unusual local symmetry of the torus, it shows that the existence of non-trivial homology cycles on T^2 lead to singularities in the circle action of the Hamiltonian vector field on T^2. These singularities do not appear on the Riemann sphere because any closed loop on S^2 is contractable, so that the singular circles above collapse to points which can be identified with the critical points of the Hamiltonian function. In fact, as we saw in Chapter 6, the only equivariant Hamiltonians on the torus are precisely those which generate translations along the homology cycles of T^2, and so we see that the Hamiltonian (3.78) generates a circle action that is singular along those cycles which are exactly the ones required for a globally equivariantly-localizable system on the torus. This is equivalent to the fact that the flow generated by V_{Σ^1} bifurcates at the saddle points of h_{Σ^1} (like the equations of motion for a pendulum), and the above shows analytically why there is no single-valued, globally-defined Riemannian geometry on the torus for which the height function h_{Σ^1} generates isometries.

The local circle action defined by the diffeomorphism (7.94) however partitions the torus into 4 open sets P_i which are the disjoint sets that remain when one removes the 2 canonical homology cycles discussed above. Each of these sets P_i is diffeomorphic to an open rectangle in $\mathbb{R}^2$ on which the Hamiltonian vector field V_{Σ^1} generates a global $\mathbb{R}^1$-action. Thus the above formalism implies that the corrections to the Duistermaat-Heckman formula for the partition function in this case is given by (7.75) evaluated on the pure singular cycles a_1 and a_2 above, and on the critical cycles b_1 and b_2 (see the previous Section). Summing the 2 contributions from the 1-form F in (7.78) along the pure homology cycles shows immediately that $\oint_{a_1} F|_{a_1} + \oint_{a_2} F|_{a_2} = 0$,

7.4 Examples   263

as anticipated. As for the integrals along the critical cycles, taking proper care of orientations induced by the contractable patches, we find that the contributions from b_1 and b_2 are the same and that the corrections can be written as

$$\delta Z_{T^2}(T) = -\frac{1}{iT \operatorname{Im} \tau} \left(e^{iT(r_2-r_1)} \int_0^\pi d\phi \, \frac{e^{-iT \operatorname{Im} \tau \cos \phi}}{\sin \phi} \right.$$
$$\left. - e^{iT(r_2+r_1)} \int_0^\pi d\phi \, \frac{e^{iT \operatorname{Im} \tau \cos \phi}}{\sin \phi} \right) \tag{7.97}$$

After a change of variables we find that the integrals in (7.97) can be expressed in terms of the exponential integral function [60]

$$\operatorname{Ei}(x) = -\int_{-x}^{\infty} dt \, \frac{e^{-t}}{t} \tag{7.98}$$

which diverges for $x \leq 0$. Here the integral denotes a Cauchy principal value integration. After some algebra we find

$$\delta Z_{T^2}(T) = -\frac{1}{iT \operatorname{Im} \tau} \left[e^{iT(r_2-r_1)} \left\{ \frac{e^{iT \operatorname{Im} \tau}}{2} \left(\operatorname{Ei}(-2iT \operatorname{Im} \tau) \right. \right. \right.$$
$$\left. - \operatorname{Ei}\left(-2iT \operatorname{Im} \tau \cos^2 \frac{y}{2}\right) \right)$$
$$\left. - \frac{e^{-iT \operatorname{Im} \tau}}{2} \left(\operatorname{Ei}\left(2iT \operatorname{Im} \tau \sin^2 \frac{\epsilon}{2}\right) - \operatorname{Ei}(2iT \operatorname{Im} \tau) \right) \right\}$$
$$- e^{iT(r_2+r_1)} \left\{ \frac{e^{-iT \operatorname{Im} \tau}}{2} \left(\operatorname{Ei}(2iT \operatorname{Im} \tau) - \operatorname{Ei}\left(2iT \operatorname{Im} \tau \cos^2 \frac{y}{2}\right) \right) \right.$$
$$\left. \left. - \frac{e^{iT \operatorname{Im} \tau}}{2} \left(\operatorname{Ei}\left(-2iT \operatorname{Im} \tau \sin^2 \frac{\epsilon}{2}\right) - \operatorname{Ei}(-2iT \operatorname{Im} \tau) \right) \right\} \right]$$
$$\tag{7.99}$$

where $y = \pi - \epsilon$ and $\epsilon \to 0$ is used to regulate the divergence of the integrals in (7.97) at $\phi = 0$ and $\phi = \pi$.

The correction term (7.99) tells us quite a bit. First of all, note that it is a sum of 4 terms which can be identified with the contributions from the critical points of the Hamiltonian h_{Σ^1}. However, these terms are resummed, since the above correction terms take into account the full loop corrections to the Duistermaat-Heckman formula. Next, the terms involving ϵ are divergent, and the overall divergence of $\delta Z_{T^2}(T)$ is anticipated from Kirwan's theorem, which says that the full saddle-point series for this Hamiltonian diverges. The exponential integral function can be expanded as the series [60]

264 7. Beyond the Semi-Classical Approximation

$$\mathrm{Ei}(x) = \gamma + \log x + \sum_{n=1}^{\infty} \frac{x^n}{nn!} \tag{7.100}$$

for x small, where γ is the Euler-Mascheroni constant. Thus the divergent pieces in (7.99) can be explicitly expanded in powers of $\frac{1}{T}$, giving a much simpler way to read off the coefficients of the loop-expansion (note the enormous complexity of the series coefficients in (7.3) for this Hamiltonian – a direct signal of the messiness of its stationary-phase series). Finally, the finite terms (those independent of the regulator ϵ), can be evaluated for $T = -i$ and $\tau = 1 + i$, and we find $\delta Z_{T^2} = 123.086$. In Section 3.5 we saw that the exact value of the partition function for this dynamical system was 2117.12, while the Duistermaat-Heckman formula gave $Z_0 = 1849.327$. Thus $Z_0 + \delta Z_{T^2} = 1972.41$, which is a better approximation to the partition function than the Duistermaat-Heckman formula. Of course, given the large divergence of the stationary phase series, we do not expect that the finite contributions in (7.99) will give the exact result for the partition function, but we certainly do get much closer. As the function χ^1 which generates the set of preferred coordinates is by no means unique, perhaps a refined definition of it could lead to a better approximation $Z_0 + \delta Z$. Then, however, we lose a lot of the geometrical interpretation of the corrections that we gave in the last Section.

The second set of examples we consider here are the potential problems (5.273) defined on the plane $\mathbb{R}^2$, where $U(q)$ is a C^∞ potential which is a non-degenerate function. In this case the equation (7.76) becomes

$$p\frac{\partial \chi^1}{\partial q} = -U'(q)\frac{\partial \chi^1}{\partial p} \tag{7.101}$$

which is solved by

$$\chi^1(q,p) = p^2/2 - U(q) \tag{7.102}$$

Then proceeding as above the local coordinates $(\bar{x}, \bar{y})$ in which the Hamiltonian vector field generates translations are

$$\bar{x}(q,p) = \frac{1}{pU'(q)}\left(qU'(q) - p^2\right) \quad , \quad \bar{y}(q,p) = \sqrt{\frac{p^2}{2} + U(q)} \tag{7.103}$$

Thus here there are only critical 'cycles' given by the infinite lines

$$P = \{(0,q) \in \mathbb{R}^2\} \quad , \quad \mathcal{U}_i = \{(p,q_i) \in \mathbb{R}^2\} \tag{7.104}$$

where q_i are the extrema of the potential $U(q)$.

Since for the Darboux Hamiltonian (5.273), V^p and V^q vanish on the 'cycles' P and $\mathcal{U}_i$ respectively, we must use the renormalized version of (7.78), namely (7.85). Combining (7.85) with (7.77) we find, for an even potential function, that the corrections are

$$\delta Z_{\mathbb{R}^2}(T) = -\frac{1}{iT}\left\{\int_0^\infty dq\,\frac{e^{iTU(q)}}{U'(q)} - \left(\sum_{q_i} e^{iTU(q_i)}\right)\int_0^\infty dp\,\frac{e^{iTp^2/2}}{p}\right\} \tag{7.105}$$

and we note the manner in which the divergences are cancelled here. From this we immediately see that for the harmonic oscillator potential $U(q) = aq^2$, the corrections (7.105) vanish (note that the integration measures in (7.105) contain implicit factors of ω_{12} that maintain covariance). Similarly, it is easily verified, by a simple change of variables, that for a potential of the form $U(q) = aq + bq^2$ these correction terms vanish, again as expected. Finally, for a quartic potential $U(q) = \frac{q^2}{2} + \frac{q^4}{4}$ (the anharmonic oscillator), a numerical integration of (7.105) for $T = i$ gives $\delta Z_{\mathbb{R}^2} = -0.538$ and the Duistermaat-Heckman formula yields $Z_0 = 2\pi$. A numerical integration of the original partition function gives $Z = 4.851$, which differs from the value $Z_0 + \delta Z_{\mathbb{R}^2} = 5.745$. The corrections do not give the exact value here, but again at least they are a better approximation than the Duistermaat-Heckman formula. Again, a refinement of the preferred coordinates could lead to a better approximation. The method of the last Section has therefore "stripped" off any potentially divergent contributions to the loop-expansion and at the same time approximated the partition function in a much better way. These last few examples illustrate the applicability and the complete consistency of the geometric approach of the last Section to the saddle-point expansion. Indeed, we see that it reproduces the precise analytic features of the loop-expansion but avoids many of the cumbersome calculations in evaluating (7.3). It would be interesting to develop some of these ideas further.

We would next like to check, following the analysis of Section 5.9, if there are any conformally-invariant geometries for this dynamical system when the potential $U(q) \geq 0$ is bounded from below. In the harmonic-polar coordinates (5.274), the conformal Killing equations (7.35) can be determined by setting the right-hand sides of the Killing equations (5.276) equal to instead $(\nabla_\theta V^\theta)g_{\mu\nu} = (\partial_\theta V^\theta + \Gamma^\theta_{\theta\theta}V^\theta)g_{\mu\nu}$. After some algebra, we find that they generate the 2 equations

$$\partial_\theta \log\left(\frac{(V^\theta)^2 g_{\theta\theta}}{g_{rr}}\right) = 2\frac{g_{r\theta}}{g_{rr}}\partial_r \log V^\theta \tag{7.106}$$

$$\partial_\theta \log\left(V^\theta \frac{g_{\theta\theta}}{g_{r\theta}}\right) = \frac{g_{\theta\theta}}{g_{r\theta}}\partial_r \log V^\theta \tag{7.107}$$

(7.107) can be formally solved as

$$g_{r\theta} = -V^\theta g_{\theta\theta} \int_{\theta_0}^\theta d\theta'\,\partial_r V^{\theta'} + f(r) \tag{7.108}$$

from which we see that again single-valuedness $g_{r\theta}(r, \theta+2\pi) = g_{r\theta}(r, \theta)$ holds only when (5.279) is true, i.e. when $U(q)$ is the harmonic oscillator potential with $V^\theta = 1$. Even for the harmonic oscillator, the equations (7.106) and (7.107) only seem to admit radially-symmetric solutions $g_{\mu\nu} = g_{\mu\nu}(r)$ so that $V^\theta = 1$ is a global isometry of g. Thus, even though we lose the third equation in (5.276) which established the results of Section 5.9 using the Killing equations, we still arrive at the conclusion that there are no single-valued metric tensors obeying the conformal Lie derivative requirement for essentially all potentials which are bounded from below (and the harmonic oscillator only seems to generate isometries). Thus the conformal symmetry requirement in the case at hand does not lead to any new localizable systems.

Finally, we examine what can be learned in these cases from the vanishing of the 2-loop correction (7.19) in harmonic coordinates. In these coordinates, the connection 1-form (7.8) has components

$$\gamma_p = 0 \quad , \quad \gamma_y = \frac{dq}{dy} \qquad (7.109)$$

and the condition (7.19) reads

$$\frac{d}{dy}\gamma_y = -\gamma_y^2 \qquad (7.110)$$

There are 2 solutions to (7.110). Either $\gamma_y = 0$, in which case $U(q)$ is the harmonic oscillator potential, or $\gamma_y = (y + a)^{-1}$, where a is an integration constant. This latter solution, however, yields $q(y) = C_1 y^2 + ay + C_0$, which gives a potential $U(q)$ which is not globally defined as a C^∞-function on $\mathbb{R}^2$. Thus the *only* potential which is bounded from below that leads to a localizable partition function is that of the simple-harmonic oscillator. This example illustrates how the deep geometric analyses of this Chapter serve of use in examining the localizability properties of dynamical systems. As for these potential problems, it could prove of use in examining the localization features of other more complicated integrable systems [56].

7.5 Heuristic Generalizations to Path Integrals: Supersymmetry Breaking

The generalization of the loop expansion to functional integrals is not yet known, although some formal suggestive techniques for carrying out the full semi-classical expansion can be found in [93] and [147]. It would be of utmost interest to carry out an analysis along the lines of this Chapter for path integrals for several reasons. There the appropriate loop space expansion should again be covariantized, but this time the functional result need not be fully independent of the loop space coordinates. This is because the quantum corrections could cause anomalies for many of the symmetries of the classical theory (i.e. of the classical partition function). In particular, the larger conformal dynamical structures discussed in Section 7.2 above could

7.5 Heuristic Generalizations to Path Integrals: Supersymmetry Breaking 267

play an important role in path integral localizations which are expressed in terms of trajectories on the phase space [134]. It would be very interesting to see if these general conformal symmetries of the classical theory remain unbroken by quantum corrections in a path integral generalization. The absence of such a conformal anomaly could then lead to a generalization of the above extended localizations to path integral localization formulas. As this symmetry in the finite-dimensional case is not represented by a nilpotent operator, such as an exterior derivative, one would need some sort of generalized supersymmetry arguments to establish the localization with these sorts of symmetries. When these supersymmetries are globally present, the vanishing of higher-loop terms in the path integral loop expansion is a result of the usual non-renormalizations of 1-loop quantities in supersymmetric quantum field theories that arise from the mutual cancellations between bosonic and fermionic loops in perturbation theory (where the fermionic loops have an extra minus sign compared to the bosonic ones).

Quite generally though, one also has to keep in mind that the loop space localization formulas are rather formal. We have overlooked several formal functional aspects, such as difficulties associated with the definition of the path integral measure. There may be anomalies associated with the argument in Section 4.4 that the path integral is independent of the limiting parameter $\lambda \in \mathbb{R}$, for instance the supersymmetry may be broken in the quantum theory (e.g. by a scale anomaly in the rescaling of the phase space metric $g \to \lambda \cdot g$). The same sort of anomalies could also break the larger conformal symmetry we have found for the classical theory above. However, even if the localization formulas are not correct as they stand, it would then be interesting to uncover the reasons for that. This could then provide one with a systematic geometric method for analysing corrections to the WKB approximation.

The ideas presented in this Chapter are a small step forward in this direction. In particular, it would be interesting to generalize the construction of Section 7.3, as this is the one that is intimately connected to the integrability features of the dynamical system. The Poincaré duality interpretation there is one possible way that the construction could generalize to path integrals. For path integrals, we would expect the feature of an invariant metric tensor that cannot be extended globally to manifest itself as a local (i.e. classical) supersymmetry of the theory which is dynamically broken globally on the loop space. This has been discussed by Niemi and Palo [122] in the context of the supersymmetric non-linear sigma-model (see Chapter 8). Another place where the metric could enter into a breakdown of the localization formulas is when the localization 1-form $\psi \sim i_W g$ does not lead to a homotopically-trivial element under the (infinitesimal) supersymmetry transformation described by Q_S. Then additional input into the localization formalism should be required on a topologically non-trivial phase space to ensure that $Q_S \psi$ indeed does reside in the trivial homotopy class. These inputs could follow from an appropriate loop space extension of the correction terms $\delta Z(T)$ discussed above, which will then always reflect global properties of the quantum theory. Other directions could also entail examining the connections between equivariant localization and other ideas we have discussed in this Book. One is the Parisi-Sourlas supersymmetry that we encountered in the evaluation of the Niemi-Tirkkonen localization formula for the height function on the sphere (Section 5.5), although this feature seems to be more intimately connected to the Kähler geometry of S^2, as we showed above. The Kähler symmetries we found in Section 7.1 would be a good probe of the path integral correction formulas, and it would interesting to see if they could also be generalized to some sort of supersymmetric structure.

8. Equivariant Localization in Cohomological Field Theory

We have seen that the equivariant localization formalism is an excellent, conceptual geometric arena for studies of supersymmetric and topological field theories, and more generally of (quantum) integrability. Given that the Hamiltonians in an integrable hierarchy are functionals of action variables alone [106], the equivariant localization formalism might yield a geometric characterization of quantum integrability, and perhaps some deeper connection between quantum-integrable bosonic theories and supersymmetric quantum field theories. This is particularly interesting from the point of view of examining corrections to the localization formulas, which in the last Chapter we have seen reflect global properties of the theory. This would be of particular interest to analyse more closely, as it could then lead to a unified description of localization in the symplectic loop space, the supersymmetric loop space and in topological quantum field theory.

In this final Chapter we shall discuss some of the true field theoretical models to which the equivariant localization formalism can be applied. We shall see that the quantum field theories which fall into this framework always have, as anticipated, some large symmetry group (such as a topological gauge symmetry or a supersymmetry) that serves to provide an equivariant cohomological structure on the space of fields that can be understood as a "hidden" supersymmetry of the theory. Furthermore, the configuration spaces of these models must always admit some sort of (pre-)symplectic structure in order that the localization properties of phase space path integrals can be applied. Because of space considerations we have not attempted to give a detailed presentation here and simply present an overview of the various constructions and applications, mainly just presenting results that have been obtained. The interested reader is refered to the extensive list of references that are cited throughout this Chapter for m ore details. We shall emphasize here the connections eluded to throughout this Book between the localization formalism for dynamical systems and genuine topological quantum field theories. Exploring the connections between the topological field theories and more conventional physical quantum field theories will then demonstrate how the equivariant localization formalism for phase space path integrals serves as the correct arena for studying the (path integral) quantization of real physical systems.

8.1 Two-Dimensional Yang-Mills Theory: Equivalences Between Physical and Topological Gauge Theories

In Section 3.8 we first pointed out that, instead of circular actions, one can consider the Poisson action of some non-abelian Lie group acting on the phase space. Then the non-abelian generalizations of the equivariant localization formulas, discussed in Sections 3.8, 4.9 and 5.8, lead to richer structures in the quantum representations discussed earlier and one obtains intriguing path integral representations of the groups involved. In this Section we shall demonstrate how a formal application of the Witten localization formalism can be used to study a cohomological formulation of 2-dimensional QCD (equivalently the weak-coupling limit of 2-dimensional pure Yang-Mills theory). This leads to interesting physical and mathematical insights into the structures of these theories. We shall also discuss how these results can be generalized to topological field theory limits of other models.

First, we briefly review some of the standard lore of 2-dimensional QCD. The action for pure Yang-Mills theory on a 2-dimensional surface Σ^h of genus h is

$$S_{YM}[A] = -\frac{1}{2e^2} \int_{\Sigma^h} \mathrm{tr}\, F_A \star F_A \tag{8.1}$$

where A is a gauge connection of a trivial prinicipal G-bundle over Σ^h, F_A is its curvature 2-form (c.f. Section 2.4), and e^2 is the coupling constant of the gauge field theory. Since $\phi \equiv \star F_A$ is a scalar field in 2-dimensions, the action (8.1) depends on the metric of Σ^h only through its dependence on the area $A(\Sigma^h) = \int_{\Sigma^h} \star 1$ of the surface. A deformation of the metric can therefore be compensated by a change in the coupling constant e^2. The action (8.1) is invariant under the gauge transformations (2.72). The corresponding quantum field theory is described by the path integral

$$Z_{\Sigma^h}(e^2) \equiv \int_{\mathcal{A}} [dA]\ e^{i S_{YM}[A]} \tag{8.2}$$

where $\mathcal{A}$ is the space of gauge connections over Σ^h.

We can write the partition function in a much simpler (first order) form by treating the **g**-valued scalar field $\phi = \star F_A \in C^\infty(\Sigma^h, \mathbf{g})$ as a Lagrange multiplier to write

$$Z_{\Sigma^h}(e^2) = \int_{\mathcal{A}} [dA] \int_{C^\infty(\Sigma^h, \mathbf{g})} [d\phi]\ e^{-i \int_{\Sigma^h} \mathrm{tr}\left(i\phi F_A + \frac{e^2}{2}\phi \star \phi\right)} \tag{8.3}$$

In the weak coupling limit $e^2 \to 0$, the action in (8.4) reduces to the topological one $\int_{\Sigma^h} \mathrm{tr}\, i\phi F_A$ (i.e. one that is independent of the metric of Σ^h and which consequently determines a cohomological quantum field theory)[1].

---

[1] This sort of topological field theory is called a 'BF theory' and it is the prototype of a Schwarz-type topological gauge theory [22].

8.1 Two-Dimensional Yang-Mills Theory

The gauge invariance of the action $S[\phi, A]$ appearing in (8.3) is expressed as $S[g^{-1}\phi g, A^g] = S[\phi, A]$ where $g \in \mathbf{g}$ and ϕ transforms under the adjoint representation of the gauge group. Because of this gauge invariance, it is necessary to fix a gauge and restrict the integration in (8.3) to the equivalence classes $\mathcal{A}/\mathcal{G}$ of gauge connections modulo gauge transformations. This can done by the standard BRST gauge fixing procedure (see Appendix A for a brief account).

For this, we introduce an auxilliary, $\mathbf{g}$-valued fermion field ψ^μ, which is an anti-commuting 1-form in the adjoint representation of $\mathbf{g}$, and write (8.3) as

$$Z_{\Sigma^h}(e^2) = \frac{1}{\text{vol } C^\infty(\Sigma^h, \mathbf{g})} \int_{\mathcal{A} \otimes \Lambda^1 \mathcal{A}} [dA] \, [d\psi] \int_{C^\infty(\Sigma^h, \mathbf{g})} [d\phi]$$

$$\times \exp\left\{ - \int_{\Sigma^h} \text{tr} \left(\phi F_A - \frac{1}{2} \psi \wedge \psi \right) - i\frac{e^2}{2} \int_{\Sigma^h} \text{tr } \phi \star \phi - i \int_{\Sigma^h} \star\{Q, \Psi\} \right\} \quad (8.4)$$

In (8.4) we have introduced the usual BRST and Faddeev-Popov gauge fixing terms defined by the graded BRST commutator of a gauge fermion $\Psi = \psi^\mu \Pi_\mu(x)$ with the usual BRST charge Q. The square of Q is $Q^2 = -i\delta_\phi$ where δ_ϕ is the generator of a gauge transformation with infinitesimal parameter ϕ. Thus Q is nilpotent on the space of physical (i.e. gauge-invariant) states of the quantum field theory. The system of fields (A, ψ, ϕ) is the basic multiplet of cohomological Yang-Mills theory. The (infinitesimal) gauge invariance of (8.4) is manifested in its invariance under the infinitesimal BRST supersymmetry transformations

$$\delta A_\mu = i\epsilon\psi_\mu \quad , \quad \delta\psi_\mu = -\epsilon(\nabla_A)_\mu \phi = -\epsilon(\partial_\mu + [A_\mu, \phi]) \quad , \quad \delta\phi = 0 \quad (8.5)$$

where ϵ is an anticommuting parameter. The supersymmetry transformations (8.5) are generated by the graded BRST commutator $\delta\Phi = -i\{Q, \Phi\}$ for each field Φ in the multiplet (A, ψ, ϕ). The ghost quantum numbers ($\mathbb{Z}$-gradings) of the fields (A, ψ, ϕ) are $(0, 1, 2)$.

We shall not enter here into a discussion of the physical characteristics of 2-dimensional Yang-Mills theory. It is a super-renormalizable quantum field theory which is exactly solvable and whose simplicity therefore allows one to explore the possible structures of more complicated non-abelian gauge theories such as higher-dimensional cohomological field theories and other physical models such as 4-dimensional QCD. It can be solved using group character expansion methods [34] or by diagonalization of the functional integration in (8.3) onto the Cartan subalgebra using the elegant Weyl integral formula [29]. Here we wish to point out the observation of Witten [171] that the BRST gauge-fixed path integral (8.4) is an infinite-dimensional version of the partition function in the last line of (3.125) used for non-abelian localization. Indeed, the integration over the auxilliary fermion fields ψ acts to

8. Equivariant Localization in Cohomological Field Theory

produce a field theoretical analog of the super-1 oop space Liouville measure introduced in Chapter 4. The "Hamiltonian" here is the field strength tensor F_A while the Lagrange multiplier fields ϕ serve as the dynamical generators of the symmetric algebra $S(\mathfrak{g}^*)$ used to generate the G-equivariant cohomology. The "phase space" $\mathcal{M}$ is now the space $\mathcal{A}$ of gauge connections, and the Cartan equivariant exterior derivative

$$D = \int_{\Sigma^h} \star \left(\psi^\mu \frac{\delta}{\delta A_\mu} - i\phi^a V^{a,\mu} \frac{\delta}{\delta \psi^\mu} \right) \tag{8.6}$$

in this case coincides with the action of the BRST charge Q, i.e. $D\Phi = -\{Q, \Phi\}$. The gauge fermion Ψ thus acts as the localization 1-form λ and, by the equivariant localization principle, the integration will localize onto the field configurations where $\lambda(V^a) = V^{a,\mu}\Pi_\mu = 0$ where $V^a = V^{a,\mu}\frac{\partial}{\partial x^\mu}$ are the vector fields generating $\mathbf{g}$.

The equivalence between the first and last lines of (3.125) is the basis of the mapping between "physical" Yang-Mills theory with action (8.1) and the cohomological Yang-Mills theory with action $\int_{\Sigma^h} \operatorname{tr} i\phi F_A$ which is defined essentially by the steps which lead to the non-abelian localization principle, but now in reverse. The extrema of the action (8.1) are the classical Yang-Mills field equations $F_A = 0$. Thus the localization of the partition function will be onto the symplectic quotient $\mathcal{M}_0$ which here is the moduli space of flat gauge connections modulo gauge transformations associated with the gauge group G. This mapping between the physical gauge theory and the cohomological quantum field theory is the basis for the localization of the 2-dimensional Yang-Mills partition function. Thus the large equivariant cohomological symmetry of this theory explains its strong solvability properties that have been known for quite some time now. More generally, as m entioned at the end of Section 5.8, the equivariant localization here also applies to the basic integrable models which are related to free field theory reductions of 2-dimensional Yang-Mills theory, such as Calegoro-Moser integrable models [56].

To carry out the localization onto $\mathcal{M}_0$ explicitly, we choose a G-invariant metric g on Σ^h and take the localization 1-form in (3.125) to be

$$\lambda = \int_{\Sigma^h} \operatorname{tr} \psi \wedge \star Df = \int_{\Sigma^h} d\mathrm{vol}(g(x))\, \operatorname{tr}\, \psi^\mu D_\mu f \tag{8.7}$$

where $f = \star F_A$. The localization onto $\lambda(V) = 0$ is then identical to localization onto the solutions of the classical Yang-Mills equations. We shall not enter into the cumbersome details of the evaluation of the partition function (8.4) at weak coupling $e^2 \to 0$ (the localization limit) using the Witten non-abelian localization formalism. For details, the reader is refered to [171]. As in Section 5.8, the final integration formula can be written as a sum over the unitary irreducible representations of G,

8.1 Two-Dimensional Yang-Mills Theory

$$Z_{\Sigma^h}(e^2) = e^{-\frac{e^2}{2}\sum_i \rho_i^2} \prod_{\alpha>0} \alpha(\rho)^{2-2h} \sum_{\lambda\in\mathbb{Z}^r} (\dim \mathcal{R}_\lambda)^{2-2h} \, e^{-\frac{e^2}{2}\sum_i (\lambda+\rho)_i^2} \quad (8.8)$$

This result follows from expanding the various (G-invariant) physical quantities appearing in the localization formula in characters of the group G (c.f. Section 5.8). From a physical standpoint the localization formula (8.8) is interesting because although it expresses the exactness of a loop approximation to the partition function, it is a non-polynomial function of the coupling constant e^2. This non-polynomial dependence arises from the contributions of the unstable classical solutions to the functional integral as described in Chapter 3. Such behaviours are not readily determined using the conventional perturbative techniques of quantum field theory. Thus the mapping provided above between the physical and topological gauge theories (equivalently the generalization of the Duistermaat-Heckman integration formula to problems with non-abelian symmetries) provides an unexpected and new insight into the structure of the partition function of 2-dimensional Yang-Mills theory. This simple mapp ing provides a clearer picture of this quantum field theoretical equivalence which is analogous to the more mysterious equivalence of topological and physical gravity in 2 dimensions. From a mathematical perspective, the quantity (8.8) is the correct one to use for determining the intersection numbers of the moduli space of flat G-connections on Σ^h [78, 82, 171]. This approach to 2-dimensional Yang-Mills theory has also been studied for genus $h = 0$ in [105].

The intriguing mapping between a physical gauge theory, with propagating particle-like local degrees of freedom, and a topological field theory with only global degrees of freedom has also been applied to more complicated models. In [28], similar considerations were applied to the non-linear cousin of 2-dimensional topological Yang-Mills theory, the gauged G/G Wess-Zumino-Witten model. This model at level $k \in \mathbb{Z}$ is defined by the action

$$S_{G/G}[g, A] = -\frac{k}{8\pi} \int_{\Sigma^h} \text{tr } g^{-1}\nabla_A g \wedge \star g^{-1}\nabla_A g + \frac{k}{12\pi} \int_M \text{tr}(g^{-1}dg)^{\wedge 3}$$
$$-\frac{k}{4\pi} \int_{\Sigma^h} \text{tr}\left(A \wedge dgg^{-1} + A \wedge A^g\right) \quad (8.9)$$

where $g \in C^\infty(\Sigma^h, \mathbf{g})$ is a smooth group-valued field, M is a 3-manifold with boundary the surface Σ^h, and A is a gauge field for the diagonal G subgroup of the $G_L \times G_R$ symmetry group of the ordinary (ungauged) Wess-Zumino-Witten model defined by the action $S_G[g] = S_{G/G}[g, A = 0]$ [54]. Since the Hodge duality operator $\star$ is conformally invariant when acting on 1-forms, the action (8.9) depends only on the chosen complex structure of Σ^h. As for the Yang-Mills theory above, the geometric interpretation of the theory comes from adding to the bosonic action (8.9) the term $\Omega(\psi) = \frac{1}{2\pi}\int_{\Sigma^h} \psi_z\psi_{\bar{z}}$ quadratic in Grassmann-odd variables ψ which represents the symplectic form

274   8. Equivariant Localization in Cohomological Field Theory

$\int_{\Sigma^h} \delta A \delta A$ on the space $\mathcal{A}$ of gauge fields on Σ^h. Again the resulting theory is supersymmetric and the infinitesimal supersymmetry transformations are

$$\delta A_z = \psi_z \ , \quad \delta \psi_z = A_z^g - A_z \ ; \quad \delta A_{\bar{z}} = \psi_{\bar{z}} \ , \quad \delta \psi_{\bar{z}} = A_{\bar{z}} - (A_{\bar{z}}^g)^{-1} \quad (8.10)$$

with the supplemental condition $\delta g = 0$. Unlike its Yang-Mills theory counterpart, the square of this supersymmetry $\Delta = \delta^2$ does not generate infinitesimal gauge transformations but rather 'global' gauge transformations (generated by the cohomological elements of the gauge group which are not connected to the identity).

Thus the action (8.9) here admits a supersymmetry which does not manifest itself as the local gauge symmetry of the quantum field theory. Nonetheless, this implies a supersymmetric structure for equivariant cohomology which can be used to obtain a localization of the corresponding path integral in the usual way. The localization formula of [23] for equivariant Kähler geometry has a field theoretic realization in this model [28] and the fixed-point localization formula is the algebraic Verlinde formula for the dimension of the space of conformal blocks of the ordinary Wess-Zumino-Witten model. For example, in the case $G = SU(2)$ the localization formula gives

$$Z_{\Sigma^h}(SU(2), k) = \int_{C^\infty(\Sigma^h, \mathrm{su}(2))} [dg] \int_{\mathcal{A}} [dA] \ \mathrm{e}^{iS_{SU(2)/SU(2)}[g,A]}$$

$$= \left(\frac{2}{k+2}\right)^{2-2h} \sum_{\ell=1}^{k+1} \left(\sin \frac{\pi \ell}{k+2}\right)^{2-2h} \quad (8.11)$$

Thus the equivariant localization formalism can also be used to shed light on some of the more formal structures of 2-dimensional conformal field theories.

Perret [138] has used the path integral for a version of the ordinary Wess-Zumino-Witten model to give a field theoretical generalization of Stone's derivation for the Weyl-Kac character formula for Kac-Moody algebras (i.e. loop groups). This is done by exploiting the $G_\mathrm{L} \times G_\mathrm{R}$ Kac-Moody symmetry associated with the quantum field theory with action $S_G[g]$ as a supersymmetry of the model along the same lines as in Chapter 5 before. Let us now briefly describe Perret's derivation. The Kac-Moody group $\tilde{G}$ is a central extension of the loop group $S^1 \to \tilde{G} \to LG$ of a compact semi-simple Lie group G, and it looks locally like the direct product $LG \otimes S^1$, i.e. an element $\tilde{g} \in \tilde{G}$ looks locally like $\tilde{g} = (g(x), c)$ where $g : S^1 \to G$ and $c \in S^1$. The coherent state path integral for the character is

$$\mathrm{tr}_\lambda \ \mathrm{e}^{iH} q^{L_0} = \int_{L\tilde{G}} [d\tilde{g}] \ \mathrm{e}^{\oint_{\tilde{g}} \langle \tilde{g} | d + i(H + \tau L_0) dt | \tilde{g} \rangle} \quad (8.12)$$

where $H = \sum_i h_i H_i \in H_C$, $q = \mathrm{e}^{i\tau}$ and L_0 is the generator of rotations of the loops. The action in (8.12) depends on 2 coordinates, the coordinate

x along the loop in Lg and the time coordinate t of the path integral. The central S^1-part of the coherent states in the path integral drops out due to gauge invariance, and thus the character representation (8.12) will define a 2-dimensional quantum field theory on the torus T^2 (i.e. the quotient of loops in the loop group LG).

It can be shown [138] that the coherent state path integral (8.12) is the quantum field theory with action

$$S_{KM} = \frac{1}{2\pi} \int_{T^2} dx\, dt\, \mathrm{tr}\left(\lambda g^{-1}(\bar{\partial}+H)g + \frac{k}{2}g^{-1}\partial_x(\bar{\partial}+2H)g\right)$$
$$+ \frac{k}{12\pi}\int_M \mathrm{tr}(g^{-1}dg)^{\wedge 3} \tag{8.13}$$

Here $k \in \mathbb{Z}$ is the given central extension of the loop group, λ is a dominant weight of G, and $\bar{\partial} = \partial_t - \tau\partial_x$ so that the Cartan angle τ becomes the modular parameter of the torus. When $\lambda = H = 0$ the action (8.13) becomes the chiral Wess-Zumino-Witten model with the single Kac-Moody symmetry $g \to k(z)g$. This symmetry is still present for generic $\lambda \neq 0$, and we can gauge the action S_{KM} with respect to an arbitrary subgroup of G by replacing H with a vector field (see (8.9)). One can now evaluate the infinite-dimensional Duistermaat-Heckman integration formula for this path integral. The critical points of the action (8.13) are in one-to-one correspondence with the affine Weyl group $W_{\mathrm{aff}}(H_C) = W(H_C) \ltimes \hat{H}_C$, where $\hat{H}_C$ is the set of co-roots. Let $r(w)$ denote the rank of an element $w \in W(H_C)$, and let h be the dual Coxeter number of the Lie group G [162]. Using modular invariance of the character to fix the conventional zero-point energy (associated with the usual $SL(2,\mathbb{R})$-invariance of the conformal field theory vacuum [54]), it can be shown that the WKB localization formula for the coherent state path integral (8.12) coincides with the Weyl-Kac character formula [138] (for notation see Section 5.1)

$$\mathrm{tr}_\lambda\, e^{iH}q^{L_0} = \sum_{w \in W(H_C), \eta \in \hat{H}_C} (-1)^{r(w)}\, q^{\sum_i((\lambda+\rho)_i + (k+h)\eta_i^2 - \rho_i^2)/2(k+h)}$$
$$\times\, e^{i(\lambda+\rho+(k+h)\eta)(H^{(w)})}\, e^{-i\rho(H^{(w)})} \prod_{n>0}(1-q^n)^{-r}$$
$$\times \prod_{\alpha>0}\left(1 - e^{i\alpha(H^{(w)})}q^n\right)^{-1}\left(1 - e^{-i\alpha(H^{(w)})}q^{n-1}\right)^{-1}$$
$$\tag{8.14}$$

which arises from the expansions of various quantities defined on the torus in terms of Jacobi theta-functions. Thus infinite dimensional analogs of the localization formulas in quantum field theory can also lead to interesting generalizations of the character formulas that the topological field theories of earlier Chapters represented.

276   8. Equivariant Localization in Cohomological Field Theory

Finally, it is possible to use some of these ideas in the context of abelian localization as well. For instance, in [126] the abelian gauge theory with action

$$S_{CSP}[A] = \frac{k}{8\pi} \int_M A \wedge dA - \frac{m}{2} \int_M A \wedge \star A \qquad (8.15)$$

defined on a 3-manifold M was studied within the equivariant localization framework. The first term in (8.15) is the Chern-Simons action which defines a topological field theory[2], while the second term is the Proca mass term for the gauge field which gives a propagating degree of freedom with mass m and thus breaks the topological invariance of the quantum field theory. In a canonical formalism where $M = \Sigma^h \times \mathbb{R}^1$, one can naturally write the model (8.15) as a quantum mechanics problem on the phase space Σ^h and apply the standard abelian equivariant localization techniques to evaluate the path integral from the ensuing supersymmetry generated by the gauge invariance of (8.15) (in the Lorentz gauge $\partial^\mu A_\mu = 0$) [26]. The path integral localization formula coincides with that of a simple harmonic oscillator of frequency $\omega_h = 8\pi m/k$, i.e. $Z = 1/2\sin(T \cdot 4\pi m/k)$, indicating a mapping once again to a topological field theory using the equivariant localization framework. The infrared limit $m \to 0$ of the model leads to the usual topological quantum mechanical models associated with Chern-Simons theory [26] and the supersymmetry, which is determined by a loop space equivariant cohomology, emerges from the symplectic structure of the theory on Σ^h and could yield interesting results in the full 3-dimensional quantum field theory defined by (8.15).

8.2 Symplectic Geometry of Poincaré Supersymmetric Quantum Field Theories

In the last Section we showed how loop space equivariant localization provides a correspondence between certain physical gauge theories and topological ones which makes manifest the localization properties of the physical models. In these cases the original theory contains a large gauge symmetry which leads to a localization supersymmetry and a limit where the model becomes topological that gives the usual localization limit. It is now natural to ask what happens when the original quantum field theory is explicitly defined with a supersymmetry (i.e. one that is not "hidden"). In Section 4.2 we saw that $N = \frac{1}{2}$ supersymmetric quantum mechanics admits a loop space equivariant cohomological structure as a result of the supersymmetry which provides an alternative explanation for the well-known localization properties of this topological field theory (where the topological nature now arises because the bosonic and fermionic degrees of freedom mutually cancel each

---

[2] The Cherns-Simons action is in fact another prototype of a Schwarz-type topological field theory [22].

8.2 Poincaré Supersymmetric Quantum Field Theories

other out). From the point of view of path integration, this approach in fact led to a nice, geometric interpretation on the functional loop space of the features of this theory. It has been argued [107, 108, 133] that this interpretation can be applied to *generic* quantum field theories with Poincaré supersymmetry. In this Section we shall briefly discuss how this works and how the techniques of equivariant localization could lead to new geometrical interpretations of such models.

To start, let us quickly review some of the standard ideas in Poincaré supersymmetric quantum field theories. The idea of supersymmetry was first used to relate particles of different spins to each other (e.g. the elementary representation theory of $SU(3)$ which groups particles of the same spin into multiplets) by joining the internal (isospin) symmetries and space-time (Poincaré) symmetries into one large symmetry group (see [155] for a comprehensive introduction). This is not possible using bosonic commutation relations because then charges of internal symmetries have to commute with space-time transformations so that dynamical breaking of these symmetries (required since in nature such groupings of particles are not observed) is not possible. However, it is possible to consider anti-commutation relations, i.e. supersymmetries, and then the imposition of the Jacobi identity for the symmetry group leads to a very restricted set of commutation relations. Here we shall be concerned only with tho se anti-commutation relations satisfied by the infinitesimal supersymmetry generators $Q_\alpha^i, \bar{Q}_{\dot\alpha}^i$,

$$\{Q_\alpha^i, \bar{Q}_{\dot\beta}^j\} = 2\delta^{ij}\Sigma^\mu_{\alpha\dot\beta}P_\mu + Z^{ij}_{\alpha\beta} \quad ; \quad i,j = 1,\ldots,N \tag{8.16}$$

where $\Sigma^\mu = \gamma^\mu \mathcal{C}$ with γ^μ the Dirac matrices and $\mathcal{C}$ the charge conjugation matrix, $P_\mu = -i\partial_\mu$ is the generator of space-time translations, and Z^{ij} is an antisymmetric matrix of operators proportional to the generators of the internal symmetry group. For the rest of the relations of the super-Poincaré group, see [155]. We assume here that the spacetime has Minkowski signature.

We shall be interested in using the relations of the super-Poincaré algebra to obtain a symplectic structure on the space of fields. For this, it turns out that only the $i = j$ terms in (8.16) are relevant. It therefore suffices to consider an $N = 1$ supersymmetry with no internal Z^{ij} symmetry group terms. The most expedient way to construct supersymmetric field theories (i.e. those with actions invariant under the full super-Poincaré group) is to use a superspace formulation. We introduce 2 Weyl spinors θ^α and $\bar\theta_{\dot\alpha}$ which parametrize the infinitesimal supersymmetry transformations. Then the $N = 1$ supersymmetry generators in 4 dimensions can be written as

$$Q_\alpha = \frac{\partial}{\partial \theta^\alpha} - i\Sigma^\mu_{\alpha\dot\alpha}\bar\theta^{\dot\alpha}\partial_\mu \quad , \quad \bar{Q}_{\dot\alpha} = -\frac{\partial}{\partial \bar\theta^{\dot\alpha}} + i\Sigma^\mu_{\alpha\dot\alpha}\theta^\alpha\partial_\mu \tag{8.17}$$

Kinetic terms in the supersymmetric action are constructed from the covariant superderivatives

$$D_\alpha = \frac{\partial}{\partial \theta^\alpha} + i\Sigma^\mu_{\alpha\dot\alpha}\bar\theta^{\dot\alpha}\partial_\mu \quad , \quad \bar{D}_{\dot\alpha} = -\frac{\partial}{\partial \bar\theta^{\dot\alpha}} - i\Sigma^\mu_{\alpha\dot\alpha}\theta^\alpha\partial_\mu \tag{8.18}$$

8. Equivariant Localization in Cohomological Field Theory

It can be readily verified that with the representation (8.17) the relations of the $N = 1$ super-Poincaré algebra are satisfied. In this superspace notation, a general group element of the supersymmetry algebra is $e^{\theta^\alpha Q_\alpha + \bar{\theta}^{\dot\alpha} \bar{Q}_{\dot\alpha}}$ and, using the supersymmetry algebra along with the Baker-Campbell-Hausdorff formula, its action on a field $A(x)$ is $e^{\theta^\alpha Q_\alpha + \bar{\theta}^{\dot\alpha} \bar{Q}_{\dot\alpha}} A(x) = e^{\theta^\alpha Q_\alpha} e^{\bar{\theta}^{\dot\alpha} \bar{Q}_{\dot\alpha}} A(y)$ where $y^\mu = x^\mu + i\theta^\alpha \Sigma^\mu_{\alpha\dot\alpha} \bar{\theta}^{\dot\alpha}$ are coordinates in superspace. Thus the supersymmetry transformation parameters live in superspace and the fields of the supersymmetric field theory are defined on superspace.

To incorporate the supersymmetry algebra as the symmetry algebra of a physical system, we need some representation of it in terms of fields defined over the space-time. The lengthy algorithm to construct supermultiplets associated with a given irreducible or reducible spin representation of the super-Poincaré algebra can be found in [155]. For instance, in 4 dimensions chiral superfields (satisfying $\bar{D}_{\dot\alpha} \Phi = 0$) are given by

$$\Phi^\mu(x, \theta, \bar{\theta}) = \phi^\mu(y) + \theta^\alpha \psi^\mu_\alpha(y) + \theta^\alpha \theta_\alpha F^\mu(y) \tag{8.19}$$

where (ϕ, ψ) are spin $(0, \frac{1}{2})$ fields, and F are auxilliary fields use to close the supersymmetry representation defined by the chiral superfields. In the following we shall consider multiplets of highest spin 1. Other multiplets can be obtained by imposing some additional constraints. The most general $N = 1$ supermultiplet in 4 space-time dimensions consists of a scalar field M, 3 pseudoscalar fields C, N and D, a vector field A_μ and 2 Dirac spinor fields χ and λ. The supersymmetry charges Q^i_α and $\bar{Q}^i_{\dot\alpha}$ respectively raise and lower the (spin) helicity components of the mulitplets by $\frac{1}{2}$. The super-Poincaré algebra can be represented in the Majorana representation where $\gamma^0 = -\sigma^2 \otimes \mathbf{1}$, $\gamma^1 = -i\sigma^3 \otimes \sigma^1$, $\gamma^2 = i\sigma^1 \otimes \mathbf{1}$ and $\gamma^3 = -i\sigma^3 \otimes \sigma^3$, with σ^i the usual Pauli spin matrices. Then the 4×4 Σ-matrices on the right-hand side of (êfNsusyalg) are

$$\Sigma^0 = \begin{pmatrix} -1 & 0 \\ 0 & -1 \end{pmatrix} \quad , \quad \Sigma^1 = \begin{pmatrix} 0 & 1 \\ 1 & 0 \end{pmatrix}$$
$$\Sigma^2 = \begin{pmatrix} 0 & -i \cdot \mathbf{1} \\ i \cdot \mathbf{1} & 0 \end{pmatrix} \quad , \quad \Sigma^3 = \begin{pmatrix} 1 & 0 \\ 0 & -1 \end{pmatrix} \tag{8.20}$$

The Majorana representation selects the preferred light-cone coordinates $x^\pm = x^2 \pm x^0$ for the translation generators P_μ in (8.16) above. Different representations of the Dirac gamma-matrices would then define different preferred light-cone directions.

The general supersymmetry transformations of the complex $N = 1$ supermultiplet $V = (C; \chi; M, N, A_\mu; \lambda; D)$ and the action of the supersymmetry charges $Q^i_\alpha, \bar{Q}^i_{\dot\alpha}$ on V can be found in [133]. There it was shown that the infinitesimal supersymmetry transformations can be written in a much simpler form using the auxilliary fields

8.2 Poincaré Supersymmetric Quantum Field Theories

$$M' = M + A_3 + \partial_1 C \quad, \quad N' = N + A_1 - \partial_3 C \quad, \quad D' = D + \partial_1 A_3 - \partial_3 A_1$$

$$\lambda'_1 = \lambda_1 - \partial_3 \chi_1 \quad, \quad \lambda'_2 = \lambda_2 - \partial_1 \chi_1 \quad, \quad \lambda'_3 = 2\lambda_3 - \partial_- \chi_1$$

(8.21)

These are precisely the (non-standard) auxilliary fields introduced in [107, 108] which, as we discussed in Section 4.9, form the basis for equivariant localization in supersymmetric quantum field theories and phase space path integrals whose Hamiltonians are functionals of isometry generators. Using these auxilliary fields we now define 2 functional derivative operators on the space of fields,

$$\mathcal{D} = \int d^3x \oint_0^T dt \left\{ \chi_2 \frac{\delta}{\delta C} + iA_+ \frac{\delta}{\delta \chi_1} + iM' \frac{\delta}{\delta \chi_3} + iN' \frac{\delta}{\delta \chi_4} \right.$$
$$\left. - \lambda'_3 \frac{\delta}{\delta A_-} - \lambda'_2 \frac{\delta}{\delta A_1} - \lambda'_1 \frac{\delta}{\delta A_3} - iD' \frac{\delta}{\delta \lambda_4} \right\}$$

(8.22)

$$\mathcal{I}_{V+} = \int d^3x \oint_0^T dt \left\{ i\partial_+ C \frac{\delta}{\delta \chi_2} + \partial_+ \chi_3 \frac{\delta}{\delta M'} + \partial_+ \chi_4 \frac{\delta}{\delta N'} + \partial_+ \chi_1 \frac{\delta}{\delta A_+} \right.$$
$$\left. - \partial_+ A_3 \frac{\delta}{\delta \lambda'_1} - \partial_+ A_1 \frac{\delta}{\delta \lambda'_2} - \partial_+ A_- \frac{\delta}{\delta \lambda'_3} - \partial_+ \lambda_4 \frac{\delta}{\delta D'} \right\}$$

(8.23)

Here we have imposed the boundary conditions on the fields $\Phi(x,t)$ that they vanish at spatial infinity and that they be periodic in time $t = x^0$,

$$\lim_{|x| \to \infty} \Phi(x,t) = 0 \quad, \quad \Phi(x, t+T) = \Phi(x,t)$$

(8.24)

so that the space of fields can be thought of as a loop space.

The operators (8.22) and (8.23) are nilpotent. If we now define

$$Q_+ = \mathcal{D} + \mathcal{I}_{V+}$$

(8.25)

then it can be checked that

$$Q_+^2 = \mathcal{D} \mathcal{I}_{V+} + \mathcal{I}_{V+} \mathcal{D} = \mathcal{L}_{V+} = \int d^3x \oint_0^T dt \, i\partial_+$$

(8.26)

The operator (8.25) generates the appropriate $N = 1$ supersymmetry transformations on the field multiplet and the supersymmetry algebra (8.26) coincides with the pertinent $N = 1$ supersymmetry algebra (8.16). The above construction therefore provides a geometric representation of the superalgebra (8.16) on the space of fields of the supersymetric field theory. A different representation of Σ^μ leads to a different choice of preferred Q_α in (8.26) and a different decomposition of the fields into loop space coordinates and 1-forms

280   8. Equivariant Localization in Cohomological Field Theory

in (8.22) and (8.23). It is now possible to examine how various supersymmetric quantum field theories decompose with respect to the above equivariant cohomological structure on the space of fields. The canonical choice is the Wess-Zumino model which is defined by the sigma-model action

$$\mathcal{S}_{WZ} = \int d^3x \oint_0^T dt \int d\theta\, d\bar{\theta} \left(\frac{1}{2}\bar{D}\bar{\Phi} \cdot D\Phi + W[\Phi]\right) \tag{8.27}$$

where $W[\Phi]$ is some super-potential. With respect to the above decompositions it is possible to show that the supersymmetric action decomposes into a sum of a loop space scalar $\mathcal{H}$ and a loop space 2-form Ω, $\mathcal{S} = \mathcal{H} + \Omega$. Because of the boundary conditions on the fields of the theory, the supersymmetry charges, which geometrically generate translations in the chosen light-cone direction, are nilpotent on the spaces of fields. By separating the loop space forms of different degrees, we find that the supersymmetry of the action, $Q_+\mathcal{S} = 0$, implies separately that $\mathcal{D}\Omega = 0$ and $\mathcal{D}\mathcal{H} = -\mathcal{I}_{V+}\Omega$. Thus the supersymmetric model admits a loop space symplectic structure and the corresponding path integral can be written as a super-loop space (i.e. phase space) functional integration. Furthermore, because $Q_+^2 = 0$ on the space of fields, the symplectic potential ϑ with $\mathcal{D}\vartheta = \Omega$ obeys $\mathcal{I}_{V+}\vartheta = \mathcal{H}$, so that the action is always locally a supersymmetry variation, $\mathcal{S} = Q_+\vartheta = (\mathcal{D} + \mathcal{I}_{V+})\vartheta$.

Thus, any generic quantum field theory with Poincaré supersymmetry group admits a loop space symplectic structure and a corresponding $U(1)$ equivariant cohomology responsible for localization of the supersymmetric path integral. The key feature is an appropriate auxilliary field formalism which defines a splitting of the fields into loop space "coordinates" and their associated "differentials" [3]. Notice that in general although the fields of the supersymmetric theory always split up evenly into loop space coordinates and 1-forms, the coordinates and 1-forms involve both bosonic and fermionic fields. It is only in the simplest cases (e.g. $N = \frac{1}{2}$ supersymmetric quantum mechanics) that the pure bosonic fields are identified as coordinates and the pure fermionic ones as 1-forms. In the auxiliary field formalism outlined above, the supersymmetry of the model is encoded within the model independent loop space equivariant cohomology defined by Q_+. In this way, one obtains a geometric interpretation of general Poincaré supersymmetric quantum field theories and an explicit localization of the supersymmetric path integral onto the constant modes (zeroes of ∂_+).

We shall not present any explicit examples of the above general constructions here. They have been verified in a number of cases. To check this formalism in special instances one needs to impose certain additional constraints on the multiplets [133] (e.g. the chirality condition mentioned above). The

---

[3] Note that the problem of choosing an appropriate set of auxilliary fields is the infinite-dimensional analog of finding a preferred set of coordinates for an isometry generator (c.f. Section 5.2).

above constructions have been in this way explictly carried out in [107, 108] for $N = 1$ supersymmetric quantum mechanics (i.e. the $(0 + 1)$-dimensional Wess-Zumino model (8.27)), the Wess-Zumino model (8.27) in both 2 and 4 dimensions, and 4-dimensional $N = 1$ supersymmetric $SU(N)$ Yang-Mills theory defined by the action

$$\mathcal{S}_{YM} = \int d^4x \left(-\frac{1}{4} F^a_{\mu\nu} F^{a,\mu\nu} + \frac{i}{2} \bar{\psi} \slashed{\nabla}_A \psi \right) \tag{8.28}$$

where ψ are Majorana fermion fields in the adjoint representation of the gauge group. The model (8.28) can be reduced to a Wess-Zumino model by eliminating the unphysical degrees of freedom and representing the theory directly in terms of transverse (physical) degrees of freedom. The equivariant localization framework has also been applied to the related supersymmetries of Parisi-Sourlas stochastic quantization [136] in [108]. Palo [133] applied these constructions to the 2-dimensional supersymmetric non-linear sigma-model (i.e. the Wess-Zumino model (8.27) with a curved target space – see the next Section).

8.3 Supergeometry and the Batalin-Fradkin-Vilkovisky Formalism

The role that the Cartan exterior derivative of equivariant cohomology plays in localization resembles a Lagrangian BRST quantization in terms of the gauge-fixing of a Lagrangian field theory over $\mathcal{M}$ with a gauge field θ_μ [22]. In [112, 113] equivariant localization was interpreted in terms of the Batalin-Vilkovisky Langrangian anti-field formalism. In this formalism, a theory with first-stage reducible constraints or with open gauge algebras is quantized by introducing an antibracket [13]–[15] which naturally introduces a new supersymmetry element into the BRST quantization scheme. This formulation is especially important for the construction of the complete quantum actions for topological gauge theories (especially those of Schwarz-type) [22]. In this Section we shall sketch some of the basic conceptual and computational ideas of this formalism, which in the context of localization for dynamical systems lead to more direct connections with super symmetric and topological field theories.

We return to the simpler situation of a generic Hamiltonian system $(\mathcal{M}, \omega, H)$. We have seen that the localization prescription naturally requires a formulation of objects defined over a super-manifold (i.e. one with bosonic and Grassmann coordinates), namely the cotangent bundle $\mathcal{M}_S \equiv \mathcal{M} \otimes T^*\mathcal{M}$. In the case of the supersymmetric quantum field theories that we considered in the previous Section, our fields where defined on a superspace and the path integral localizations were carried out over a superloop space. We would now like to try to exploit the mathematical characteristics of a super-manifold and reformulate the localization concepts in the more rigorous framework of supergeometry. This is particularly important for some of the other localization features of topological field theories that we shall discuss in the next 2 Sections.

First, we shall incorporate the natural geometrical objects of the Batalin-Vilkovisky formalism into the equivariant localization framework. The local coordinates on the super-manifold M_S are denoted as $z^A = (x^\mu, \eta^\mu)$. We define a Grassmann-odd degree symplectic structure on M_S by the non-degenerate odd symplectic 2-form

$$\Omega^1 = dz^A \wedge \Omega^1_{AB} dz^B = \omega_{\mu\nu} dx^\mu \wedge d\eta^\nu + \frac{\partial \omega_{\mu\nu}}{\partial x^\lambda} \eta^\lambda d\eta^\mu \wedge d\eta^\nu \tag{8.29}$$

The 2-form (8.29) determines an odd Poisson bracket on M_S called the antibracket. It is defined by

$$\{\mathcal{A}, \mathcal{B}\}_1 = \frac{\partial \mathcal{A}}{\partial z^A} \Omega_1^{AB} \mathcal{B} \frac{\overleftarrow{\partial}}{\partial z^B} = \omega^{\mu\nu} \left(\frac{\partial \mathcal{A}}{\partial x^\mu} \mathcal{B} \frac{\overleftarrow{\partial}}{\partial \eta^\nu} - \frac{\partial \mathcal{B}}{\partial x^\mu} \mathcal{A} \frac{\overleftarrow{\partial}}{\partial \eta^\nu} \right) \\ + \frac{\partial \omega^{\mu\nu}}{\partial x^\lambda} \eta^\lambda \frac{\partial \mathcal{A}}{\partial \eta^\mu} \mathcal{B} \frac{\overleftarrow{\partial}}{\partial \eta^\nu} \tag{8.30}$$

where $\mathcal{A}(x, \eta)$ and $\mathcal{B}(x, \eta)$ are super-functions on M_S. The grading and antisymmetry properties are opposite to those of the ordinary graded Poisson bracket,

$$\{\mathcal{A}, \mathcal{B}\}_1 = -(-1)^{(p(\mathcal{A})+1)(p(\mathcal{B})+1)} \{\mathcal{B}, \mathcal{A}\}_1$$
$$\{\mathcal{A}, \mathcal{BC}\}_1 = \{\mathcal{A}, \mathcal{B}\}_1 \mathcal{C} + (-1)^{p(\mathcal{B})(p(\mathcal{A})+1)} \mathcal{B}\{\mathcal{A}, \mathcal{C}\}_1$$
$$\{\mathcal{A}, \{\mathcal{B}, \mathcal{C}\}_1\}_1 - (-1)^{(p(\mathcal{A})+1)(p(\mathcal{C})+1)} \{\mathcal{B}, \{\mathcal{A}, \mathcal{C}\}_1\}_1 = \{\{\mathcal{A}, \mathcal{B}\}_1, \mathcal{C}\}_1 \tag{8.31}$$

where $p(\mathcal{A})$ is the Grassmann degree of the super-function $\mathcal{A}(x, \eta)$ with the property $p(\{\mathcal{A}, \mathcal{B}\}) = p(\mathcal{A}) + p(\mathcal{B}) + 1$. In particular, the super-coordinate antibrackets are

$$\{x^\mu, x^\nu\}_1 = 0 \quad , \quad \{x^\mu, \eta^\nu\}_1 = -\{\eta^\nu, x^\mu\}_1 = \omega^{\mu\nu}$$
$$\{\eta^\mu, \eta^\nu\}_1 = -\{\eta^\nu, \eta^\mu\}_1 = \frac{\partial \omega^{\mu\nu}}{\partial x^\lambda} \eta^\lambda \tag{8.32}$$

We define a mapping on $C^\infty(\mathcal{M}) \to C^\infty(M_S)$ using the super-function $\omega(z) = \frac{1}{2} \omega_{\mu\nu}(x) \eta^\mu \eta^\nu$ by

$$f(x) \to \mathcal{Q}_f(z) = \{f, \omega\}_1 = \frac{\partial f}{\partial x^\mu} \eta^\mu \tag{8.33}$$

Then the antibracket coincides with the original Poisson bracket of the phase space, $\{\mathcal{A}, f\}_1 = \{\mathcal{A}, f\}_\omega$. In particular, the dynamical systems $(\mathcal{M}, \omega, H)$

8.3 Supergeometry and the Batalin-Fradkin-Vilkovisky Formalism 283

and $(M_S, \Omega^1, \mathcal{Q}_H)$ determine a bi-Hamiltonian pair. The corresponding equations of motion are

$$\dot{x}^\mu = \{\!\{x^\mu, \mathcal{Q}_H\}\!\}_1 = \{x^\mu, H\}_\omega = V^\mu \quad , \quad \dot{\eta}^\mu = \{\!\{\eta^\mu, \mathcal{Q}_H\}\!\}_1 = \partial_\nu V^\mu \eta^\nu \quad (8.34)$$

Generally, it is readily seen that the operation $\{\!\{\omega, \cdot\}\!\}_1$ acts as exterior differentiation d, $\{\!\{H, \cdot\}\!\}_1$ acts like interior multiplication i_V with respect to the Hamiltonian vector field V, and $\{\!\{\mathcal{Q}_H, \cdot\}\!\}_1$ acts as the Lie derivative $\mathcal{L}_V$ along V. The antibracket provides an equivalent supersymmetric generalization of the ordinary Hamiltonian dynamics.

The key feature is that the supersymmetry of the odd Hamiltonian system $(M_S, \Omega^1, \mathcal{Q}_H)$ is equivalent to the equivariant cohomology determined by the equivariant exterior derviative $D_V = d + i_V$. If $\mathcal{M}$ admits an invariant Riemannian metric tensor g (the equivariant localization constraints), then the super-function

$$\mathcal{I}_H = \frac{1}{2} g_{\mu\nu} V^\mu \eta^\nu \qquad (8.35)$$

is an integral of motion for the Hamiltonian system $(M_S, \Omega^1, \mathcal{Q}_H)$, i.e. $\{\!\{\mathcal{Q}_H, \mathcal{I}_H\}\!\}_1 = 0$. Furthermore, $\mathcal{I}_H$ determines the usual bi-Hamiltonian structure on $\mathcal{M}$ because $\{\!\{\omega, \mathcal{I}_H\}\!\}_1 = \frac{1}{2}(\Omega_V)_{\mu\nu}\eta^\mu\eta^\nu$ and $\{\!\{H, \mathcal{I}_H\}\!\}_1 = K_V$. With these observations one can easily now establish the (classical) equivariant localization principle. The usual localization integral (2.128) can be written as an integral over the super-manifold M_S,

$$\mathcal{Z}(s) = \frac{1}{(iT)^n} \int_{M_S} d^{4n}z \; e^{iT(H+\omega) - s\{\!\{H-\omega, \mathcal{I}_H\}\!\}_1} \qquad (8.36)$$

where as always the classical partition function is $Z(T) = \mathcal{Z}(0)$. The volume form $d^{4n}z = d^{2n}x \, d^{2n}\eta$ is invariant under the equivariant transformations of D_V and $\mathcal{L}_V$ determined by the anti-brackets. Furthermore, we have

$$\{\!\{H - \omega, \; e^{iT(H+\omega) - s\{\!\{H-\omega, \mathcal{I}_H\}\!\}_1}\}\!\}_1 = \{\!\{\mathcal{Q}_H, \; e^{iT(H+\omega) - s\{\!\{H-\omega, \mathcal{I}_H\}\!\}_1}\}\!\}_1 = 0$$

$$\{\!\{\mathcal{Q}_H, \mathcal{I}_H \; e^{iT(H+\omega) - s\{\!\{H-\omega, \mathcal{I}_H\}\!\}_1}\}\!\}_1 = 0$$

(8.37)

The first 2 vanishing conditions just represent the invariance of the integrand in (8.36) under the actions of the operators D_V and $\mathcal{L}_V$, so that the usual equivariant cohomological structure for localization in (8.36) is manifested in the supersymmetry of the antibracket formalism. With the identities (8.37), it is straightforward to establish that $\frac{d}{ds}\mathcal{Z}(s) = 0$, and hence the localization principle (i.e. the Duistermaat-Heckman theorem).

Thus the lifting of the original Hamiltonian system to the odd one defined over a supermanifold has provided another supersymmetric way to interpret the localization, this time in terms of the presence of supersymmetric bi-Hamiltonian dynamics with even and odd symplectic structures which is the usual Batalin-Vilkovisky procedure for the evaluation of BRST gauge-fixed

284   8. Equivariant Localization in Cohomological Field Theory

path integrals. The representation (8.36) of the canonical localization integral formally coincides with the representation of differential forms in the case where the original space $\mathcal{M}$ is a supermanifold. In [150], Schwarz and Zaboronsky derived some general localization formulas for integrals over a finite-dimensional supermanifold $\mathcal{M}$ where the integrand is invariant under the action of an odd vector field W. Using the supergeometry of $\mathcal{M}$, they formulated sufficient conditions which generalize those above under which the integral localizes onto the zero locus of the c-number part $W(\eta = 0)$ of W. Their theorems quite naturally generalize the usual equivariant localization principles and could apply to physical models such as those where the Batalin-Vilkovisky formalism is applicable [149] or in the dimensional reduction mechanism of the Parisi-Sourlas model [136]. A generalization to more general (non-linear) supermanifolds can be found in [175].

Nersessian has also demonstrated how to incorporate the anti-bracket structure into the other models of equivariant cohomology (other than the Cartan model – see Appendix B) in [115] and how the usual equivariant characteristic class representations of the localization formulas appear in the Batalin-Vilkovisky formalism in [113]. The superspace structure of cohomological field theories in this context has been studied by Niemi and Tirkkonen in [129]. They discussed the role of the BRST model of equivariant cohomology in non-abelian localization (see Appendix B) and topological field theories and showed how an appropriate superfield formulation can be used to relate these equivariant cohomological structures to the Batalin-Fradkin-Vilkovisky Hamiltonian quantization of constrained systems with first stage reducible constraints. This suggests a geometric (superspace) picture of the localization properties of some topological quantum field theories such as 4-dimensional topological Yang-Mills theory (defined by the action $\int \mathrm{tr}\, F_A \wedge F_A$). This picture is similar to those of the Poincaré supersymmetric theories described in the previous Section in that the BRST charge of the cohomological field theory can be taken to generate translations in the η-direction in superspace and then the connection between equivariant cohomology and Batalin-Fradkin-Vilkovisky quantization of 4-dimensional topological Yang-Mills theory becomes transparent. These superfield formalisms therefore describe both equivariant cohomology in the symplectic setting relevant for localization and the BRST structure of cohomological field theories. This seems to imply the existence of a unified description of localization in the symplectic loop space, the supersymmetric loop space of the last Section, and in cohomological field theory. Indeed, it is conjectured that all lower dimensional integrable models are obtainable as dimensional reductions of 4-dimensional self-dual Yang-Mills theory (i.e. $F_A = \star F_A$) which is intimately connected to topological Yang-Mills theory.

Finally, the incorporation of the Batalin-Vilkovisky formalism into loop space localization has been discussed recently by Miettinen in [103]. For path integral quantization, one needs an even Hamiltonian and symplectic struc-

8.3 Supergeometry and the Batalin-Fradkin-Vilkovisky Formalism

ture. This can be done provided that $\mathcal{M}$ has on it a Riemannian structure. An even symplectic structure on M_S is given by the super-symplectic 2-form

$$\Omega^i = \frac{1}{2}\left(\omega^i_{\mu\nu} + R_{\mu\nu\lambda\rho}\eta^\lambda\eta^\rho\right)dx^\mu \wedge dx^\nu + \frac{1}{2}g_{\mu\nu}D_g\eta^\mu \wedge D_g\eta^\nu \qquad (8.38)$$

where

$$D_g\eta^\mu = d\eta^\mu + \Gamma^\mu_{\nu\lambda}\eta^\nu dx^\lambda \qquad (8.39)$$

is the covariant derivative on $\mathcal{M}$, and the subscript $i = 0, 2$ labels the Hamiltonian systems $(\mathcal{M}, \omega^0 = \omega, H_0 = H)$ and $(\mathcal{M}, \omega^2 = \Omega_V, H_2 = K_V)$. The corresponding symplectic 1-forms, with $\Omega^i = d\Theta^i$, are then

$$\Theta^0 = \theta_\mu dx^\mu + g_{\mu\nu}\eta^\mu D_g\eta^\nu \quad , \quad \Theta^2 = g_{\mu\nu}V^\nu dx^\mu + g_{\mu\nu}\eta^\mu D_g\eta^\nu \qquad (8.40)$$

The 2-forms Ω^i determine the even Poisson brackets on M_S

$$[\mathcal{A}, \mathcal{B}]_i = (\nabla_\mu \mathcal{A})[\omega^i_{\mu\nu} + R_{\mu\nu\lambda\rho}\eta^\lambda\eta^\rho]^{-1}\nabla_\nu\mathcal{B} + g^{\mu\nu}\frac{\partial\mathcal{A}}{\partial\eta^\mu}\mathcal{B}\frac{\overleftarrow{\partial}}{\partial\eta^\nu} \qquad (8.41)$$

where

$$\nabla_\mu = \partial_\mu - \Gamma^\lambda_{\mu\nu}\eta^\nu\frac{\partial}{\partial\eta^\lambda} \qquad (8.42)$$

Then the equations of motion for the odd and even Poisson brackets on M_S coincide, $[z^A, \mathcal{H}_i]_i = \{z^A, \mathcal{Q}_H\}_1$, where $\mathcal{H}_i = H_i + \Omega_V$. Thus the odd and even Poisson brackets provide bi-Hamiltonian structures also on the super-symplectic manifold M_S and therefore the Hamiltonian system is also integrable on M_S.

Given this integrability feature on M_S, we can now examine the corresponding localizations [150]. We write the partition functions $\mathcal{Z}_i(T) = \mathrm{str}\| e^{-iT\mathcal{H}_i}\|$ as path integrals over a super-loop space corresponding to LM_S by absorbing the Liouville measure factors associated with Ω^i into the argument of the action in the usual way (c.f. Chapter 4). Then the Hamiltonian systems $(M_S, \Omega^i, \mathcal{H}_i)$ have the quantum actions [103]

$$S_0 = \int_0^T dt \left(\theta_\mu \dot{x}^\mu - H + \frac{1}{2}\eta^\mu g_{\mu\nu}\frac{D_g\eta^\nu}{dt} + \frac{1}{2}(\Omega_V)_{\mu\nu}\eta^\mu\eta^\nu + \frac{1}{2}\omega_{\mu\nu}\lambda^\mu\lambda^\nu \right.$$

$$\left. + \frac{1}{2}R_{\mu\nu\lambda\rho}\lambda^\mu\lambda^\nu\eta^\lambda\eta^\rho + \frac{1}{2}g_{\mu\nu}F^\mu F^\nu\right)$$

$$S_2 = \int_0^T dt \left(g_{\mu\nu}V^\nu \dot{x}^\mu - K_V + \frac{1}{2}\eta^\mu g_{\mu\nu}\frac{D_g\eta^\nu}{dt} + \frac{1}{2}(\Omega_V)_{\mu\nu}\eta^\mu\eta^\nu \right.$$

$$\left. + \frac{1}{2}(\Omega_V)_{\mu\nu}\lambda^\mu\lambda^\nu + \frac{1}{2}R_{\mu\nu\lambda\rho}\lambda^\mu\lambda^\nu\eta^\lambda\eta^\rho + \frac{1}{2}g_{\mu\nu}F^\mu F^\nu\right) \qquad (8.43)$$

where the quantum partition functions are

$$\mathcal{Z}_i(T) = \int_{LM_S \otimes L\Lambda^1 M_S} [d^{2n}x]\, [d^{2n}\eta]\, [d^{2n}F]\, [d^{2n}\lambda]\, e^{iS_i[x,\eta;F,\lambda]} \qquad (8.44)$$

and we have introduced auxilliary anticommuting variables $\lambda^\mu \sim dx^\mu$ and bosonic variables $F^\mu \sim d\eta^\mu$ to exponentiate the determinant factors in the usual way. If $H = 0$ (the topological limit), then the action $S_0 + \int_0^T dt\, \frac{1}{2}g_{\mu\nu}\dot{x}^\mu\dot{x}^\nu$ is that of the $(0+1)$-dimensional $N=1$ supersymmetric non-linear sigma-model, i.e. the action of $N = 1$ DeRham supersymmetric quantum mechanics in background gravitational and gauge fields[4].

One can now develop the standard machinery to evaluate these path integrals using super-loop space equivariant cohomology. This has been done explicitly in [122]. The super-loop space equivariant exterior derivative is

$$Q = \int_0^T dt\, \left(\lambda^\mu \frac{\delta}{\delta x^\mu} + F^\mu \frac{\delta}{\delta \eta^\mu} + (\dot{x}^\mu - V^\mu)\frac{\delta}{\delta \lambda^\mu} + (\dot{\eta}^\mu + \partial^\mu V^\nu \eta_\nu)\frac{\delta}{\delta F^\mu} \right)$$
(8.45)

with $Q^2 = \mathcal{L}_S = \mathcal{L}_{\dot{x}} - \mathcal{L}_V$ where S is the classical action associated with the original Hamiltonian system $(\mathcal{M},\omega,H)$. Notice that a canonical conjugation $Q \to e^{-\Phi} Q\, e^{\Phi} \equiv \tilde{Q}$ does not alter the cohomology groups of the derivative operator Q. Choosing the loop space functional

$$\Phi = \int_0^T dt\, \Gamma^\lambda_{\mu\nu} \eta_\lambda \lambda^\nu \frac{\delta}{\delta F^\mu} \qquad (8.46)$$

the (topologically equivalent) operator $\tilde{Q}$ can be explicitly worked out (see [122]). With regards to this supersymmetry charge, the pertinent action S_0 in (8.43) can be obtained from the 2-dimensional $N = 1$ supersymmetric sigma-model by partial localization of it to a 1-dimensional model. This is done by breaking its (left-right) $(1,1)$ supersymmetry explicitly by the Hamiltonian flow. In this procedure the usual boson kinetic term (in light-cone coordinates) $g_{\mu\nu}\partial_+\phi^\mu\partial_-\phi^\nu$ drops out. The path integral $\mathcal{Z}_0(T)$ can be evaluated by adding an explicit gauge-fixing term $\tilde{Q}\psi$ to the action for an appropriate gauge fermion ψ. Taking $\psi = \int_0^T dt\, (g_{\mu\nu}F^\mu\eta^\nu + \frac{s}{2}g_{\mu\nu}(\dot{x}^\mu - V^\mu)\lambda^\nu)$ localizes

---

[4] The action for $N = 1$ supersymmetric quantum mechanics can be obtained from the Wess-Zumino model action (8.27) by integrating out the auxilliary field F^μ resulting from the chiral superfields (8.19) and integrating over the θ coordinates of the superspace. The action for $N = \frac{1}{2}$ supersymmetric quantum mechanics discussed in Section 4.2 can be obtained from the $N = 1$ model by setting $\lambda^\mu = \eta^\mu$ above. Notice that the Riemann curvature term then drops o ut because of its symmetry properties.

the path integral in the limit $s \to \infty$ onto the T-periodic classical trajectories of the original action S,

$$\mathcal{Z}_0(T) = \sum_{x(t) \in L\mathcal{M}_S} \text{sgn}[\det \|\delta^2 S(x(t))\|] \, e^{iS[x(t)]} \tag{8.47}$$

On the other hand, selecting $\psi = \int_0^T dt \, (g_{\mu\nu} F^\mu \eta^\nu + \frac{s}{2} g_{\mu\nu} \dot{x}^\mu \lambda^\nu)$ we find that the path integral localizes onto an ordinary integral over equivariant characteristic classes of the phase space $\mathcal{M}$,

$$\mathcal{Z}_0(T) = \int_{\mathcal{M}} \text{ch}_V(-iT\omega) \wedge E_V(R) \tag{8.48}$$

The equality of these 2 expressions for the quantum partition function $\mathcal{Z}_0(T)$ can be thought of as an equivariant, loop space generalization of the relation (3.71) in Morse theory between the Gauss-Bonnet-Chern and Poincaré-Hopf theorems for the representation of the Euler characteristic $\chi(\mathcal{M})$ of the manifold $\mathcal{M}$. Indeed, in the limit $H, \theta_\mu \to 0$ the quantities (8.47) and (8.48) reproduce exactly the relation (3.71). These relations have been used to study the set of Hamiltonian systems which satisfy the Arnold conjecture on the space of T-periodic classical trajectories for (time-dependent) classical Hamiltonians [119, 122]. Thus, the path integrals associated with the supermanifolds defined by the Batalin-Vilkovisky formalism for dynamical systems lead to loop space and equivariant generalizations of other familiar topological invariants. Furthermore, these models are closely related to supersymmetric non-linear sigma-mode ls which ties together the "hidden" supersymmetry of the given Hamiltonian system with the Poincaré supersymmetry of the localizable quantum field theories. These ideas lead us naturally into the final topic of this Book which emphasizes these sorts of relations between equivariant cohomology and topological quantum field theories. The discussion of this Section then shows how this next topic is related to the equivariant localization formalism for dynamical systems.

8.4 Equivariant Euler Numbers, Thom Classes and the Mathai-Quillen Formalism

We have now almost completed the connections between the equivariant localization formalism, cohomological field theories and their relations to physical systems, thus uniting most of the ideas presented in this Book. The last Section showed how the localization formalism connects phase space path integrals of dynamical systems to some basic topological field theory models (namely supersymmetric sigma-models). Conversely, in Section 8.2 we demonstrated that arbitrary field theoretical models of these types could be placed into the loop space equivariant localization framework so that there

288   8. Equivariant Localization in Cohomological Field Theory

is a sort of equivalence between dynamical systems and field theory models in this geometric context. The discussion of Section 8.1 then illustrated certain genuine, geometrical equivalences between physical and topological gauge theories which demonstrates the power of the formalisms of both topological field theory and equivariant localization of path integrals in describing the quantum characteristics of physical systems.

There is one final step for this connection which is the Atiyah-Jeffrey geometric interpretation of generic cohomological field theories [10] which is based on the Mathai-Quillen construction of Gaussian-shaped Thom forms [99]. This construction is the natural arena for the study of the localization properties of topological field theories, and in its infinite dimensional versions it can be used to build up topological gauge models. This approach, although based on equivariant cohomology, is rather different in spirit than the equivariant localization formalisms we have discussed thus far and we therefore only very briefly highlight the details for the sake of completeness. At the end of this Chapter, we will discuss a bit the connections with the other ideas of this Book. More detailed reviews of the Mathai-Quillen formalism in topological field theory can be found in [27, 29, 34].

The basic idea behind the Mathai-Quillen formalism is the relation (3.71) between the Poincaré-Hopf and Gauss-Bonnet-Chern representations of the Euler characteristic. It represents the localization of an explicit differential form representative of the Euler class of a vector bundle onto the zero locus of some section of that bundle. The original idea for the application and generalization of this relation to cohomological field theories traces back to Witten's connection between supersymmetric quantum mechanics and Morse theory [166]. Let us start with a simple example in this context. Given the local coordinates (x, η) on the cotangent bundle $\mathcal{M} \otimes T^*\mathcal{M}$ of some phase space $\mathcal{M}$, we denote $p_\mu \equiv \frac{\partial}{\partial x^\mu}$ and $\bar{\eta}_\mu \equiv \frac{\partial}{\partial \eta^\mu}$. In the spirit of the previous Section, we then interpret $(x, \bar{\eta})$ as local coordinates on a supermanifold $S^*\mathcal{M}$ and $\eta^\mu \sim dx^\mu, p_\mu \sim d\bar{\eta}_\mu$ as the local basis for the cotangent bundle of $S^*\mathcal{M}$.

The nilpotent exterior derivative operator on $S^*\mathcal{M}$ can be written as

$$d = \eta^\mu \frac{\partial}{\partial x^\mu} + p_\mu \frac{\partial}{\partial \bar{\eta}_\mu} \tag{8.49}$$

The invertible conjugation $d \to e^{-\Phi} d\, e^{\Phi}$ produces another linear derivation which generates the same cohomology as d. If $\mathcal{M}$ has metric g, then we can select $\Phi = -\Gamma^\lambda_{\mu\nu} \eta^\nu \bar{\eta}_\lambda \frac{\partial}{\partial p_\mu}$ (as in (8.46)) so that (8.49) conjugates to

$$d = \eta^\mu \frac{\partial}{\partial x^\mu} + (p_\mu + \Gamma^\lambda_{\mu\nu} \eta^\nu \bar{\eta}_\lambda) \frac{\partial}{\partial \bar{\eta}_\mu} + (\Gamma^\lambda_{\mu\nu} p_\lambda \eta^\nu - \frac{1}{2} R^\lambda_{\mu\rho\nu} \eta^\nu \eta^\rho \bar{\eta}_\lambda) \frac{\partial}{\partial p_\mu} \tag{8.50}$$

The action of (8.50) on the local coordinates of the cotangent bundle of the supermanifold $S^*\mathcal{M}$

8.4 The Mathai-Quillen Formalism

$$dx^\mu = \eta^\mu \quad , \quad dp_\mu = \Gamma^\lambda_{\mu\nu} p_\lambda \eta^\nu - \frac{1}{2} R^\lambda_{\mu\rho\nu} \eta^\nu \eta^\rho \bar\eta_\lambda$$
$$d\eta^\mu = 0 \quad , \quad d\bar\eta_\mu = p_\mu + \Gamma^\lambda_{\mu\nu} \eta^\nu \bar\eta_\lambda \tag{8.51}$$

coincides with the standard infinitesimal transformation laws of $N = 1$ DeRham supersymmetric quantum mechanics.

We now consider the following integral

$$\mathcal{Z}^* = \int_{S^*\mathcal{M} \otimes T^*(S^*\mathcal{M})} d^{2n}x\, d^{2n}p\, d^{2n}\eta\, d^{2n}\bar\eta\ e^{d\psi[x,p;\eta,\bar\eta]} \tag{8.52}$$

Since $d^2 = 0$ the integral (8.52) is formally independent of the function ψ on the super-manifold $S^*\mathcal{M} \otimes T^*(S^*\mathcal{M})$, i.e. $\frac{\delta \mathcal{Z}^*}{\delta \psi} = 0$. We can therefore evaluate (8.52) in 2 equivalent ways. First, we introduce a Hamiltonian vector field V on $\mathcal{M}$ and take $\psi = \psi_V = \frac{1}{2} V^\mu \bar\eta_\mu$ so that

$$d\psi_V = p_\mu V^\mu + \eta^\mu \nabla_\mu V^\nu \bar\eta_\nu \tag{8.53}$$

The integration in (8.52) can then be carried out explicitly. The integration over p_μ produces a delta-function $\delta(V)$ localizing the integral onto the zero locus $\mathcal{M}_V$ of the vector field V. The integration over the Grassmann coordinates in (8.53) yields a determinant of ∇V. Computing the relevant Jacobian for the transformation $x \to V(x)$ (c.f. Section 2.6), we arrive finally at

$$\mathcal{Z}^* = \sum_{p \in \mathcal{M}_V} \operatorname{sgn} \det \mathcal{H}(p) \tag{8.54}$$

Next, we take $\psi = \psi_g = g^{\mu\nu} p_\mu \bar\eta_\nu$ so that

$$d\psi_g = g^{\mu\nu} p_\mu p_\nu - \frac{1}{2} R^\lambda_{\mu\rho\nu} \eta^\nu \eta^\rho \bar\eta_\lambda g^{\mu\sigma} \bar\eta_\sigma \tag{8.55}$$

Evaluating the Gaussian integrals over p_μ and $\bar\eta_\mu$ in (8.52) leads to

$$\mathcal{Z}^* = \int_{\mathcal{M} \otimes T^*\mathcal{M}} d^{2n}x\, d^{2n}\eta\ \operatorname{Pfaff}\left(\frac{1}{2} R^\mu_{\nu\lambda\rho} \eta^\lambda \eta^\rho\right) \tag{8.56}$$

which we recognize as the Euler class of the tangent bundle $T\mathcal{M}$. Thus the equality of (8.54) and (8.56) leads immediately to the relation (3.71).

The exterior derivative operator (8.50) produces the Mathai-Quillen representative of the Euler class of the tangent bundle of $\mathcal{M}$. The above derivation is a special case of a more general construction of explicit differential form representatives for the Euler numbers of vector bundles $E \to \mathcal{M}$. These representatives are so-called Gaussian-shaped Thom forms whose constructions are best understood within the framework of equivariant cohomology. The idea is that one realizes the vector bundle $E \to \mathcal{M}$ as its associated principal G-bundle $P \times W$ (with W the standard fiber space of E) and constructs a particular representative (the Thom class) of the G-equivariant cohomology of $P \times W$. Given a section $V : \mathcal{M} \to E$, the regularized Euler class $E_V(F_A)$ (with F_A the curvature of a connection A of the bundle) is then the pullback of the Thom class to $\mathcal{M}$ under this section. It can be expressed as

$$E_V(F_A) = \int d^m\eta \; e^{-\frac{1}{2}\|V\|^2 + \frac{1}{2}\eta_\mu F_A^{\mu\nu}\eta_\nu + i\nabla_A V^\mu \eta_\mu} \quad (8.57)$$

where $m = \dim \mathcal{M}$ and η_μ are Grassmann variables. The norm $\|V\|^2$ is with respect to a fixed fiber metric on E and ∇_A is a compatible connection. We shall not go into the details of the construction of Thom classes using equivariant cohomology, but refer the reader to [27, 34, 81] for lucid accounts of this formalism. The important features of the Mathai-Quillen representative (8.57) are as follows. For general V, integrating out the Grassmann variables shows that $E_V(F_A)$ is a $2m$-form, and the fact that it is closed follows from the invariance of the exponent in (8.57) under the supersymmetry transformations

$$\delta V^\mu = \nabla_A V^\mu \quad , \quad \delta\eta^\mu = iV^\mu \quad (8.58)$$

with the additional condition $\delta x^\mu = \eta^\mu$ when (8.57) is integrated over $x \in \mathcal{M}$. Note that setting $V = 0$ in (8.57) and integrating out the Grassmann coordinates we see that it coincides with the usual Euler characteristic class $E(F_A) = \text{Pfaff}(F_A)$. Since the associated Thom class is closed, (8.57) is independent of the chosen section V, i.e. $E_V(F_A)$ is cohomologous to $E(F_A)$ for any V. This means that the Euler characteristic $\chi(E \to \mathcal{M}) = \int_\mathcal{M} E_V(F_A)$ can be evaluated by rescaling V by $s \in \mathbb{R}$ and localizing the integral in the limit $s \to \infty$ onto the zeroes of the section V (note the curvature term in this limit does not contribute), thus reproducing in this way the standard relation between the Poincaré-Hopf and Gauss-Bonnet-Chern theorems for generic vector bundles over $\mathcal{M}$. Thus the Thom class not only yields a representative of the Euler class of a vector bundle, but it also produces the Poincaré-dual form of t he zero locus of a given section of the bundle. The use of equivariant cohomology and localization techniques therefore also reproduce some classical results from geometry and topology.

Niemi and Palo [121] have shown how to construct equivariant generalizations of the Mathai-Quillen formalism. For this one considers the usual Cartan equivariant exterior derivative D_V and the associated Lie derivative $\mathcal{L}_V = D_V^2$ on the super-manifold $S^*\mathcal{M}$. If the Christoffel connection satisfies $\mathcal{L}_V \Gamma = 0$, then the conjugation by Φ introduced above of these operators produces an action on the local coordinates of the cotangent bundle of $S^*\mathcal{M}$ for $\mathcal{L}_V$ which generates the usual covariant coordinate transformation laws with respect to coordinate change defined by the Hamiltonian vector field V. Thus the integral

$$\mathcal{Z}_V^* = \int_{S^*\mathcal{M} \otimes T^*(S^*\mathcal{M})} d^{2n}x \, d^{2n}p \, d^{2n}\eta \, d^{2n}\bar{\eta} \; e^{i\phi \otimes (H+\omega) + D_V \psi} \quad (8.59)$$

8.5 The Mathai-Quillen Formalism for Infinite-Dimensional Vector Bundles   291

is formally independent of any generally covariant function ψ on the cotangent bundle of $S^*\mathcal{M}$. This is again just the equivariant localization principle. The measure in (8.59) is the invariant Liouville measure on the extended phase space. Evaluating (8.59) using the 2 choices for ψ mentioned above, we arrive at the relation

$$\sum_{p\in\mathcal{M}_V} e^{i\phi\otimes H(p)} \operatorname{sgn}\det \mathcal{H}(p)$$
$$= \int_{\mathcal{M}\otimes T^*\mathcal{M}} d^{2n}x\, d^{2n}\eta\, e^{i\phi\otimes (H+\omega)} \operatorname{Pfaff}\left[\nabla_\nu V^\mu + \frac{1}{2}R^\mu_{\nu\lambda\rho}\eta^\lambda\eta^\rho\right] \quad (8.60)$$

which can be recognized as an equivariant generalization of (3.71). Thus an appropriate equivariantization leads to a Mathai-Quillen representative for the *equivariant* Euler number of an equivariant vector bundle. In the limit $V, \phi \to 0$, (8.60) reduces to the usual relation. The non-degenerate cases are also possible to treat in this way [121].

8.5 The Mathai-Quillen Formalism for Infinite-Dimensional Vector Bundles

In this final Section of this Book, we shall discuss briefly the explicit connection to cohomological field theories. This will make explicit the relations of localization quite generically to topological field theory that we have mentioned through out this Book. As originally pointed out by Atiyah and Jeffrey [10], although the Euler number itself does not make sense for an infinite dimensional vector bundle, the Mathai-Quillen form $E_V(F_A)$ can be used to define regularized Euler numbers $\chi_V(E \to \mathcal{M}) = \int_\mathcal{M} E_V(F_A)$ of such bundles for those choices of V whose zero locus is finite-dimensional so that the localization makes these quantities well-defined. Although these numbers are not independent of V as in the finite-dimensional cases, they are naturally associated with $\mathcal{M}$ for certain choices. The functional integrals which arise in this way are equivalent to ordinary finite-dimensional integrals and represent the fundamental property of topological field theories, i.e. that t heir path integrals represent characteristic classes. This has been noted throughout this book as our central theme, and indeed most topological field theories can be obtained or interpreted in terms of the infinite-dimensional Mathai-Quillen formalism [34].

The simplest example is the regularized Euler number of the loop space $L\mathcal{M} \to \mathcal{M}$ over a manifold $\mathcal{M}$. The canonical vector field associated with this bundle is $\dot{x}^\mu(t)$. Now the loop space version of the integral (8.52) is a path integral over an extended superloop space, and we replace the exterior derivative d there by the equivariant, loop space one $Q_{\dot{x}}$ on $L(S^*\mathcal{M})$. The Lie derivative $\mathcal{L}_{\dot{x}} = Q_{\dot{x}}^2$ as before is the generator of time translations, and employing the standard conjugation above the path integral can be localized using any single-valued functional ψ on $L(S^*\mathcal{M})$. This follows from the equivariant localization principle for the model independent S^1-action. Choosing the natural loop space extensions of the functionals ψ used in the finite-dimensional calculations above, we arrive at precisely the same results (8.54) and (8.56) for the Euler characteristic of $\mathcal{M}$. The path integral analog of (8.57) with $V = \dot{x}$ yields the action of $N = 1$ DeRham supersymmetric quantum mechanics. In Section 4.2 we saw that the path integral for $N = \frac{1}{2}$ Dirac supersymmetric quantum mechanics localized onto constant modes and yielded the index of the twisted spin complex of $\mathcal{M}$. In the present case the localization of this Witten index onto constant loops yields the index of the DeRham complex of $\mathcal{M}$ (i.e. the Euler characteristic). This was the fundamental observation of Witten [166] and was one of the main ingredients in the birth of topological field theory. If the target space manifold has a Kähler structure then the sigma-model actually has 2 independent (holomorphic and anti-holomorphic) supersymmetries. Restricting the computation of the supersymmetric quantum mechanics partition function to the anti-holomorphic sector of the Hilbert space as described earlier leads to the representation of the index of the Dolbeault complex in terms of the Todd class.

Equivariant generalizations of this simple example are likewise possible. In (8.59) the path integration now involves the action S rather than the Hamiltonian, and D_V gets replaced by Q_S in the usual routine of Chapter 4. Now we conjugate the relevant operators and find that the localization priniciple requires the localization functionals ψ to be generally covariant and single-valued. The resulting path integrations yield precisely the computation at the end of Section 8.3 above. For Hamiltonians which generate circle actions, the right-hand side of (8.47) coincides with the left-hand side of (8.60) because of the structure of the set $L\mathcal{M}_S$ discussed at the beginning of Section 4.6. Thus in this case we again obtain the ordinary finite-dimensional relation (8.60). These relations play a deeper role when the Hamiltonian depends explicitly on time t. Then the right-hand side of (8.47) represents a regularized measure of the number of T- periodic classical trajectories of the given dynamical system [119, 122]. Thus the classical dynamics of a physical system can be characterized in this way via the localization properties of supersymmetric non-linear sigma-models. In [124], these relations were related to a functional Euler character in the quantum cohomology defined by the topological non-linear sigma-model and also to a loop space generalization of the Lefschetz fixed point theorem.

Besides supersymmetric quantum mechanics the localization features of more complicated topological gauge theories can be studied by the computing the Euler numbers of vector bundles over the infinite-dimensional space $\mathcal{A}/\mathcal{G}$ of gauge connections modulo gauge transformations of a principal G-bundle. One can either start with a given topological field theory and analyse its localization characteristics using the techniques of this Chapter, or conversely

8.5 The Mathai-Quillen Formalism for Infinite-Dimensional Vector Bundles

by applying the Mathai-Quillen formalism to some vector bundle over $\mathcal{A}/\mathcal{G}$ and reconstructing the action of the corresponding topological gauge theory from there. The resulting path integrals always compute sorts of intersection numbers on moduli space. A discussion of these models is beyond the scope of this book and we refer to [34] for an extensive discussion of the theories which can be viewed in this way. The basic example is Donaldson theory [22] which is the prime example of a cohomological field theory and is used to calcula te intersection numbers of moduli spaces of instantons for the study of 4-manifolds. Topological Yang-Mills theory in 4-dimensions is another interesting application of this formalism. The field theoretic generalization of supersymmetric quantum mechanics, i.e. the topological sigma-model [22], is the appropriate setting for studying the quantum symmetries of string theory and more generally super-conformal field theories. The Mathai-Quillen formalism applied to 2-dimensional topological gravity could presumably shed light on its equivalence with physical gravity in 2 dimensions. The coupling of the topological sigma-model to topological gravity can be interpreted as topological string theory and studied using these methods. Finally, viewing 2-dimensional Yang-Mills theory as a topological field theory (see Section 8.1 above) leads in this way to a localization onto the rather complicated Hurwitz space of branched covers of the Riemann surface. This construction has been exploited recently as a candida te for a string theoretical realization of 2-dimensional Yang-Mills theory [34].

Thus, the Mathai-Quillen formalism serves as the natural arena for the localization properties of cohomological field theories. However, the connection between the localization formalisms of the earlier Chapters of this Book (i.e. the stationary-phase formula) and the constructive Mathai-Quillen formalism above has yet to be completely clarified, as the latter relies on quite different cohomological symmetries than the ordinary BRST supersymmetries responsible for equivariant localization [121]. Recall these models all possess a Grassmann-odd symmetry δ that defines a supersymmetry transformation (8.58) which resembles the usual BRST supersymmetries of equivariant localization. It is possible to argue [29, 34] that the δ-action is not free and that the path integral receives contributions from some arbitrarily small δ-invariant tubular neighbourhood of the fixed point set of δ. The integration over the directions normal to this fixed point set can be calculated in a stationary-phase approximation. One readily sees from (8.58) that the fixed point set of δ is the precisely the moduli space $\mathcal{M}_V$ described by the zero locus of V and its tangents ψ satisfying the linearized equation $\nabla_A V(\psi) = 0$. In this way the topological field theory path integral reduces to an integration of differential forms over $\mathcal{M}_V$. It remains though to still obtain a more precise connection between these BRST fixed points, localization, and the interpretation of the geometrical and topological features of path integrals in terms of the Mathai-Quillen formalism which shows how such infinite-dimensional integrations are *a priori* designed to represent finite-dimensional integrals. The antibracket formalism developed in Section 8.3 above is a key stepping stone between the Mathai-Quillen localization features of topological field theory path integrals, and the path integral localizations of generic Hamiltonian systems. The supersymmetric formulation of equivariant cohomology developed in [129], and its connections with 4-dimensional topological Yang-Mill s theory, could serve as another approach to this connection. This might give a more direct connection between localization and some of the more modern theories of quantum integrability [35], such as R-matrix formulations and the Yang-Baxter equation. This has been discussed somewhat in [56]. These connections are all important and should be found in order to have full understandings of the structures of topological and integrable quantum field theories, and hence generic physical models, from the point of view of loop space equivariant localization.

9. Appendix A: BRST Quantization

BRST quantization was first introduced in the quantization of Yang-Mills theory as a useful device for proving the renormalizability of non-abelian gauge theories in 4 dimensions. It was shown that a global fermionic symmetry was present after Yang-Mills gauge fixing which incorporated the original gauge invariance of the model and ultimately led to straightforward derivations of the Ward identities associated with the gauge symmetry in both quantum electrodynamics and quantum chromodynamics. New impetus came when the BRST theory was applied to the quantization of Hamiltonian systems with first class constraints [69]. For completeness, in this Appendix we shall outline the essential features of the BRST quantization scheme of which the loop space localization principle can be thought of as a special instance.

Consider any physical system with symmetry operators K^a that (possibly locally) generate a closed Lie algebra $\mathbf{g}$,

$$[K^a, K^b] = f^{abc} K^c \tag{9.1}$$

Introduce Faddeev-Popov ghost and anti-ghost fields θ^a, $\bar{\theta}^a \sim \frac{\partial}{\partial \theta^a}$ which are anticommuting Grassmann variables that transform in the adjoint representation of $\mathbf{g}$. They have the canonical anticommutator

$$[\bar{\theta}^a, \theta^b]_+ = \delta^{ab} \tag{9.2}$$

We define the ghost number operator as

$$U = \theta^a \bar{\theta}^a \tag{9.3}$$

whose eigenvalues are integers running from 0 to dim $\mathbf{g}$.

We now introduce the operator

$$Q = \theta^a K^a - \frac{1}{2} f^{abc} \theta^a \theta^b \bar{\theta}^c \tag{9.4}$$

In the physics literature the operator Q is known as the BRST charge, while in the mathematics literature it is the Lie algebra coboundary operator that computes the cohomology of the Lie algebra $\mathbf{g}$ with values in the representation defined by the operators K^a. The crucial property of Q is that it is nilpotent, $Q^2 = 0$, which can be seen from (9.1) and the identity

$$f^{abc}f^{cde} + f^{bdc}f^{cae} + f^{dac}f^{cbe} = 0 \tag{9.5}$$

which follows from (9.1) via the Jacobi identity for the Lie bracket. Let $\mathcal{H}^k$ be the Hilbert space of states of ghost number k, i.e. $U\Psi = k \cdot \Psi$ for $\Psi \in \mathcal{H}^k$. We say that a state $\Psi \in \mathcal{H}^k$ is BRST invariant if it is annihilated by Q, $Q\Psi = 0$, where in general the action of Q on any state raises the ghost number by 1. Any other state $\Psi' = \Psi + Q\chi$ of ghost number k is regarded as equivalent to $\Psi \in \mathcal{H}^k$ for any other state $\chi \in \mathcal{H}^{k-1}$. The space of Q-equivalence classes of ghost number k is called the BRST-cohomology in the physics literature. Mathematically, it forms the k-th cohomology group $H^k(\mathbf{g}; R)$ of the Lie algebra $\mathbf{g}$ with values in the representation R carried by the symmetry operators K^a.

Of particular interest from a physical standpoint are the BRST-invariant states of ghost number 0. From (9.3) it follows that a state Ψ of ghost number 0 must be annihilated by all of the anti-ghost fields $\bar{\theta}^a$, so that the action of Q on such a state is

$$Q\Psi = \theta^a K^a \Psi \quad , \quad \Psi \in \mathcal{H}^0 \tag{9.6}$$

The anticommutation relations (9.2) imply that a state annihilated by all $\bar{\theta}^a$ cannot be annihilated by any of the ghost fields θ^a, and so the condition $Q\Psi = 0$ is equivalent to

$$K^a \Psi = 0 \quad , \quad a = 1, \ldots, \dim \mathbf{g} \quad ; \quad \Psi \in H^0(\mathbf{g}; R) \tag{9.7}$$

Therefore a state Ψ of ghost number 0 is BRST-invariant if and only if it is $\mathbf{g}$-invariant, and thus the cohomology group $H^0(\mathbf{g}; R)$ coincides with the space of $\mathbf{g}$-invariant states that do not contain any ghosts, i.e. the physical states.

In a gauge theory with gauge group G, the partition function must be evaluated as always with gauge-fixing functions g^a, $a = 1, \ldots, \dim G$, which specify representatives of the gauge equivalence classes of the theory and restrict the functional integration to a subspace U_0 of the original configuration space of the field theory defined by the zeroes of the functions g^a. Then the path integral can be written symbolically as

$$\int_{U_0} e^{iS} = \text{vol}(G) \int_{U_0} \prod_{a=1}^{\dim G} \delta(g^a) \det \|V^b(g^c)\| \, e^{iS} \tag{9.8}$$

where V^a are as usual vector fields associated with an orthonormal basis $\{X^a\}$ of $\mathbf{g}$ (i.e. $\text{tr}(X^a X^b) = \delta^{ab}$). Here S is the classical G-invariant gauge field action and the volume factor $\text{vol}(G)$ is infinite for a local gauge field theory. Modulo this infinite factor, the right-hand side of (9.8) is what is taken as the definition of the quantum gauge theory partition function. Introducing Faddeev-Popov ghost fields and additional auxilliary fields ϕ^a, we can absorb the additional factors on the right-hand side of (9.8) into the exponential to write

$$\int_{U_0} e^{iS} = \text{vol}(G) \int_{U_0} e^{iS_q} \tag{9.9}$$

where

$$S_q = S + \phi^a g^a + \bar{\theta}^a V^a(g^b)\theta^b \tag{9.10}$$

is the gauge-fixed, quantum action.

The BRST-symmetry of this model is defined by the following differential,

$$s(\Phi) = V^a(\Phi)\theta^a \quad , \quad s(\phi^a) = 0 \quad ; \quad s(\theta^a) = -\frac{1}{2}f^{abc}\theta^b\theta^c \quad , \quad s(\bar{\theta}^a) = -\phi^a \tag{9.11}$$

where Φ is any scalar-valued functional of the gauge fields of the theory. With this definition we have $s^2 = 0$, $s(S) = 0$ and the quantum action (9.10) can be written as

$$S_q = S + s(-g^a\bar{\theta}^a) \tag{9.12}$$

Thus s is a BRST operator that determines an $N = \dim G$ supersymmetry of the gauge-fixed field theory, and the statement that the partition function (9.8) is independent of the choice of gauge-fixing functions g^a is equivalent to the fact that the path integral depends only on the BRST-cohomology class of the action S, not on its particular representative. Thus the BRST-supersymmetry here represents the local gauge symmetry of the theory. The gauge variation of any functional $\mathcal{O}$ of the fields of the theory is then represented as a graded commutator

$$\{s, \mathcal{O}\} = s\mathcal{O} - (-1)^p\mathcal{O}s \tag{9.13}$$

with the fermionic charge s, where p is the ghost-degree of $\mathcal{O}$. The physical (i.e. gauge-invariant) Hilbert space of the gauge theory is the space of BRST-cohomology classes of ghost number 0.

(9.8) and (9.12) demonstrate the explicit relationship between equivariant localization of path integrals and BRST quantization (see the localization principle in Section 4.4). In the next Appendix we shall make this connection a bit more explicit. As we have mentioned, BRST-cohomology is the fundamental structure in topological field theories [22]. By definition, a topological action is a Witten-type action if the the classical action S is BRST-exact, while it is a Schwarz-type action if the gauge-fixed, quantum action S_q is BRST-exact (but not the classical one). In the case of the localization formalism for phase space path integrals, the BRST-operator is identified with the loop space equivariant exterior derivative of the underlying equivariant cohomological structure, and the "Hilbert space" of physical states consists of loop space functionals which are invariant under the flows of the loop space Hamiltonian vector field. This BRST-supersymmetry is always the symmetry that is responsible for localization in these models. The BRST formalism can also be applied to Hamiltonian systems with first-class constraints, i.e. those whose constraint functions K^a generate a Poisson subalgebra representation of a Lie algebra (9.1). The supersymmetric states then represent the observables which respect the constraints of the dynamical system (as in a gauge theory).

10. Appendix B: Other Models of Equivariant Cohomology

In this Appendix we shall briefly outline some of the other standard models for the G-equivariant cohomology of a differentiable manifold $\mathcal{M}$ and compare them with the Cartan model which was used extensively throughout this Book. We shall also discuss how these other models apply to the derivation of some of the more general localization formulas which were just briefly sketched in Section 4.9, as well as their importance to other ideas in topological quantum field theory.

10.1 The Topological Definition

As with ordinary cohomology, equivariant cohomology has a somewhat direct interpretation in terms of topological characteristics of the manifold $\mathcal{M}$ (and in this case also the Lie group G) [9]. This can be used to develop an axiomatic formulation of equivariant cohomology which in the usual way provides properties that uniquely characterize the cohomology groups [34]. This topological definition resides heavily in the topology of the Lie group G through the notion of a classifying space [73]. A classical theorem of topology tells us that to G we can associate a very special space E_G which is characterized by the fact that it is contractible and that G acts on it without fixed points. The space E_G is called the universal G-bundle. The classifying space B_G for G-bundles is then defined as the base space of a universal bundle whose total space is E_G. The space B_G is unique up to homotopy and E_G is unique up to equivariant homotopy (i.e. smooth continuous equivariant deformations of the space). The bundle $E_G \to B_G$ has 2 remarkable universal properties. The first one is that any given principal G-bundle $E \to \mathcal{M}$ over a manifold $\mathcal{M}$ has an isomorphic copy sitting inside $E_G \to B_G$. The isomorphism classes of principal G-bundles are therefore in one-to-one correspondence with the homotopy classes of maps $f : \mathcal{M} \to B_G$. The second property is that all natural ways of measuring the topology of $E \to \mathcal{M}$ can be obtained from $H^*(B_G)$.

For example, when $G = \bigoplus_{i=1}^n \mathbb{Z}$, we have $E_{\mathbb{Z}^n} = \mathbb{R}^n$ and $B_{\mathbb{Z}^n} = (S^1)^n$, while for $G = U(1)$ we get $E_{U(1)} = S(H) = \bigcup_{n=0}^\infty S^{2n+1}$ (the Hilbert sphere) and $B_{U(1)} = \mathbb{C}P^\infty = \bigcup_{n=0}^\infty \mathbb{C}P^n$. In gauge theories G is the group of local gauge transformations, so that E_G is the space $\mathcal{A}$ of Yang-Mills potentials while $B_G = \mathcal{A}/G$ is the space of gauge orbits. In string theory G is the semi-direct product of the diffeomorphism and Weyl groups of a Riemann surface Σ^h of genus h, E_G is the space of metrics on Σ^h, and B_G is the moduli space $\mathcal{M}_{\Sigma^h}$ of Σ^h. From this point of view, one can define topological field theory and topological string theory as the study of $H^*(B_G)$ and related cohomologies using the language of local quantum field theory.

Given a smooth G-action on a manifold $\mathcal{M}$, we thus have 2 spaces E_G and $\mathcal{M}$ on which G acts. Thus G also acts on the Cartesian product space $\mathcal{M} \times E_G$ via the diagonal action

$$G \times (\mathcal{M} \times E_G) \to \mathcal{M} \times E_G$$
$$(g, x, e) \to (g \cdot x, g \cdot e) \tag{10.1}$$

Like the G-action on E_G, this action is also free and thus the quotient space

$$\mathcal{M}_G = (\mathcal{M} \times E_G)/G \tag{10.2}$$

is a smooth manifold called the homotopy quotient of $\mathcal{M}$ by G. Since E_G is contractible, $\mathcal{M} \times E_G$ is homotopic to $\mathcal{M}$. Furthermore, if the G-action on $\mathcal{M}$ is free then $\mathcal{M}_G$ is homotopic to $\mathcal{M}/G$, so that both spaces have the same (ordinary) cohomology groups. In the general case we can regard $\mathcal{M}_G$ as a bundle over B_G with fiber $\mathcal{M}$. These observations motivate the topological definition of equivariant cohomology as

$$H^k_{G,\text{top}}(\mathcal{M}) = H^k(\mathcal{M}_G) \tag{10.3}$$

Notice that if $\mathcal{M}$ is the space consisting of a single point, then $\mathcal{M}_G \simeq E_G/G = B_G$. Thus

$$H^k_{G,\text{top}}(\text{pt}) = H^k(B_G) \tag{10.4}$$

and so the G-equivariant cohomology of a point is the ordinary cohomology of the classifying space B_G. This latter cohomology can be quite complicated [9], and the topological definition therefore shows that the equivariant cohomology measures much more than simply the cohomology of a manifold modulo a group action on it. It is this feature that makes the non-abelian localization formalisms of topological field theories and integrable models very powerful techniques.

10.2 The Weil Model

The topological definition of the G-equivariant cohomology above can be reformulated in terms of nilpotent differential operators [9, 81]. In this formulation, the equivariant cohomology is obtained in a more algebraic way by exploiting differential properties of the Lie algebra **g** of the group G.

More precisely, to describe the exterior differential calculus of the symmetric algebra $S(\mathbf{g}^*)$, we introduce the Weil algebra

$$W(\mathbf{g}) = S(\mathbf{g}^*) \otimes \Lambda \mathbf{g}^* \tag{10.5}$$

As in Section 2.6, (10.5) algebraically describes the exterior bundle of $S(\mathbf{g}^*)$ where the exterior algebra $\Lambda \mathbf{g}^*$ consists of multilinear antisymmetric forms on $\mathbf{g}$ and it is generated by an anti-commuting basis of Grassmann numbers θ^a, $a = 1, \ldots, \dim G$ (i.e. the 1-forms $\theta^a \sim d\phi^a$). The Weil algebra has the usual $\mathbb{Z}$-grading (ghost number), i.e. the generators ϕ^a of $S(\mathbf{g}^*)$ have degree 2 while the generators θ^a of $\Lambda \mathbf{g}^*$ have degree 1. Both of these sets of generators are dual to the same fixed basis $\{X^a\}_{a=1}^{\dim G}$ of $\mathbf{g}$.

There are 2 differential operators of interest acting on $W(\mathbf{g})$. The first is the "abelian" exterior derivative on $\Lambda \mathbf{g}^*$

$$d_0 = \phi^a \otimes I_a \tag{10.6}$$

where $I_a = \frac{\partial}{\partial \theta^a}$ is the interior multiplication of degree -1 on $W(\mathbf{g})$. (10.6) identifies ϕ^a as the superpartner of θ^a and its non-vanishing actions are

$$d_0 \theta^a = \phi^a \quad , \quad d_0 \frac{\partial}{\partial \phi^a} = -I_a \tag{10.7}$$

Next, we introduce the linear derivations of degree 0

$$L_a = -f^{abc} \left(\phi^b \frac{\partial}{\partial \phi^c} + \theta^b I_c \right) \tag{10.8}$$

which generate the coadjoint action of G on $W(\mathbf{g})$ explicitly by

$$L_a \phi^b = f^{abc} \phi^c \quad , \quad L_a \theta^b = f^{abc} \theta^c \tag{10.9}$$

and which yield a representation of G on $W(\mathbf{g})$,

$$[L_a, L_b] = f^{abc} L_c \tag{10.10}$$

Using (10.8), we define our second differential operator, the coboundary operator

$$d_{\mathbf{g}} = \theta^a L_a + \frac{1}{2} f^{abc} \theta^a \theta^b I_c \tag{10.11}$$

which computes the $W(\mathbf{g})$-valued Lie algebra cohomology of $\mathbf{g}$, i.e. it is the BRST operator associated with the constraint operators L_a acting on $W(\mathbf{g})$ (see (9.4)).

The sum of (10.6) and (10.11) is known as the Weil differential,

$$d_{\mathcal{W}} = d_0 + d_{\mathbf{g}} \tag{10.12}$$

whose action on the generators of $W(\mathbf{g})$ is

$$d_W \theta^a = \phi^a - \frac{1}{2}f^{abc}\theta^b\theta^c \quad , \quad d_W \phi^a = -f^{abc}\theta^b\phi^c \tag{10.13}$$

These 3 differential operators are all nilpotent derivations of degree 1,

$$d_W^2 = d_0^2 = d_{\mathbf{g}}^2 = 0 \tag{10.14}$$

and they act as exterior derivatives on $W(\mathbf{g})$. (10.12) makes the Weil algebra into an exterior differential algebra. However, the cohomology of the Weil differential d_W on $W(\mathbf{g})$ is trivial. This can be seen by redefining the basis of $W(\mathbf{g})$ by the shift $\phi^a \to \phi^a - \frac{1}{2}f^{abc}\theta^b\theta^c$. In the new basis, $\phi^a = d_W\theta^a$ are exact and the cohomology of d_W coincides with that of d_0 on $W(\mathbf{g})$ so that

$$H^k(W(\mathbf{g}), d_W) = \mathbb{R} \tag{10.15}$$

The cohomology can be made non-trivial using the 2 derivations I_a and L_a on $W(\mathbf{g})$ introduced above. We notice first of all the analogy between (10.13) and the algebra of connections and curvatures. The first relation in (10.13) is the definition of the curvature of a connection 1-form $A \sim \theta^a$ on a principal G-bundle $E \to \mathcal{M}$, while the second one is the Bianchi identity for the curvature 2-form $F \sim \phi^a$, i.e.

$$dA = F - \frac{1}{2}[A \stackrel{\wedge}{,} A] \quad , \quad dF = -[A \stackrel{\wedge}{,} F] \tag{10.16}$$

Here we recall that the characteristic classes of $E \to \mathcal{M}$ are constructed from $A \in \Lambda^1(E, \mathbf{g})$, which can be regarded as a map $A : \mathbf{g}^* \to \Lambda^1 E$, and from the field strength $F \in \Lambda^2(E, \mathbf{g})$, which can be regarded as a map $F : \mathbf{g}^* \to \Lambda^2 E$. These maps generate a differential algebra homomorphism $(W(\mathbf{g}), d_W) \to (\Lambda E, d)$ which is called the Chern-Weil homomorphism. This homomorphism is unique and it maps the algebraic connection and curvature (θ^a, ϕ^a) to the geometric ones (A, F) [99]. In this setting, a connection on a principal G-bundle $E \to \mathcal{M}$ is just the same thing as a homomorphism $W(\mathbf{g}) \to \Lambda E$. Thus the Weil algebra is an algebraic analog of the universal G-bundle E_G. Like E_G, it possesses universal properties, and therefore it provides a universal model of connections on G-bundles. In particular, the contractibility of E_G is the analog of the triviality (10.15) of the cohomology of the Weil algebra.

Pursuing this analogy between E_G and $W(\mathbf{g})$, we can find non-trivial and universal cohomology classes by considering the so-called "basic forms" [9]. First, we note that the operators I_a and L_a above are the algebraic analogues of the interior multiplication and Lie derivative of differential forms with respect to the infinitesimal generators ϕ^a of the G-action on $W(\mathbf{g})$. Indeed, the operator (10.8) can be expressed in terms of the Weil differential as

$$L_a = I_a d_W + d_W I_a \equiv [d_{\mathbf{g}}, I_a]_+ \tag{10.17}$$

and furthermore we have

10.2 The Weil Model

$$[I_a, I_b] = 0 \quad , \quad [L_a, I_b] = f^{abc} I_c \tag{10.18}$$

Thus the derivation L_a has the natural structure of a Lie derivative on $W(\mathbf{g})$ that commutes with all the derivatives above,

$$[d_W, L_a] = [d_\mathbf{g}, L_a] = [d_0, L_a] = 0 \tag{10.19}$$

In particular, the (anti-)commutation relations above among d_W, L_a and I_a are all independent of the choice of basis of $\mathbf{g}$. As we saw in Section 2.3, these relations also reflect the differential geometric situation on a manifold $\mathcal{M}$ with a G-action on it, and the Chern-Weil homomorphism above maps $(d_W, L_a, I_a) \to (d, \mathcal{L}_{V^a}, i_{V^a})$ between the differential algebras $W(\mathbf{g}) \to \Lambda\mathcal{M}$.

We can finally define the Weil model for equivariant cohomology. For this we consider the tensor product $W(\mathbf{g}) \otimes \Lambda\mathcal{M}$ of the Weil algebra with the exterior algebra over the manifold $\mathcal{M}$. The replacement $\Lambda\mathcal{M} \to W(\mathbf{g}) \otimes \Lambda\mathcal{M}$ for the description of the equivariant cohomology is the algebraic equivalent of the replacement $\mathcal{M} \to E_G \times \mathcal{M}$ in Section B.1 above. The basic subalgebra $(W(\mathbf{g}) \otimes \Lambda\mathcal{M})_{\text{basic}}$ of $W(\mathbf{g}) \otimes \Lambda\mathcal{M}$ consists of those forms which have no vertical component (i.e. the horizontal forms) and which are G-invariant (i.e. have no vertical variation). These 2 conditions mean, respectively, that the basic forms are those annihilated by all the operators $I_a \otimes \mathbf{1} + \mathbf{1} \otimes i_{V^a}$ and $L_a \otimes \mathbf{1} + \mathbf{1} \otimes \mathcal{L}_{V^a}$ (recall that $i_V \alpha$ is the component of α along the (vertical) vector field $V \in T\mathcal{M}$), so that

$$(W(\mathbf{g}) \otimes \Lambda\mathcal{M})_{\text{basic}} = \left(\bigcap_{a=1}^{\dim G} \ker\left(I_a \otimes \mathbf{1} + \mathbf{1} \otimes i_{V^a}\right) \right)$$
$$\cap \left(\bigcap_{b=1}^{\dim G} \ker\left(L_a \otimes \mathbf{1} + \mathbf{1} \otimes \mathcal{L}_{V^a}\right) \right) \tag{10.20}$$

This subalgebra is stable under the action of the extended DeRham exterior derivative

$$d_T = d_W \otimes \mathbf{1} + \mathbf{1} \otimes d \tag{10.21}$$

and the cohomology of (10.21) on (10.20) is the algebraic definition of the G-equivariant cohomology of $\mathcal{M}$,

$$H^k_{G,\text{alg}}(\mathcal{M}) = H^k((W(\mathbf{g}) \otimes \Lambda\mathcal{M})_{\text{basic}}, d_T) \tag{10.22}$$

The Chern-Weil homomorphism $W(\mathbf{g}) \to \Lambda E$ with $E = E_G$ then reduces to an isomorphism of cohomology groups if G is compact and connected [9] and we have

$$H^k_{G,\text{alg}}(\mathcal{M}) \simeq H^k_{G,\text{top}}(\mathcal{M}) \tag{10.23}$$

so that the algebraic and topological definitions of equivariant cohomology are equivalent.

We close this Section with a remark concerning the basic subcomplex $B_\mathbf{g}$ of $W(\mathbf{g})$ ($V^a = 0$ in (10.20)). Since d_W vanishes on this basic subcomplex, we have $H^k(B_\mathbf{g}, d_W) = B_\mathbf{g}$. Horizontality in this case means θ^a-independence, so that the basic forms in $W(\mathbf{g})$ lie in only $S(\mathbf{g}^*)$. G-invariance in this case translates into invariance under the coadjoint action of G on $\mathbf{g}^*$. Thus

$$H^k(B_\mathbf{g}, d_W) = B_\mathbf{g} = S(\mathbf{g}^*)^G \qquad (10.24)$$

and so the basic subalgebra of the Weil algebra coincides with the algebra of invariant polynomial functions on the Lie algebra $\mathbf{g}$, i.e. $B_\mathbf{g}$ is the algebra of corresponding Casimir invariants. It is known [9] that if G is a compact connected Lie group, then $H^k(B_G) = H^k(B_\mathbf{g}, d_W)$, and so comparing (10.24) with (10.4) we find that the G-equivariant cohomology of a point is simply the algebra of G-invariant polynomials on $\mathbf{g}$. This is in agreement with what we found in Section 2.3 from the Cartan model. In the next Section we shall indeed find that this correspondence is no accident.

10.3 The BRST Model

The final model for the G-equivariant cohomology of a manifold $\mathcal{M}$ interpolates between the Cartan and Weil models, and it therefore relates the Cartan model to the topological characteristics of this cohomology theory. It also ties in with the BRST quantization ideas that are directly related to the localization formalisms and it is the model of equivariant cohomology which arises naturally in the physical context of topological field theories. The (unrestricted) BRST algebra B of topological models on quotient spaces is the same as that for the Weil model as a vector space, $B = W(\mathbf{g}) \otimes \Lambda \mathcal{M}$ [81, 132]. Now, however, the differential on it is the BRST operator

$$\delta = d_W \otimes \mathbf{1} + \mathbf{1} \otimes d + \theta^a \otimes \mathcal{L}_{V^a} - \phi^a \otimes i_{V^a} \qquad (10.25)$$

which is a nilpotent graded derivation of degree 1 on B. The BRST operator is the natural nilpotent extension of the Cartan equivariant exterior derivative $D_\mathbf{g}$ defined in (2.59). The Weil differential d_W in (10.25) takes care of the non-nilpotency (2.61) of the Cartan model derivative.

The Kalkman parametric model for the equivariant cohomology is defined by the Kalkman differential [81]

$$\delta_t = e^{t\theta^a \otimes i_{V^a}} d_T\, e^{-t\theta^a \otimes i_{V^a}} = d_T + t\theta^a \otimes \mathcal{L}_{V^a} - t\phi^a \otimes i_{V^a} + \frac{1}{2}t(1-t)f^{abc}\theta^a\theta^b \otimes i_{V^c} \qquad (10.26)$$

where $t \in [0,1]$. Notice that $\delta_0 = d_T$ and $\delta_1 = \delta$, so that (10.26) interpolates between the differentials of the Weil and BRST models. Furthermore, it satisfies

$$[\delta_t, I_a \otimes \mathbf{1} + (1-t)\mathbf{1} \otimes i_{V^a}] = L_a \otimes \mathbf{1} + \mathbf{1} \otimes \mathcal{L}_{V^a} \qquad (10.27)$$

so that we obtain in this way a family of Lie super-algebras acting on $W(\mathbf{g}) \otimes \Lambda \mathcal{M}$. Notice also that

$$d_T = e^{-\frac{1}{2} f^{abc} \theta^a \theta^b \frac{\partial}{\partial \phi^c}} (d + d_0) \, e^{\frac{1}{2} f^{abc} \theta^a \theta^b \frac{\partial}{\partial \phi^c}} \quad (10.28)$$

and thus the cohomology of δ on B coincides with the DeRham cohomology of d on $\Lambda \mathcal{M}$.

Thus the BRST operator δ does not capture the G-equivariant cohomology of $\mathcal{M}$, because the $W(\mathbf{g})$ part of (10.25) can be conjugated to the cohomologically trivial operator (10.6) (equivalently the cohomology of d_W is trivial). We have to accompany δ with a restriction of its domain in the same way that (2.59) computes the equivariant cohomology when restricted to the G-invariant subspace (2.40). The appropriate restriction is to the θ^a-independent and G-invariant subalgebra of B. This reduction maps B to $\Lambda_G \mathcal{M}$ and δ to $D_\mathbf{g}$, so that the mapping $\theta^a \to 0$, $B \to S(\mathbf{g}^*) \otimes \Lambda \mathcal{M}$ induces the isomorphism of complexes

$$(B_{\text{basic}}, d_W) \simeq (\Lambda_G \mathcal{M}, D_\mathbf{g}) \quad (10.29)$$

between the Weil and Cartan models. This restriction can be formulated by introducing another nilpotent operator $\mathcal{W}$ whose kernel is the desired G-invariant and θ^a-independent subalgebra [129, 132].

For this, we introduce another copy $\bar{W}(\mathbf{g})$ of the Weil algebra. It is generated by $\bar{\phi}^a$ and $\bar{\theta}^a$ which are the $\mathbf{g}^*$-valued coefficients corresponding, respectively, to θ^a-independence (generated by I_a) and G-invariance (generated by $L_a \otimes 1 + 1 \otimes \mathcal{L}_{V^a}$). A nilpotent operator with kernel $\Lambda_G \mathcal{M}$ is

$$\mathcal{W} = d_{\bar{\mathbf{g}}} \otimes \mathbf{1} \otimes \mathbf{1} + \bar{\theta}^a \otimes (L_a \otimes \mathbf{1} + \mathbf{1} \otimes \mathcal{L}_{V^a}) - \bar{\phi}^a \otimes I_a \otimes \mathbf{1} \quad (10.30)$$

where $d_{\bar{\mathbf{g}}}$ is the Lie algebra coboundary operator (10.11) on $\bar{W}(\mathbf{g})$ which makes the overall combination in (10.30) nilpotent. The action of δ on $\bar{W}(\mathbf{g})$ is taken as that of the abelian differential (10.6) so that δ commutes with $\mathcal{W}$. As the cohomology of $d_{\bar{0}}$ on $\bar{W}(\mathbf{g})$ is trivial, this alteration of δ does not affect its cohomology. The equivariant BRST operator is therefore

$$\delta = \mathbf{1} \otimes d_W \otimes \mathbf{1} + \mathbf{1} \otimes \mathbf{1} \otimes d + d_{\bar{0}} \otimes \mathbf{1} \otimes \mathbf{1} + \mathbf{1} \otimes \theta^a \otimes \mathcal{L}_{V^a} - \mathbf{1} \otimes \phi^a \otimes i_{V^a} \quad (10.31)$$

and δ and $\mathcal{W}$ satisfy the nilpotent algebra

$$\delta^2 = [\mathcal{W}, \delta]_+ = \mathcal{W}^2 = 0 \quad (10.32)$$

The G-equivariant cohomology of $\mathcal{M}$ is isomorphic to the cohomology of δ on the subalgebra of $\bar{W}(\mathbf{g}) \otimes W(\mathbf{g}) \otimes \Lambda \mathcal{M}$ which is annihilated by $\mathcal{W}$. This defines the BRST model of equivariant cohomology.

We remark that it is also possible to formulate the restriction onto the basic subcomplex in the Weil model using the nilpotent operator

$$\mathcal{W}_W = d_{\bar{\mathbf{g}}} \otimes \mathbf{1} \otimes \mathbf{1} + \bar{\theta}^a \otimes (L_a \otimes \mathbf{1} + \mathbf{1} \otimes \mathcal{L}_{V^a}) - \bar{\phi}^a \otimes (I_a \otimes \mathbf{1} + \mathbf{1} \otimes i_{V^a}) \quad (10.33)$$

and the corresponding extension of the Weil differential is

$$\bar{d}_{\mathcal{W}} = \mathbf{1} \otimes d_T + d_{\bar{0}} \otimes \mathbf{1} \otimes \mathbf{1} \tag{10.34}$$

The operators (10.33) and (10.34) are similarity transforms of (10.30) and (10.31), respectively, just as in (10.28). They therefore obey the algebra (10.32) as well, as required for the appropriate restriction process above.

It is this model that is relevant for the construction of non-abelian generalizations of the Duistermaat-Heckman integration formula, such as the Witten localization formalism (see Section 3.8). Modelling the equivariant cohomological structure as above is the correct way to incorporate the idea of equivariant integration that we discussed earlier. In these models the generators ϕ^a of the symmetric algebra $S(\mathbf{g}^*)$ generate dynamics of their own and only after they are fully incorporated as above can one define properly the required equivariant localization. All 4 equivariant derivations discussed in this Book – the Cartan, Weil, Kalkman and BRST differentials, have been related by Nersessian [114, 115] to the geometric anti-bracket structure of the Batalin-Vilkovisky formalism (see Section 8.3). In these formalisms, one constructs antibrackets for $W(\mathbf{g})$ in addition to the usual ones for the cotangent bundle $\mathcal{M} \otimes \Lambda^1 \mathcal{M}$.

10.4 Loop Space Extensions

Loop space generalizations of the constructions above have been presented by Tirkkonen in [161] in the context of the dynamical localizations of Section 4.9. For this, we introduce superpartners $\theta^a(t)$ for the multipliers $\phi^a(t)$ and make them dynamical by adding a kinetic term for the Grassmann coordinates θ^a to the action,

$$S \to S_T = S + \int_0^T dt \, \theta^a(t) \dot{\theta}^a(t) \tag{10.35}$$

The circle action generator in (4.151) is also extended to include θ^a, i.e.

$$V_{S^1} \to V_{S^1} + \int_0^T dt \, \dot{\theta}^a(t) \frac{\delta}{\delta \theta^a(t)} \tag{10.36}$$

Thus the path integral is now formulated over a superloop space with local coordinates (x, θ) (recall from (10.7) that θ^a are coordinates and $\phi^a = d_0 \theta^a$ are the corresponding 1-forms in the Weil algebra). To exploit the loop space isometry generated by the semi-direct product $LG \otimes S^1$, we need to construct the corresponding equivariant operators. Because the fields ϕ^a and θ^a are dynamical and are therefore an important part of the path integration, this has to be done in terms of the BRST model above.

10.4 Loop Space Extensions

In the corresponding equivariant BRST operator, the part corresponding to the LG-action in (4.151) generated by V_{LG} on $L\mathcal{M}$ is just lifted from the corresponding equivariant BRST operator (10.31) for the G-action on $\mathcal{M}$. The part associated with the circle action in (4.151) as generated by (10.36) on the superloop space is given by the Cartan model operator $Q_{\dot{x}} = d_L - i_{\dot{x}}$. Since $Q_{\dot{x}}^2 = \int_0^T dt \, \frac{d}{dt}$, the operator $Q_{\dot{x}}$ is nilpotent on the loop space $L\mathcal{M}$. The total equivariant BRST operator for the semi-direct product $LG \otimes S^1$ then combines (10.31) lifted to the loop space with $Q_{\dot{x}}$,

$$Q_T = d_L + d_W^{(L)} + \int_0^T dt \left(\theta^a \mathcal{L}_{V^a} - \phi^a i_{V^a} + \dot{x}^\mu \frac{\delta}{\delta \eta^\mu} + \dot{\theta}^a \frac{\delta}{\delta \phi^a} \right) \quad (10.37)$$

where $d_W^{(L)}$ is the Weil differential (10.12) lifted to the superloop space in the standard way. Furthermore, to the restriction operator (10.30) we add a piece corresponding to the S^1-action [161]

$$\mathcal{W}_T = \mathcal{W}_L + \int_0^T dt \, \dot{\theta}^a(t) \frac{\delta}{\delta \phi^a(t)} \quad (10.38)$$

with $\mathcal{W}_L$ the lifting of the restriction operator (10.30) to the superloop space. The algebra of Q_T and $\mathcal{W}_T$ is then

$$Q_T^2 = \int_0^T dt \, \frac{d}{dt} \quad , \quad \mathcal{W}_T^2 = [Q_T, \mathcal{W}_T]_+ = 0 \quad (10.39)$$

The function $F(\phi) = \frac{1}{2}(\phi^a)^2$ which is effectively added to the action as discussed in Section 4.9 can now be interpreted as a symplectic 2-form on the Weil algebra $W(\mathbf{g})$ (as ϕ^a are interpreted as super 1-forms). The superloop space symplectic 2-form is then

$$\Omega_T = \Omega + \int_0^T dt \, \frac{1}{2} (\phi^a(t))^2 \quad (10.40)$$

and the equivariance properties of the phase space path integral are summarized by the equations

$$Q_T(S_T + \Omega_T) = \mathcal{W}_T(S_T + \Omega_T) = 0 \quad (10.41)$$

The first vanishing condition in (10.41) identifies the superloop space action S_T as the moment map for the action of the semi-direct product $LG \otimes S^1$ on $L\mathcal{M}$, while the second one states that this group action is symplectic. To carry out the Niemi-Tirkkonen localization procedure over the superloop space with the BRST operator Q_T, we generalize the gauge fermion field (4.123) to the Weil algebra,

$$\psi_T = \int_0^T dt \left(g_{\mu\nu}\dot{x}^\mu \eta^\nu + \hat{\phi}^a \hat{\theta}^a \right) \qquad (10.42)$$

where we have left the zero mode part of $\phi^a \theta^a$ out of ψ_T. We can now carry out the same localization procedure which led to the Niemi-Tirkkonen formula, and thus arrive at the localization formula (4.153) [161].

The above construction is a non-abelian generalization of the model-independent auxilliary field formalism which we discussed in Section 4.9. In the abelian case where we can *a priori* fix any function $F(\phi)$ for the localization, instead of modifying the loop space symplectic structure as in (10.40), which in the model-independent formalism appears as the functional determinant $\det \|\partial_t\|$, we shift the gauge fermion field (10.42) as

$$\psi_T \to \psi_T - \int_0^T dt \left(\theta_\mu \eta^\mu - \left\{ \xi(\phi) - \frac{1}{2}\bar{\phi} \right\} \bar{\theta} \right) \qquad (10.43)$$

The standard Niemi-Tirkkonen localization procedure then leads to the localization formula (4.149) [128]. For the generic, non-abelian group actions discussed above, the situation for loop space localization is different, and even the second Weil algebra copy $\bar{W}(\mathbf{g})$ is made dynamical. For discussions of the relations between the BRST model of equivariant cohomology and Witten-type topological field theories, see [27, 34, 129].

References

1. R. Abraham and J. E. Marsden, *Foundations of Mechanics*, Addison-Wesley (New York) (1978)
2. A. Alekseev and S. L. Shatashvili: Path Integral Quantization of the Coadjoint Orbits of the Virasoro Group and 2D Gravity, *Nucl. Phys.* **B323** (1989), 719
3. A. Alekseev, L. D. Faddeev and S. L. Shatashvili: Quantization of the Symplectic Orbits of Compact Lie Groups by means of the Functional Integral, *J. Geom. Phys.* **5** (1989), 391
4. O. Alvarez, I. M. Singer and P. Windey: Quantum Mechanics and the Geometry of the Weyl Character Formula, *Nucl. Phys.* **B337** (1990), 467
5. L. Alvarez-Gaumé: Supersymmetry and the Atiyah-Singer Index Theorem, *Commun. Math. Phys.* **90** (1983), 161
6. V. I. Arnold, *Mathematical Methods of Classical Mechanics*, Springer-Verlag (Berlin) (1978)
7. V. I. Arnold and S. P. Novikov, *Dynamical Systems*, Springer-Verlag (Berlin) (1990)
8. M. F. Atiyah: Circular Symmetry and Stationary Phase Approximation, *Asterisque* **131** (1985), 43
9. M. F. Atiyah and R. Bott: The Moment Map and Equivariant Cohomology, *Topology* **23** (1984), 1
10. M. F. Atiyah and L. C. Jeffrey: Topological Lagrangians and Cohomology, *J. Geom. Phys.* **7** (1990), 120
11. M. Audin, *The Topology of Torus Actions on Symplectic Manifolds*, Birkhäuser (Basel) (1991)
12. A. P. Balachandran, G. Marmo, B.-S. Skagerstam and A. Stern, *Classical Topology and Quantum States*, World Scientific (Singapore) (1991)
13. I. A. Batalin and E. S. Fradkin: A Generalized Canonical Formalism and Quantization of Reducible Gauge Theories, *Phys. Lett.* **B122** (1983), 157
14. I. A. Batalin and G. A. Vilkovisky: Relativistic S-matrix of Dynamical Systems with Boson and Fermion Constraints, *Phys. Lett.* **B69** (1977), 309
15. I. A. Batalin and G. A. Vilkovisky: Quantization of Gauge Theories with Linearly Dependent Generators, *Phys. Rev.* **D28** (1983), 2567
16. L. Baulieu: Perturbative Gauge Theories, *Phys. Rep.* **129** (1985), 1
17. F. A. Berezin, *The Method of Second Quantization*, Nauka (Moscow) (1986)
18. M. Bergeron, G. W. Semenoff and R. J. Szabo: Canonical BF-type Topological Field Theory and Fractional Statistics of Strings, *Nucl. Phys.* **B437** (1995), 695
19. N. Berline and M. Vergne: Classes Characteristiques Equivariantes, *C. R. Acad. Sci. Paris* **295** (1982), 539
20. N. Berline and M. Vergne: Zeros d'un Champ de Vecteurs et Classes Characteristiques Equivariantes, *Duke Math. J.* **50** (1983), 539
21. N. Berline, E. Getzler and M. Vergne, *Heat Kernels and Dirac Operators*, Springer-Verlag (Berlin) (1991)

22. D. Birmingham, M. Blau, M. Rakowski and G. Thompson: Topological Field Theory, *Phys. Rep.* **209** (1991), 129
23. J.-M. Bismut: Index Theorem and Equivariant Cohomology on the Loop Space, *Commun. Math. Phys.* **98** (1985), 213
24. J.-M. Bismut: Localization Formulas, Superconnections and the Index Theorem for Families, *Commun. Math. Phys.* **103** (1986), 127
25. J.-M. Bismut: Equivariant Bott-Chern Currents and the Ray-Singer Analytic Torsion, *Math. Ann.* **287** (1990), 495
26. M. Blau: Chern-Simons Quantum Mechanics, Supersymmetry and Symplectic Invariants, *Intern. J. Mod. Phys.* **A6** (1991), 365
27. M. Blau: The Mathai-Quillen Formalism and Topological Field Theory, *J. Geom. Phys.* **11** (1993), 95
28. M. Blau and G. Thompson: Equivariant Kähler Geometry and Localization in the G/G Model, *Nucl. Phys.* **B439** (1995), 367
29. M. Blau and G. Thompson: Localization and Diagonalization, *J. Math. Phys.* **36** (1995), 2192
30. M. Blau, E. Keski-Vakkuri and A. J. Niemi: Path Integrals and Geometry of Trajectories, *Phys. Lett.* **B246** (1990), 92
31. M. Böhm and G. Junker: Path Integration over Compact and Non-compact Rotation Groups, *J. Math. Phys.* **28** (1987), 1978
32. R. Bott and L. W. Tu, *Differential Forms in Algebraic Topology*, Springer-Verlag (New York) (1986)
33. H. Cartan: La Transgression dans un Group de Lie et dans un Fibre Principal, in *Colloque de Topologie*, CBRM (Bruxelles) (1950), 57
34. S. Cordes, G. Moore and S. Ramgoolam: Lectures on $2D$ Yang-Mills Theory, Equivariant Cohomology and Topological Field Theories, *Nucl. Phys.* **B41** (Proc. Suppl.) (1995), 184; in *Fluctuating Geometries in Statistical Mechanics and Field Theory*, eds. F. David, P. Ginsparg and J. Zinn-Justin, North-Holland (Amsterdam) (1996)
35. A. Das, *Integrable Models*, World Scientific (Singapore) (1989)
36. P. Di Francesco, P. Ginsparg and J.-B. Zuber: $2D$ Gravity and Random Matrices, *Phys. Rep.* **254** (1995), 1
37. W. Dittrich and M. Reuter, *Classical and Quantum Dynamics*, Springer-Verlag (Berlin) (1992)
38. J. S. Dowker: When is the 'Sum over Classical Paths' Exact?, *J. Phys.* **A3** (1970), 451
39. J. J. Duistermaat and G. J. Heckman: On the Variation in the Cohomology of the Symplectic Form of the Reduced Phase Space, *Invent. Math.* **69** (1982), 259; **72** (1983), 153
40. H. M. Dykstra, J. D. Lykken and E. J. Raiten: Exact Path Integrals by Equivariant Localization, *Phys. Lett.* **B302** (1993), 223
41. T. Eguchi, P. B. Gilkey and A. J. Hanson: Gravitation, Gauge Theories and Differential Geometry, *Phys. Rep.* **66** (1980), 213
42. L. P. Eisenhart, *Riemannian Geometry*, Princeton University Press (Princeton) (1949)
43. L. P. Eisenhart, *Continuous Groups of Transformations*, Dover (New York) (1961)
44. E. Ercolessi, G. Morandi, F. Napoli and P. Pieri: Path Integrals for Spinning Particles, Stationary Phase and the Duistermaat-Heckman Theorem, *J. Math. Phys.* **37** (1996), 535
45. L. D. Faddeev: Introduction to Functional Methods, in *Methods in Field Theory*, eds. R. Balian and J. Zinn-Justin, North-Holland (Amsterdam) (1976), 1

References   311

46. R. P. Feynman: Space-time Approach to Nonrelativistic Quantum Mechanics, *Rev. Mod. Phys.* **20** (1948), 367
47. J. P. Francoise: Canonical Partition Functions of Hamiltonian Systems and the Stationary Phase Formula, *Commun. Math. Phys.* **117** (1988), 34
48. D. Friedan and P. Windey: Supersymmetric Derivation of the Atiyah-Singer Index Theorem, *Nucl. Phys.* **B235** [FS11] (1984), 395
49. K. Fujii, T. Kashiwa and S. Sakoda: Coherent States over Grassmann Manifolds and the WKB-exactness in Path Integrals, *J. Math. Phys.* **37** (1996), 567
50. K. Funahashi, T. Kashiwa, S. Nima and S. Sakoda: More About Path Integral for Spin, *Nucl. Phys.* **B453** (1995), 508
51. K. Funahashi, T. Kashiwa, S. Sakoda and K. Fujii: Coherent States, Path Integrals and Semiclassical Approximation, *J. Math. Phys.* **36** (1995), 3232
52. K. Funahashi, T. Kashiwa, S. Sakoda and K. Fujii: Exactness in the WKB Approximation for some Homogeneous Spaces, *J. Math. Phys.* **36** (1995), 4590
53. H. Georgi, *Lie Algebras in Particle Physics*, Addison-Wesley (Redwood City) (1982)
54. P. Ginsparg: Applied Conformal Field Theory, in *Fields, Strings and Critical Phenomena*, eds. E. Brézin and J. Zinn-Justin, Elsevier Science (Amsterdam) (1989), 1
55. H. Goldstein, *Classical Mechanics*, Addison-Wesley (New York) (1950)
56. A. Gorsky and N. Nekrasov: Hamiltonian Systems of Calegero-type and Two-dimensional Yang-Mills Theory, *Nucl. Phys.* **B414** (1994), 213
57. E. Gozzi: Hidden BRS Invariance in Classical Mechanics, *Phys. Lett.* **B201** (1988), 525
58. E. Gozzi and M. Reuter: Classical Mechanics as a Topological Field Theory, *Phys. Lett.* **B240** (1990), 137
59. E. Gozzi, M. Reuter and W. D. Thacker: Hidden BRS Invariance in Classical Mechanics 2, *Phys. Rev.* **D40** (1989), 3363
60. I. S. Gradsteyn and I. M. Ryzhnik, *Table of Integrals, Series and Products*, Academic (San Diego) (1981)
61. M. B. Green, J. H. Schwarz and E. Witten, *Superstring Theory*, Cambridge University Press (Cambridge) (1987)
62. C. Grosche and F. Steiner: How to Calculate Path Integrals in Quantum Mechanics, *J. Math. Phys.* **36** (1995), 2354
63. V. Guilleman and E. Prato: Heckman, Kostant and Steinberg Formulas for Symplectic Manifolds, *Adv. Math.* **82** (1990), 160
64. V. Guilleman and S. Sternberg: Geometric Asymptotics, *AMS Math. Surveys* **14** (1977)
65. V. Guilleman and S. Sternberg, *Symplectic Techniques in Physics*, Cambridge University Press (Cambridge) (1984)
66. S.-O. Hahn, P. Oh and M.-H. Kim: Quantum Mechanics of Integrable Spins on Coadjoint Orbits, *J. Korean Phys. Soc.* **29** (1996), 409
67. Harish-Chandra: Differential Operators on a Semisimple Lie Algebra, *Am. J. Math.* **79** (1957), 87
68. S. Helgason, *Differential Geometry, Lie Groups and Symmetric Spaces*, Academic (New York) (1978)
69. M. Henneaux and C. Teitelboim, *Quantization of Gauge Systems*, Princeton University Press (Princeton) (1992)
70. A. Hietamäki and A. J. Niemi: Index Theorems and Loop Space Geometry, *Phys. Lett.* **B288** (1992), 321
71. A. Hietamäki, A. Yu. Morozov, A. J. Niemi and K. Palo: Geometry of $N = \frac{1}{2}$ Supersymmetry and the Atiyah-Singer Index Theorem, *Phys. Lett.* **B263** (1991), 417

72. L. Hörmander, *The Analysis of Linear Partial Differential Operators*, Springer-Verlag (Berlin) (1983)
73. D. Husemoller, *Fiber Bundles*, Springer-Verlag (Berlin) (1990)
74. Y. Imayoshi and M. Taniguchi, *An Introduction to Teichmüller Spaces*, Springer-Verlag (Tokyo) (1992)
75. C. Itzykson and J.-B. Zuber, *Quantum Field Theory*, McGraw-Hill (New York) (1980)
76. C. Itzykson and J.-B. Zuber: The Planar Approximation, *J. Math. Phys.* **21** (1980), 411
77. R. Jackiw: Field Theoretic Investigations in Current Algebra, in *Current Algebra and Anomalies*, eds. S. B. Treiman, R. Jackiw, B. Zumino and E. Witten, Princeton University Press (Princeton) (1985)
78. L. C. Jeffrey and F. Kirwan: Localization for Non-abelian Group Actions, *Topology* **34** (1995), 291
79. K. Johnson: Functional Integrals for Spin, *Ann. Phys.* **192** (1989), 104
80. J. D. S. Jones and S. B. Petrack: The Fixed Point Theorem in Equivariant Cohomology, *Trans. Am. Math. Soc.* **322** (1990), 35
81. J. Kalkman: BRST Model for Equivariant Cohomology and Representatives for the Equivariant Thom Class, *Commun. Math. Phys.* **153** (1993), 447
82. J. Kalkman: Residues in Non-abelian Localization, Cambridge preprint (1994)
83. T. Kärki: Path Integral Localization of the Laplacian on Lie Groups and Selberg's Trace Formula, Helsinki preprint HIP-1997-71-TH (1997)
84. T. Kärki and A. J. Niemi: On the Duistermaat-Heckman Formula and Integrable Models, in *Proc. XXVII Intern. Ahrenshoop Symp. on the Theory of Elementary Particles*, eds. D. Lüst and G. Weigt, DESY 94-053 (1994), 175
85. E. Keski-Vakkuri, A. J. Niemi, G. W. Semenoff and O. Tirkkonen: Topological Quantum Theories and Integrable Models, *Phys. Rev.* **D44** (1991), 3899
86. M.-H. Kim and P. Oh: Integrable Systems on Flag Manifold and Coherent State Path Integral, *Mod. Phys. Lett.* **A10** (1995), 1847
87. A. A. Kirillov, *Elements of the Theory of Representations*, Springer-Verlag (Berlin) (1976)
88. F. Kirwan: Morse Functions for which the Stationary Phase Approximation is Exact, *Topology* **26** (1987), 37
89. J. R. Klauder: Coherent States without Groups, *Mod. Phys. Lett.* **A8** (1993), 1735
90. J. R. Klauder and B.-S. Skagerstam, *Coherent States*, World Scientific (Singapore) (1985)
91. H. Kleinert, *Path Integrals in Quantum Mechanics, Statistics and Polymer Physics*, World Scientific (Singapore) (1990)
92. I. I. Kogan, A. Yu. Morozov, G. W. Semenoff and N. Weiss: Area Law and Continuum Limit in Induced QCD, *Nucl. Phys.* **B395** (1993), 547
93. F. Langouche, D. Roekaerts and E. Tirapegui, *Functional Integration and Semiclassical Expansions*, Reidel (Dordrecht) (1982)
94. A. Lichernowicz, *Geometry of Groups of Transformations*, Noordhoff (Leyden) (1977)
95. R. Loudon: One-dimensional Hydrogen Atom, *Am. J. Phys.* **27** (1959), 649
96. M. S. Marinov: Path Integrals on Homogeneous Manifolds, *J. Math. Phys.* **36** (1995), 2458
97. J. Marsden and A. Weinstein: Reduction of Symplectic Manifolds with Symmetry, *Rep. Math. Phys.* **5** (1974), 121
98. W. S. Massey, *Algebraic Topology: An Introduction*, Springer-Verlag (New York) (1977)

References

99. V. Mathai and D. Quillen: Superconnections, Thom Classes and Equivariant Differential Forms, *Topology* **25** (1986), 85
100. D. McMullan and I. Tsutsui: On the Emergence of Gauge Structures and Generalized Spin when Quantizing on a Coset Space, *Ann. Phys.* **237** (1995), 269
101. A. Messiah, *Quantum Mechanics*, Wiley (New York) (1976)
102. M. Miettinen: On Localization and Regularization, *J. Math. Phys.* **37** (1996), 3141
103. M. Miettinen: Antibrackets, Supersymmetric σ-model and Localization, *Phys. Lett.* **B388** (1996), 309
104. J. W. Milnor and J. D. Stasheff, *Characteristic Classes*, Princeton University Press (Princeton) (1974)
105. J. A. Minahan and A. P. Polychronakos: Classical Solutions for Two Dimensional QCD on the Sphere, *Nucl. Phys.* **B422** (1994), 172
106. A. Yu. Morozov: Matrix Models as Integrable Systems, in *Particles and Fields '94*, eds. L. Pelletier and L. Vinet (to appear) (1996)
107. A. Yu. Morozov, A. J. Niemi and K. Palo: Supersymmetry and Loop Space Geometry, *Phys. Lett.* **B271** (1991), 365
108. A. Yu. Morozov, A. J. Niemi and K. Palo: Supersymplectic Geometry of Supersymmetric Quantum Field Theories, *Nucl. Phys.* **B377** (1992), 295
109. D. Mumford, *Tata Lectures on Theta*, Birkhäuser (Basel) (1983)
110. R. Narasimhan, *Compact Riemann Surfaces*, Birkhäuser (Basel) (1992)
111. C. Nash, *Differential Topology and Quantum Field Theory*, Academic (San Diego) (1991)
112. A. P. Nersessian: Antibrackets and Localization of Path Integrals, *JETP Lett.* **58** (1993), 64
113. A. P. Nersessian: Equivariant Localization, *NATO ASI Series* **B331** (1994), 353
114. A. P. Nersessian: From Antibracket to Equivariant Characteristic Classes, Joint Institute for Nuclear Research preprint JINR E2-94-377 (1994)
115. A. P. Nersessian: Antibrackets and Non-abelian Equivariant Cohomology, *Mod. Phys. Lett* **A10** (1995), 3043
116. H. Nicolai: Supersymmetry and Functional Integration Measures, *Nucl. Phys.* **B176** (1980), 419
117. H. B. Nielsen and D. Rohrlich: A Path Integral to Quantize Spin, *Nucl. Phys.* **B299** (1988), 471
118. A. J. Niemi: Pedagogical Introduction to BRST, *Phys. Rep.* **184** (1989), 147
119. A. J. Niemi: On the Number of Periodic Classical Trajectories in a Hamiltonian System, *Phys. Lett.* **B355** (1995), 501
120. A. J. Niemi and K. Palo: On Quantum Integrability and the Lefschetz Number, *Mod. Phys. Lett.* **A8** (1993), 2311
121. A. J. Niemi and K. Palo: Equivariant Morse Theory and Quantum Integrability, Uppsala preprint UU-ITP 10/94 (1994)
122. A. J. Niemi and K. Palo: On Supersymmetric Nonlinear σ-models and Classical Dynamical Systems, *Intern. J. Mod. Phys.* **A11** (1996), 1101
123. A. J. Niemi and P. Pasanen: Orbit Geometry, Group Representations and Topological Quantum Field Theories, *Phys. Lett.* **B253** (1991), 349
124. A. J. Niemi and P. Pasanen: Topological σ-model, Hamiltonian Dynamics and Loop Space Lefschetz Number, *Phys. Lett.* **B386** (1996), 123
125. A. J. Niemi and G. W. Semenoff: Fermion Number Fractionization in Quantum Field Theory, *Phys. Rep.* **135** (1986), 99
126. A. J. Niemi and V. V. Sreedhar: On the Infrared Limit of the Chern-Simons-Proca Theory, *Phys. Lett.* **B336** (1994), 381

127. A. J. Niemi and O. Tirkkonen: Cohomological Partition Functions for a Class of Bosonic Theories, *Phys. Lett.* **B293** (1992), 339
128. A. J. Niemi and O. Tirkkonen: On Exact Evaluation of Path Integrals, *Ann. Phys.* **235** (1994), 318
129. A. J. Niemi and O. Tirkkonen: Equivariance, BRST Symmetry and Superspace, *J. Math. Phys.* **35** (1994), 6418
130. P. Oh: Classical and Quantum Mechanics of Non-abelian Chern-Simons Particles, *Nucl. Phys.* **B462** (1996), 551
131. P. Oh and M.-H. Kim: Action-angle Variables for Complex Projective Space and Semiclassical Exactness, *Mod. Phys. Lett.* **A9** (1994), 3339
132. S. Ouvry, R. Stora and P. van Baal: Algebraic Characterization of Topological Yang-Mills Theory, *Phys. Lett.* **B220** (1989), 1590
133. K. Palo: Symplectic Geometry of Supersymmetry and Non-linear Sigma Model, *Phys. Lett.* **B321** (1994), 61
134. L. D. Paniak, G. W. Semenoff and R. J. Szabo: Conformal Motions and the Duistermaat-Heckman Integration Formula, *Phys. Lett.* **B372** (1996), 236
135. P.-É. Paradan: Action Hamiltoniene d'un Tore et Formule de Localisation en Cohomologies Equivariante, *C. R. Acad. Sci. Paris* **324** (1997), 491
136. G. Parisi and N. Sourlas: Random Magnetic Fields, Supersymmetry and Negative Dimensions, *Phys. Rev. Lett.* **43** (1979), 744
137. A. M. Perelomov, *Generalized Coherent States and their Applications*, Springer-Verlag (Berlin) (1986)
138. R. E. Perret: Path Integral Derivation of the Characters for Kac-Moody Groups, *Nucl. Phys.* **B356** (1991), 229
139. R. F. Picken: The Propagator for Quantum Mechanics on a Group Manifold from an Infinite-dimensional Analogue of the Duistermaat-Heckman Integration Formula, *J. Phys.* **A22** (1989), 2285
140. R. F. Picken: The Duistermaat-Heckman Integration Formula on Flag Manifolds, *J. Math. Phys.* **31** (1990), 616
141. E. Prato and S. Wu: Duistermaat-Heckman Measures in a Non-compact Setting, Princeton preprint (1993)
142. A. M. Pressley and G. B. Segal, *Loop Groups*, Oxford University Press (Oxford) (1986)
143. S. G. Rajeev, S. K. Rama and S. Sen: Symplectic Manifolds, Coherent States and Semiclassical Approximation, *J. Math. Phys.* **35** (1994), 2259
144. F. Rief, *Fundamentals of Statistical and Thermal Physics*, McGraw-Hill (New York) (1965)
145. M. Schlichenmaier, *An Introduction to Riemann Surfaces, Algebraic Curves and Moduli Spaces*, Springer-Verlag (Berlin) (1989)
146. L. S. Schulman: A Path Integral for Spin, *Phys. Rev.* **176** (1968), 1558
147. L. S. Schulman, *Techniques and Applications of Path Integration*, Wiley (New York) (1981)
148. A. S. Schwarz: The Partition Function of a Degenerate Quadratic Functional and the Ray-Singer Invariants, *Lett. Math. Phys.* **2** (1978), 247
149. A. S. Schwarz: Semiclassical Approximation in Batalin-Vilkovisky Formalism, *Commun. Math. Phys.* **158** (1993), 373
150. A. S. Schwarz and O. Zaboronsky: Supersymmetry and Localization, *Commun. Math. Phys.* **183** (1997), 463
151. G. W. Semenoff: Quantum Adiabatic Phases, Effective Actions and Anomalies, in *Topological and Geometrical Methods in Field Theory*, eds. J. Hietarinta and J. Westerholm, World Scientific (Singapore) (1988)

152. G. W. Semenoff: Chern-Simons Gauge Theory and the Spin-statistics Connection in Two-dimensional Quantum Mechanics, in *Physics, Geometry and Topology*, ed. H. C. Lee, Plenum (New York) (1991)
153. G. W. Semenoff and R. J. Szabo: Equivariant Localization, Spin Systems and Topological Quantum Theory on Riemann Surfaces, *Mod. Phys. Lett.* **A9** (1994), 2705
154. S. L. Shatashvili: Correlation Functions in the Itzykson-Zuber Model, *Commun. Math. Phys.* **154** (1993), 421
155. M. Sohnius: Introducing Supersymmetry, *Phys. Rep.* **128** (1985), Nos. 2,3
156. M. Stone: Supersymmetry and the Quantum Mechanics of Spin, *Nucl. Phys.* **B314** (1989), 557
157. R. J. Szabo: Geometrical Aspects of Localization Theory, Ph.D. Thesis, University of British Columbia (unpublished) (1995)
158. R. J. Szabo and G. W. Semenoff: Eta-invariants and Fermion Number in Finite Volume, *Phys. Lett.* **B284** (1992), 317
159. R. J. Szabo and G. W. Semenoff: Phase Space Isometries and Equivariant Localization of Path Integrals in Two Dimensions, *Nucl. Phys.* **B421** (1994), 391
160. O. Tirkkonen: Quantum Integrability and Localization Formulas, *Theor. Math. Phys.* **95** (1993), 395
161. O. Tirkkonen: Equivariant BRST and Localization, *Theor. Math. Phys.* **98** (1994), 492
162. V. S. Varadarajan, *Lie Groups, Lie Algebras and their Representations*, Springer-Verlag (New York) (1984)
163. M. Vergne: A Note on the Jeffrey-Kirwan-Witten Localization Formula, *Topology* **34** (1996), 243
164. S. Weinberg, *Gravitation and Cosmology*, Wiley (New York) (1972)
165. A. Weinstein: Cohomology of Symplectomorphism Groups and Critical Values of Hamiltonians, *Math. Z.* **201** (1989), 75
166. E. Witten: Supersymmetry and Morse Theory, *J. Diff. Geom.* **17** (1982), 661
167. E. Witten: Constraints on Supersymmetry Breaking, *Nucl. Phys.* **B202** (1982), 253
168. E. Witten: Non-abelian Bosonization in Two Dimensions, *Commun. Math. Phys.* **92** (1984), 455
169. E. Witten: Topological Quantum Field Theory, *Commun. Math. Phys.* **117** (1988), 353
170. E. Witten: Introduction to Cohomological Field Theory, *Intern. J. Mod. Phys.* **A6** (1991), 2775
171. E. Witten: Two Dimensional Gauge Theories Revisited, *J. Geom. Phys.* **9** (1992), 303
172. N. Woodhouse, *Geometric Quantization*, Clarendon Press (Oxford) (1980)
173. S. Wu: An Integration Formula for the Square of Moment Maps of Circle Actions, *Lett. Math. Phys.* **29** (1993), 311
174. Y. Yasui and W. Ogura: Vortex Filament in Three-manifold and the Duistermaat-Heckman Formula, *Phys. Lett.* **A210** (1996), 258
175. O. Zaboronsky: Dimensional Reduction in Supersymmetric Field Theories, University of California preprint (1996)
176. J. Zinn-Justin, *Quantum Field Theory and Critical Phenomena*, Clarendon Press (Oxford) (1989)

Printing: Druckhaus Beltz, Hemsbach
Binding: Buchbinderei Schäffer, Grünstadt